G. Anger

Inverse Problems in Differential Equations

Mathematische Lehrbücher und Monographien
Herausgegeben von der Akademie der Wissenschaften der DDR
Karl-Weierstraß-Institut für Mathematik

II. Abteilung
Mathematische Monographien
Band 79

Inverse Problems in Differential Equations

von G. Anger

Inverse Problems in Differential Equations

by Gottfried Anger

With 56 figures

Akademie-Verlag Berlin
1990

Verfasser:

Prof. Dr. Gottfried Anger

Martin-Luther-Universität Halle
Sektion Mathematik

Kluwer Plenum ISBN 0-306-43164-5
ISBN 3-05-500657-7
ISSN 0076-5430

Erschienen im Akademie-Verlag Berlin, Leipziger Str. 3—4, Berlin, DDR-1086

Lizenznummer: 202 · 100/406/89
Printed in the German Democratic Republic
Gesamtherstellung: VEB Druckhaus „Maxim Gorki“, Altenburg, 7400
Lektor: Dr. Reinhard Höppner
LSV 1035
Bestellnummer: 7639629 (9198)
05600

Preface

In re mathematica ars proponendi quaestionem pluris facienda est quam solvendi.
GEORG CANTOR (1845–1918), Thesis, Berlin 1867

Most mathematical problems in science, technology and medicine are inverse problems. Studying such problems is the only complete way of analyzing experimental results. Often these problems concern the determination of properties of some inaccessible regions from observation on the boundary of that region, as in geophysics and medicine. Further, the automatization of physical processes leads necessarily to inverse problems, which absolutely must be solved. The practical importance of inverse and improperly posed problems is such that they may be considered among the pressing problems of current mathematical research.

The main problem in mathematical physics is to study the information content of an inverse problem, i.e., to find out which internal parameters of a system inaccessible to measurement can be uniquely determined by measurements. The book gives a complete analysis both of these problems and of the solution of some inverse problems, in which the information content of the coefficients and the source term of a given differential equation is not too large. In order for the book to be accessible to engineers and physicists as well as to mathematicians, a large part of this book uses classical mathematics only. But if the set of all solutions of a special inverse problem is to be studied from a systematic point of view, the results of algebra (equivalence classes), topology (adapted notions of convergence) and functional analysis (convex sets) have to be applied. Such systematic investigations constitute the great strength of mathematics. In Chapter 1 we study some special inverse problems on the real line. Many problems in applications like mechanics, computed tomography, and so on, can be reduced to an Abel integral equation for which a complete theory exists. The book opens with such problems. The difficulties arising in inverse problems with regard to differential equations can be characterized in a comparatively simple way. Chapter 2 is devoted to actual differential equations including the basic equations for carrier transport in semiconductors. In this chapter we formulate many inverse problems, but only some special problems can be studied in this book. Added references give further information concerning these problems. Classical and modern potential theory is one base of our book. In Chapter 3 one can find results on potential theory. Such problems arise if the source term is to be determined. Since in inverse theory the set of all solutions of a special inverse problem is to be studied, an integration theory without any restrictions on the boundary of a domain is necessary.

Such an integration theory is sketched in this chapter. Chapter 4 deals with the solution of the Dirichlet problem and its adjoint mapping describing the set of all solutions of an inverse source problem. In that direction approximation theorems play an important role. In Chapter 5 one can find a systematic approach for studying inverse source problems for elliptic and parabolic differential equations of second order. In Chapters 6 and 7 some special inverse problems are studied. The coefficients and the source term to be studied depend on the solution u or one variable only. In Chapter 8 some global uniqueness theorems are proved for linear differential equations by means of results of modern potential theory. There is an extensive bibliography on inverse and improperly posed problems (340 titles).

The objective of this book is to elucidate the fundamental mathematical structures of inverse problems. Sound study of an inverse problem requires knowledge in optimization (G. Duvaut and J. L. Lions [1], R. Kluge [1], E. Zeidler [1]), in numerical analysis (J. Baumeister [1], B. Hofmann [1], F. Natterer [1]) as well as in stochastics (J. V. Beck et al. [1], J. L. Doob [1], A. Tarantola [1]). Due to limitation in space, both these results and further details on inverse problems cannot be presented in this book.

The author is very grateful to Prof. R. P. Gilbert (Newark, Delaware) for encouraging him to write a book on inverse problems in differential equations. He also wishes to thank Prof. G. Fanselau † (Potsdam), Prof. E. W. Grafarend (Stuttgart), Prof. B. Hofmann (Karl-Marx-Stadt), Prof. H. Jochmann (Potsdam), Prof. R. Kluge (Berlin), Prof. H. Moritz (Graz), Prof. R. Rösler (Freiberg, Sachsen), Dr. L. Ballani (Potsdam), Dr. M. A. Brodsky (Moscow), Dr. E. Kleine (Berlin) and Dr. W. Webers (Potsdam) for many common discussions on inverse problems. Further, the author is very grateful to his wife Brigitte Anger for typing the manuscript, to his son Nils Anger and to Mr. Johannes Gottlieb for many discussions on special problems in physics and for reading the manuscript. Finally he wishes to express his gratitude to Mr. Werner Voigt for reading the English text and to Dr. R. Höppner, of the Akademie-Verlag, for his support in preparing the manuscript.

Berlin, June 1988

Contents

On the interpretation of nature . . . 9

1. Some special inverse problems on the real line . . . 14
   1.1. Some remarks on the set of real numbers . . . 14
   1.2. The Abel integral equation . . . 14
   1.3. Some special differential equations of the first order . . . 26
   1.4. Differential equations of the second order . . . 32
   1.5. Inverse spectral problems . . . 35
   1.6. Differential equations of higher order . . . 36
   1.7. Consequences for the theory of inverse problems . . . 37

2. Actual differential equations in applications . . . 40
   2.1. The Laplace equation and elliptic differential equations of second order . 40
   2.2. Heat equation and parabolic differential equations of second order . . . 49
   2.3. Wave equation and hyperbolic differential equations of second order . . 58
   2.4. Systems of partial differential equations . . . 63
   2.5. Equations of higher order . . . 64

3. Potentials . . . 66
   3.1. Classical potentials . . . 66
   3.2. Some uniqueness theorems for classical potentials . . . 75
   3.3. Some remarks on integration theory . . . 80
   3.4. Potentials, continuous potentials, $\mathbb{C}$-capacity . . . 91

4. The space of continuous functions in inverse problems . . . 105
   4.1. A new transformation . . . 105
   4.2. The balayage principle as projection on the dual space of the space of continuous functions . . . 109
   4.3. Approximation theorems . . . 115
   4.4. A general form of the balayage principle . . . 125

5. Inverse source problems . . . 132
   5.1. Some special inverse source problems in $\mathbb{R}^1$ . . . 132
   5.2. Some special extremal elements of $\mathfrak{B}(\delta_{x^0})$ . . . 136
   5.3. General theorems on inverse source problems . . . 144
   5.4. The geometrical interpretation of extremal measures . . . 149
   5.5. The case of a harmonic density. Analytic continuation of special potentials 160
   5.6. An inverse source problem for the heat equation . . . 174

6. Some one-dimensional inverse problems in differential equations . . . 190
   6.1. Explicitly solvable inverse problems in heat conduction . . . 190

6.2. A nonlinear inverse problem in which the coefficients depend on the gradient of the solution . . . 194

7. Some inverse problems for nonlinear equations . . . 204
7.1. A uniqueness theorem for determining conductivity from overspecified data 204
7.2. A nonlinear parabolic inverse coefficient problem . . . 210

8. Global uniqueness theorems for linear differential equations . . . 216

9. Some historical remarks . . . 222

Conclusions . . . 225

References . . . 227

Symbol Index . . . 250

Subject Index . . . 252

On the interpretation of nature

For practical applications, the interpretation of experimental findings is of fundamental importance. Many physical parameters f do not lend themselves to direct measurement and the effect g which they induce is then the only known criterion; that is, often g is the only measurable value. Between f and g there exist relations based on physical laws which may be formally expressed as equations of the first kind: $Af = g$, with f to be determined by calculation. In doing so, an essential step consists in modelling the process resulting in a symbolic description of A. The basic equations of mathematical physics are derived by assuming the knowledge of the material parameters f at each point of the system considered. Such parameters are known, for instance, for the atmosphere, for interplanetary space and for materials for which experiments of high level exist, while in geophysics the earth's parameters are to be determined by measurements on the surface of the earth.

The calculation of Af with given f is called a direct problem. Typically, this does not offer any major difficulties, as it is possible to use the constructive methods of mathematics. The determination of f from the relation $Af = g$, with given g, is called an inverse problem. The solution of inverse problems is the only way to completely investigate experimental findings. Equations of the first kind are basically different from the more commonly occuring equations of the second kind: $Af + f = g$ arising, e.g., in spectral theory. Equations of the first kind often have a great number of solutions f with $Af = g$. Each equation of the first kind describes a particular physical process and a uniform theory for such equations cannot be expected.

The first inverse problem leading to an integral equation of the first kind was solved by N. H. Abel in 1823. Suppose we slide a particle up a frictionless hill with initial energy E and measure the time $T(E)$ required for it to return. If we vary E and measure $T(E)$, can we determine the shape of the hill? If s is the arc length of the hill and $h(s)$ with $h(0) = 0$ is its height, then both the arc length s and the energy V are coordinates suitable to describe the process as $V(s) = mg_0h(s)$. If V is used as an independent variable, and $V \mapsto s(V)$ is a strictly increasing function, the given problem can be uniquely solved by applying the laws of point mechanics. If, however, the hill has maxima and minima, the problem cannot be solved uniquely (J. B. Keller (1976)). In this case it even has non-denumerably many solutions due to the many-valued mapping $V \mapsto s(V)$ (G. Anger (1987)). Similar conditions apply to the determination of the shape of a mountain by means of light (electromagnetic waves). The back of a mountain cannot be identified from an observation point. Further, it is not possible to look into the interior of an obstacle by means of electromagnetic waves. These examples raise the question of the information content of a physical process or of the relevant mathematical model: which values f do lend themselves to unique and stable determination using the mapping A from $Af = g$?

Inverse problems occurring in optics can be comparatively easily understood and formulated. In this field the direct problem consists in the calculation of the emission, the scattering and the propagation of radiation on the basis of known parameters, both of sources and of diffusers. On the other hand, the inverse problem consists in the determination of the parameters of sources, diffusers, or propagation media by means of the radiation received by a detector. We are well aware of the intuitive solution of the inverse problem: we derive the size, the shape, the surface finish and the structure of an object from the scattering and absorption of light perceived by the eye. In doing so, we have to take into consideration that the human eye has an extremely complex structure. The brain is able to synthesize information in the form of visible light to identify objects, but if optic information beyond the visual impression, or the analysis of optic information by means of electrooptic automatic instruments is required, intuition has to be replaced by mathematical reconstruction, that means by the solution of inverse problems (H. P. Baltes (1980)). This is true for practically any physical and biological process. The highest level of development has been reached with regard to inverse spectral problems as they are closely related to equations of the second kind. A first theorem of uniqueness has been proved by V. Ambarzumian (1929). In the present book, only a few results along these lines are briefly outlined, as a number of monographs exist which deal with this issue.

Another inverse problem that has been under investigation for more than 100 years is the inverse source problem in gravimetry. It consists in the determination of the earth's density $\varrho(x)$, $s \in \mathbb{R}^3$, by measuring the gravity $g = \operatorname{grad} u$ on the earth's surface, with the gravitational field u satisfying the Poisson equation $-\Delta u = \varrho(x)$. From the knowledge of u for the whole space follows the uniqueness of ϱ. In this inverse problem, u and grad u are known just with regard to the earth's surface, while away from the surface of the earth the solution of the Dirichlet problem or of the problem of the oblique derivative has to be used to determine them. The lack of information on u inside the earth results in ambiguity with regard to the inverse source problem. This fact was already known to G. G. Stokes (1867), G. V. Schiaparelli (1875), C. Neumann (1906), P. Pizetti and G. Lauricella (1907–1911), as well as to G. Herglotz (1914). First uniqueness results (for special mass distributions) were obtained by G. Herglotz (1914), P. S. Novikov (1938), and I. M. Rapoport (1940). As far as the information content is concerned, one can as a first step introduce and study the set of all mass distributions producing the same potential on the earth. The integrals of these mass distributions are equal for harmonic functions (G. Lauricella (1911)). It has been modern potential theory, one of the most efficient theories in mathematics (O. Perron (1923),

R. Remak (1924), N. Wiener (1924), M. Brelot (1903–1987), J. L. Doob, G. Choquet, H. Bauer, G. Anger) which, since about the year 1960 has made it possible to consider the characterization of the set of all mass distributions having the same potential (G. Anger (1973–1976)) from a systematic point of view. Additional information is needed to uniquely choose an element out of this set in order to arrive at a solution. This may be the information that the density ϱ is constant but such assumptions are not true for geophysics or for medicine. It is more essential to consider various coupled fields. In geophysics coupling of gravitational and seismic fields (A. S. Alekseev and B. A. Bubnov (1981, 1984)), as well as coupling of gravitational and magnetic fields (D. Zidarov (1968–1984), M. S. Zhdanov (1981, 1988)) are of particular interest. In addition, satellites introduce the possibility of measuring various kinds of radiation.

We are now observing just the beginning of such investigations. The results outlined in this book ought to be extended in the direction indicated, i.e., in modelling more consideration should be given to the complexity of physical processes. Owing to the present lack of information, we are now unable to determine the mass distributions which produce the earth's potential. The above considerations apply also to sources of electrostatic fields. Nevertheless, profound structural investigations may classify the sources (measures), which, outside a body, produce the same potential. The set of these sources is convex and weakly compact. Thus, results of functional analysis are available for concrete issues in physics. The attempt may then be made in a laboratory to find, out of the set of solutions, a solution relevant for a given use. This aspect is gaining increasing importance in applied mathematics. Owing to the existing lack of information, only partial processes can be solved in complex systems by means of mathematics.

In a sense, the results outlined above apply also to inverse source problems for other differential equations, as well as to most identification problems. With regard to the latter, certain coefficients of a differential equation are to be determined on the basis of additional information. If, however, the information content of individual coefficients or of the right-hand side of a differential equation is not too large, these values can be uniquely calculated on the basis of additional information (Chapters 6 and 7). This is true for the diffusion coefficient $k = k(u)$ in equation (7.1.1), for the density $\varrho = \varrho(|\text{grad } u|)$ in equation (6.2.18), for the right-hand side $f = f(u)$ in equation (7.2.1), and for one-dimensional inverse problems. These special problems are a small part of the set of all identification problems.

The difficulties arising in inverse problems with regard to differential equations can be characterized in a comparatively simple way. To do so, we consider the simplest differential equation

$$u' = f, \quad u(a) = 0.$$

The solution u depends on the integral of f. Different f may have the same value $u(b) = g$, $a < b$. If f is linear the relation between f and g is one-to-one. The function f is uniquely determined, provided u is known for the whole interval. However, this information is not available in practice.

As another example we consider the differential equation

$$u' - \big(q(x) + \varphi(x)\big)\,u = 0, \qquad u(a) = u_0, \qquad \varphi(x) = 0 \text{ for } x \notin [\alpha, \beta] \subset [a, b],$$

with the integral of φ being zero. The function φ does not influence the solution u for $x \notin [\alpha, \beta]$. These so-called "non-radiating" physical parameters are of principal importance for the solution of equations of the first kind $Af = g$, as they do not lead to a unique solution f. They cause particular difficulties in problems arising in geophysics and in medicine where a unique solution has to be found. If $\varphi = 0$ and $q = q_0 =$ const, then q_0 can be determined upon one measurement of $u(x^0)$. The relation between $u(x^0)$ and q_0 is strictly non-linear, which is another difficulty in inverse problems.

In both physics and biology, microstructures are gaining increasing importance. At present, supercomputers render possible the precalculation of systems comprising no more than 1000 particles (Dupuis (1986)). But one cubic centimetre of a solid body contains 10^{23} atoms, so, modelling of the properties of microstructures still presents great difficulties. Therefore, one has to continue to use models based on the continuum, in particular models formulated as differential equations. It is not possible to calculate the density $\varrho(x)$ in each point of a body from finitely many measurements performed outside the body. In the case of the earth one needs non-denumerably many values for this purpose. The same situation occurs in the coefficients $a_\alpha(x)$ of a differential equation, if the information content is too large. For this reason the information content has to be reduced with regard to ϱ or a_α, respectively. This can be done by integration (averaging) of certain subsets. As has been said earlier, different ϱ or a_α may have the same average value.

The present state of inverse theory may be evaluated as follows: each particular physical or biological problem is of a complex nature; the models used in mathematics comprise only a small number of material parameters. When faced with a multiplicity of inverse problems, one tries to determine just one parameter by measurement outside an open set. This approach leads to a limitation of the problem, which in general can be dealt with in laboratories only. In laboratories most parameters can be kept constant, and different experiments can be controlled (G. Duvaut and J. L. Lions (1976)). Every inverse problem requires a particular approach. In designing instruments one finds certain additional degrees of freedom as far as geometrical dimensions and materials are concerned. The great success in design engineering is due to this fact. Design perception corresponds to deductive perception in mathematics. On the other hand, interpretive sciences such as geophysics, soil mechanics, meteorology, space research, biology, medicine, etc., are facing particular difficulties. There, one tries to determine a great number of inner parameters from a small number of measurements. Lack of information often results in ambiguity, so different inner parameters produce the same measured values. This ambiguity can only be eliminated by means of additional information to be obtained from experiment. For inverse problems we have the following principles: only in exceptional cases can an inverse problem be decided in a purely mathematical way. In such cases, the solution often depends discontinuously on measured values (A. N. Tikhonov (1943, 1963), V. K. Ivanov

(1962, 1965), M. M. Lavrent'ev (1962)). In case of ambiguity, only the set of all solutions can be introduced and studied as a first step (information content of an inverse problem). One can try to choose a solution relevant for the given use on the basis of laboratory experiments. This result has important consequences for scientific work. Success may be achieved only through an interdisciplinary approach (J. Polking (1987)). For the complex systems of nature, the scientist largely depends on practical experience. In the future, intense fundamental research on inverse problems will be imperative. The determination of the information content of mathematical models, that is the possibility of unique and stable determination of inner parameters, will be most important. The solution of the direct problem is the presupposition of a successful approach to the inverse problem.

We have repeatedly pointed out that an equation of the first kind $Af = g$ may have non-denumerably many solutions f with the same image g. Here Cantor's Continuum Hypothesis plays a role. More than a hundred years ago, G. Cantor of the University Halle-Wittenberg, tried to prove that any infinite set of real numbers may be mapped one-to-one either on the set $\mathbb{N}$ of natural numbers, or on the set $\mathbb{R}$ of real numbers. K. Gödel (1931) and P. Cohen (1963) pointed out that the existing mathematical axioms are not sufficient to prove or to refute this statement. Similar assertions are valid for any inverse problem. Such problems do not lend themselves to decision without substantial additional suppositions gained from experiments. It has repeatedly been stressed that with regard to such undecidable problems profound structural investigations may be needed, and these may be of primary importance for the future.

It may be that results in inverse theory are most important for simulation. By means of these results, simulation can use the information content of the special model considered more effectively.

1. Some special inverse problems on the real line

1.1. Some remarks on the set of real numbers

In mathematics and its applications the set $\mathbb{R}$ of all real numbers x plays a fundamental role. The real numbers can be constructed from the set $\mathbb{N}$ of all natural numbers. For simplicity it is easier to state the properties of the real numbers in the form of axioms. The first group of axioms specifies the algebraic properties for $x + y$ and $x \cdot y$, the second group of axioms the order properties $x \leqq y$. The third axiom gives the important property of *order-completeness*, i.e., if $\mathcal{A}$ is any non-empty set of real numbers which is bounded above, the set of majorants of $\mathcal{A}$ has a least element. This means that there exists a unique real number $s =: \sup \mathcal{A}$ — the *supremum* of $\mathcal{A}$ — satisfying $x \leqq s$ for all $x \in \mathcal{A}$, and if $x \leqq u$ for all $x \in \mathcal{A}$, then $s \leqq u$. If $s \in \mathcal{A}$ then s is called the *maximum* of $\mathcal{A}$ and denoted by $s =: \max \mathcal{A}$. The third axiom distinguishes the set of real numbers from all sets whose objects satisfy the first group and the second group of axioms. It can be shown that these three axioms do in fact characterize the real numbers, in that any set of objects satisfying them is mathematically indistinguishable from the set of real numbers. From the third axiom the *Archimedian property* of the real numbers follows: if x and y are positive real numbers, there exists a natural number n such that $y < nx$. G. Cantor proved that the set of real numbers is non-denumerable (uncountable). A set $\mathcal{A} \subset \mathbb{R}$ is called *denumerable* (countable), if and only if there exists a one-to-one mapping from $\mathbb{N}$ onto $\mathcal{A}$.

1.2. The Abel integral equation

The *Abel integral equation*

$$\int_a^\eta (\eta - \xi)^{\alpha-1} f(\xi)\, \mathrm{d}\xi = g(\eta), \qquad \xi, \eta \in [a, b], \qquad 0 < \alpha < 1, \tag{1}$$

is an equation of the first kind. Its solution is given by

$$f(\xi) = \frac{\sin \pi\alpha}{\pi} \frac{\mathrm{d}}{\mathrm{d}\xi} \int_a^\xi (\xi - \eta)^{-\alpha}\, g(\eta)\, \mathrm{d}\eta\,. \tag{2}$$

In applications most often one has $\alpha = 1/2$. *Equations of the first kind* can formally be written in the form

$$Af = g. \tag{3}$$

Here g is given by the external parameters of the special problem considered, and f is determined by the internal parameters, inaccessible to measurement. Most problems in science and technology can be written as equation of the first kind.

The Abel integral equation plays an important role in applications, for instance, in mechanics, seismics, stereology, heat conduction, enzyme kinetics, spectroscopy of gas discharges, determination of the refractive index of optical fibres, determination of a potential from a period of oscillation, scattering theory, fusions theory, etc. (Abel [1], Cormack [1], Fenyö and Stolle [1], Gorenflo [1], Gorenflo and Vessella [4], Keller [1], Lasch and Kühnau [1]).

Equation (1) was explicitly solved by N. H. Abel in 1823. This equation is studied on an infinite-dimensional space, for instance, on $\mathbf{L}^p([a, b])$ with $p = 1$. The space $\mathbf{L}^p([a, b])$ consists of all functions whose p-th power is integrable over the closed interval

$$[a, b] := \{x : a \leqq x \leqq b, x \in \mathbb{R}\}.$$

We often have to consider the extended real line

$$\overline{\mathbb{R}} := \{y : -\infty \leqq y \leqq +\infty\} \quad \text{and} \quad \mathbb{R}^+ := \{y : 0 \leqq y < +\infty\}.$$

The kernel

$$K(\eta, \xi) := \begin{cases} (\eta - \xi)^{\alpha-1} & \text{if} \quad \xi < \eta, \\ 0 & \text{if} \quad \eta < \xi, \\ +\infty & \text{if} \quad \xi = \eta \end{cases} \tag{4}$$

is a mapping

$$K: [a, b] \times [a, b] \mapsto \overline{\mathbb{R}}.$$

Then (1) can be written as

$$(Af)(\eta) := \int_a^b K(\eta, \xi) f(\xi) \, d\xi. \tag{5}$$

In order to get all solutions of special physical problems by means of a special equation it must be studied on an infinite-dimensional space. Of course, measurements can be made only at a finite number of points. One of the most serious problems in mathematics consists in proving properties from finite sets to the continuum, i.e., to non-denumerable sets. Such proofs are possible for compact sets or for power series. A set is *compact* if and only if each infinite subset contains a sequence converging to an element of this set (Dunford and Schwartz [1]). A set is called *relatively compact* if its closure is compact.

It is well known that a function $f: \mathbb{C} \mapsto \mathbb{C}$ represented by a power series is zero on a denumerable and bounded set, the function must be identically zero. Here $\mathbb{C}$ is the set of complex numbers. Further, the set $\mathbb{R}$ is often used in statistics. In order to conclude to $\mathbb{R}$ from a finite number of measurements an a priori information concerning the special form of the distribution function, for instance, the Gaussian distribution, must be known. If one has too little information of the special problem under consideration the statement of Cantor's Continuum Hypothesis holds: neither provable, nor refutable. This result was proved by K. Gödel in 1930–1936 and P. Cohen in 1963 (COHEN [1] and DAHN [1]). Roughly speaking *Cantor's Continuum Hypothesis* is the following: to each infinite subset $B \subset \mathbb{R}$ there exists a one-to-one mapping onto $\mathbb{N}$ or $\mathbb{R}$.

The compact sets are applied for studying *compact operators,* often called *compact mappings* or *compact transformations.* A compact operator maps any bounded set of a functional space into a relatively compact set. Most solutions of boundary-value or initial-value problems can be represented by compact integral operators. Therefore, these operators can be approximated by operators defined on finite-dimensional spaces. This fact is important for the numerical treatment of such problems.

Studying equations of the type (3) on infinite-dimensional spaces one can immediately see difficulties arising in inverse theory: the non-uniqueness or the instability of (3). A mapping A, for which the inverse A^{-1} exists, is called *instable* if A^{-1} is discontinous. If A is a compact one-to-one mapping on an infinite-dimensional space, then A^{-1} must be discontinuous. This follows from the following facts: if A^{-1} is continuous then the mapping $A^{-1}A = I$ is compact since it maps any bounded set into a relatively compact set. But the identity operator I is not compact on an infinite-dimensional space. Therefore A^{-1} must be discontinuous. In a concrete case see formula (11). In order to get useful results for solution $f = A^{-1}g$ regularization methods have been developed (BECK et al. [1], ENGL et al. [1], GROETSCH [1], HOFMANN [1], IVANOV et al. [1], MOROZOV [1, 2], TIKHONOV and ARSENIN [1]).

For linear equations of the first kind often infinitely many solutions $f \neq 0$ of the homogeneous equation $Af = 0$ exist (see Sect. 1.3). One reason for the property $Af = 0$ is that insufficient information is available for the problem under consideration. In order to prove properties for solutions of the first kind from a finite number of points to the continuum additional a priori information is necessary (HOFMANN [1]). Such additional conditions must be verified by experience gained in experiments. Roughly speaking, one can characterize the results of mathematics in the area of analysis as follows: the study of metric spaces on which *equations of the second kind* $Af + f = g$ are studied. A finite number of solutions of the homogeneous equation $Af + f = 0$ exists for compact linear operators A. This theory was developed by I. Fredholm (1900), D. Hilbert (1904), E. Schmidt (1905), F. Riesz (1918), J. v. Neumann (1932) and many other mathematicians (DUNFORD and SCHWARTZ [1]).

In the following we shall consider some special applied problems which can be reduced to an Abel integral equation. Further, the Abel integral equation is important as a special case of the more general Volterra equation (FENYÖ and STOLLE [1]).

Example 1.1 (Mechanical problem). The problem consists in determining the shape of a hill from travel time. Suppose we slide a particle up a frictionless hill with initial energy E, and measure the travel time $T(E)$ required for it to return. If we vary E and measure $T(E)$, can we determine the shape of the hill from it (ABEL [1], GORENFLO and VESSELLA [1], KELLER [1]) (see Fig. 1.1)?

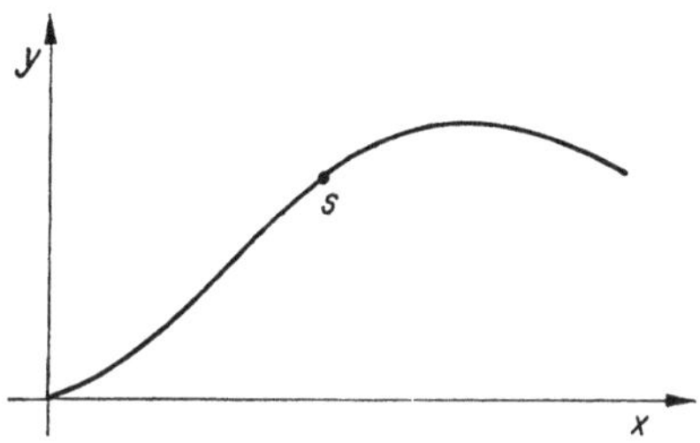

Fig. 1.1

This problem was formulated and solved by Abel. To analyze it we shall first formulate and solve the direct problem: given the shape of the hill, find the travel time $T(E)$. Let s denote the arc length along the hill and let the height of the hill be $h(s)$ with $h(0) = 0$. We denote by m the mass of the particle, by g_0 the acceleration of gravity, by $s(t)$ the position of the particle at time t. Then the equation of motion of the particle is

$$m \frac{\mathrm{d}^2 s(t)}{\mathrm{d}t^2} = -\frac{\mathrm{d}V(s)}{\mathrm{d}s}, \tag{6}$$

where $V(s) = mg_0h(s)$ is the potential energy of the particle at s. We shall measure s and t from the initial position,

$$s(0) = 0, \qquad \frac{\mathrm{d}s(0)}{\mathrm{d}t} = v_0 > 0.$$

We multiply (6) by $\mathrm{d}s/\mathrm{d}t$ and integrate to get the energy equation

$$\frac{m}{2}\left(\frac{\mathrm{d}s}{\mathrm{d}t}\right)^2 + V(s) = E. \tag{7}$$

E is the total energy, given by $E = mv_0^2/2$. From (7) it follows that

$$\left(\frac{\mathrm{d}t}{\mathrm{d}s}\right)^2 = \frac{m}{2}\frac{1}{E - V(s)}$$

and

$$t = \left(\frac{m}{2}\right)^{1/2} \int_0^s \frac{\mathrm{d}s'}{(E - V(s'))^{1/2}}. \tag{8}$$

Let $s_1(E)$ be the smallest value of $s \geqq 0$ for which $E - V(s) = 0$. The corresponding t, which we call $T(E)/2$, is given by setting $s = s_1(E)$. Then

$$T(E) = (2m)^{1/2} \int_0^{s_1(E)} \frac{\mathrm{d}s'}{\big(E - V(s')\big)^{1/2}}. \tag{9}$$

The inverse problem consists in the following: given the travel time $T(E)$ for $E \geqq 0$, find the shape of the hill, i.e., find the potential $V(s)$ for $s > 0$. Equation (8) is a nonlinear integral equation for V. Let us introduce V as integration variable (see Fig. 1.2).

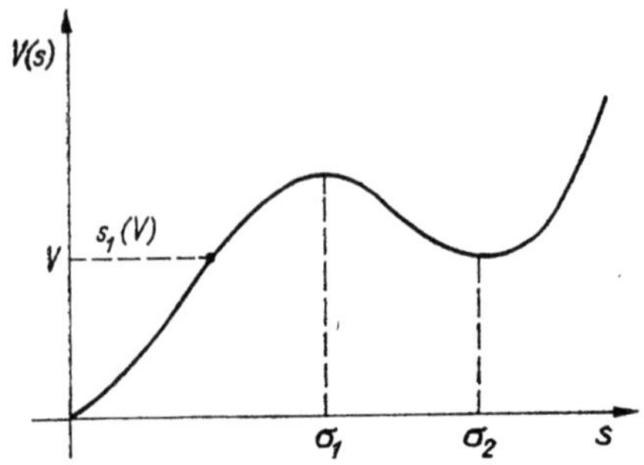

Fig. 1.2

This change of integration variable must be made separately in each interval within $V(s)$ is monotone. Let $s = s_1(V)$ denote the inverse of $V(s)$ within this interval, $0 \leqq E \leqq V(\sigma_1)$. Then

$$T(E) = (2m)^{1/2} \int_0^E (E - V)^{-1/2} \frac{\mathrm{d}s_1(V)}{\mathrm{d}V} \mathrm{d}V, \qquad 0 \leqq E \leqq V(\sigma_1). \tag{10}$$

Equation (10) is a linear equation of the first kind for $\mathrm{d}s_1/\mathrm{d}V$. The solution of (10) is given by (see Theorem 1.1)

$$\frac{\mathrm{d}s_1(V)}{\mathrm{d}V} = \frac{T(0)}{(2mV)^{1/2}} + \frac{1}{(2m)^{1/2}\,\pi} \int_0^V (V - E)^{-1/2}\, T'(E)\, \mathrm{d}E. \tag{11}$$

Integrating (11) from 0 to V we get in the interval $0 \leqq s \leqq \sigma_1$

$$s_1(V) = \frac{1}{(2m)^{1/2}\,\pi} \int_0^V (V - E)^{-1/2}\, T(E)\, \mathrm{d}E, \qquad 0 \leqq V \leqq V(\sigma_1). \tag{12}$$

Remark 1.1. The equations of Example 1.1 also hold for a particle in a potential well $V(s)$. Then the period $P(E)$ of oscillation of a particle with energy E is to be determined (direct problem). The inverse problem consists in determining the potential $V(s)$ from the period of oscillation $P(E)$ (Keller [1]).

Example 1.2 (Seismic sounding). We suppose that (i) the earth is an elastic body with spherical symmetry, (ii) the velocity of propagation v of a seismic ray depends only on the depth of which it is an increasing function of depth, (iii) the propagation of a seismic ray obeys the laws of geometric optics. Under this assumptions the velocity of seismic waves in the earth's interior is to be determined from measurements of their travel times at the surface. See also equation (15) (Cassinis [1], Carrion [1], Fitzgibbon [1], Gorenflo and Vessella [1], Grant and West [1]).

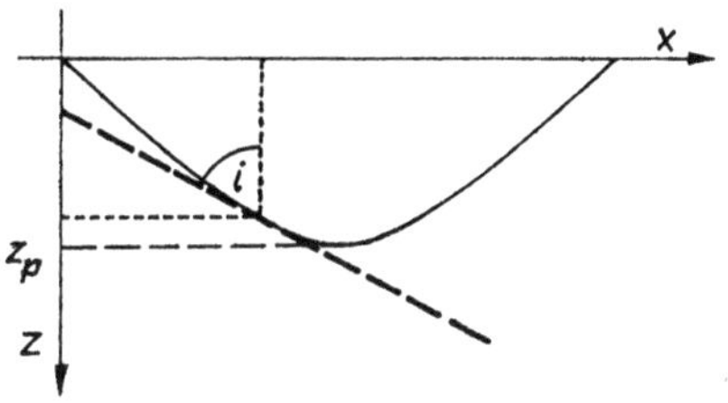

Fig. 1.3

For simplicity we consider a plane model. The velocity increases continuously with depth (see Fig. 1.3). Let i be the angle between the seismic ray and the vertical z-axis, $v(z)$ the velocity of the propagation of the seismic ray at a point with the z-coordinate equal to z, z_p the maximum depth of seismic ray, $u = 1/v$,

$$p = \frac{\sin i}{v}, \qquad u(z_p) = p.$$

Further, let $z(u)$ be the inverse function of $u(z)$, $u_0 = 1/v(0)$. Then by Snell's law one gets for the travel time $\tau(p)$

$$\tau(p) = 2 \int_p^{u_0} \frac{uz(u)}{(u^2 - p^2)^{1/2}} \, \mathrm{d}u. \tag{13}$$

This integral is by the change of variables

$$\xi = 1 - \left(\frac{u}{u_0}\right)^2, \qquad \eta = 1 - \left(\frac{p}{u_0}\right)^2$$

an equation of type (1). For ambiguities arising in seismic interpretation see Grant and West [1], Keller [1], Gerver and Markushevich [1].

Example 1.3 (Computed tomography in medicine). It consists in the reconstruction of densities from their projection along lines (Anderson and Newsan [1], Cormack [1], Natterer [1], Tikhonov et al. [1]).

We denote with I_0 the intensity of a ray travelling along L as it emanates from an exterior source, the intensity past the object is assumed to be (see Fig. 1.4)

$$I = I_0 \exp\left(-\int_L f(P) \, \mathrm{d}s\right) \tag{14}$$

From I/I_0 one can determine the line integral.

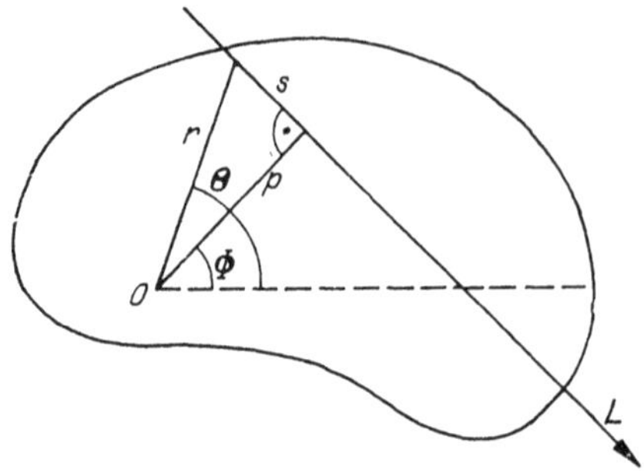

Fig. 1.4

First consider in $\mathbb{R}^2$ a circularly symmetrical function $f(r)$. Then if $\hat{f}(p)$, called *Radon transform* of f, is the line integral along a line which is a perpendicular distance from the origin, it is easy to show that

$$\hat{f}(p) := \int_L f(P)\,\mathrm{d}s = 2\int_p^\infty \frac{rf(r)\,\mathrm{d}r}{(r^2-p^2)^{1/2}}. \tag{15}$$

To find $f(r)$ one needs only $\hat{f}(p)$ for $p \geqq r$. This remains true for non-circularly symmetrical functions. Let the line along which $\hat{f}$ is computed be specified by (p, ϕ) as in Fig. 1.4. Then if one writes

$$f(r, \theta) = \sum_{k=0}^\infty f_k(r)\,\mathrm{e}^{\mathrm{i}k\theta}, \qquad \hat{f}(p, \phi) = \sum_{k=0}^\infty \hat{f}_k(p)\,\mathrm{e}^{\mathrm{i}k\phi}, \tag{16}$$

one can show that (Cormack [1])

$$\hat{f}_k(p) = 2\int_p^\infty \frac{f_k(r)\,T_k(r/p)\,\mathrm{d}r}{(r^2-p^2)^{1/2}}, \tag{17}$$

where T_k is the *Tschebicheff polynomial* of the first kind which is orthogonal in $]-1, 1[$. It has the form

$$\begin{aligned} T_k(x) &= \frac{1}{2}\left[\left(x+\mathrm{i}\sqrt{1-x^2}\right)^k + \left(x-\mathrm{i}\sqrt{1-x^2}\right)^k\right] \\ &= x^k - \binom{k}{2} x^{k-2}(1-x^2) + \binom{k}{4} x^{k-4}(1-x^2)^2 - + \cdots. \end{aligned} \tag{18}$$

The solution of (15) is given by (Cormack [1])

$$f(r) = -\frac{1}{\pi}\int_r^\infty \frac{\hat{f}'(p)\,\mathrm{d}p}{(p^2-r^2)^{1/2}} \tag{19}$$

and the solution of (17) by

$$f_k(r) = -\frac{1}{\pi}\int_r^\infty \frac{\hat{f}_k'(p)\,T_k(p/r)\,\mathrm{d}p}{(p^2-r^2)^{1/2}}. \tag{20}$$

The inversion formula for the Radon transform in $\mathbb{R}^n$ can be viewed as a decomposition of a function f into plane waves (JOHN [1, 2]). One can find a further development of these ideas in integral geometry. The planes are to be replaced by a family of surfaces (ROMANOV [1, 2]). Further, magnetic resonance imaging becomes more and more important (BOERNER et al. [1], GRÜNBAUM [1]). This method does not use X-ray examinations.

Let us prove formulas (2), (19) and (20) (FENYÖ and STOLLE [1], GORENFLO and VESSELLA [1, 2], TONELLI [1]).

Theorem 1.1 (Existence and uniqueness of Abel's equation).

(i) *If there exists a solution of* (1) *it must have the form*

$$\int_a^x f(\xi)\,\mathrm{d}\xi = \frac{\sin \pi\alpha}{\pi} \int_a^x (x-\eta)^{-\alpha}\, g(\eta)\,\mathrm{d}\eta. \tag{21}$$

If f is continuous we get

$$f(\xi) = \frac{\sin \pi\alpha}{\pi} \frac{\mathrm{d}}{\mathrm{d}\xi} \int_a^\xi (\xi-\eta)^{-\alpha}\, g(\eta)\,\mathrm{d}\eta \tag{22}$$

or, if a and $g(a)$ are finite and g' exists

$$f(\xi) = \frac{\sin \pi\alpha}{\pi} \left[g(a)\,(\xi-a)^{-\alpha} + \int_a^\xi (\xi-\eta)^{-\alpha}\, g'(\eta)\,\mathrm{d}\eta \right]. \tag{23}$$

(ii) *There exists at most one solution of* (1).

(iii) *If g is the difference of two monotone functions continuous from the right (i.e., $g(\xi) = \lim_{k\to\xi^+} g(x)$) then equation* (1) *has a solution in $\mathbf{L}^1([a, b])$, being then given by formula* (22) *or* (23), *where the equality holds in the sense of $\mathbf{L}^1([a, b])$.*

Proof. (i) The following formula is the base in our considerations (FENYÖ and STOLLE [1]):

$$\int_0^1 \lambda^{-\alpha}(1-\lambda)^{\alpha-1}\,\mathrm{d}\lambda = \frac{\pi}{\sin \pi\alpha}. \tag{24}$$

Multiplying both sides of (1) by $(y-\eta)^{-\alpha}$, where $y \in\,]a, b[$ is fixed, and then integrating over the open interval $]a, b[$ we obtain

$$\int_a^y \left\{ (y-\eta)^{-\alpha} \int_a^\eta (\eta-\xi)^{\alpha-1} f(\xi)\,\mathrm{d}\xi \right\} \mathrm{d}\eta = \int_a^y (y-\eta)^{-\alpha}\, g(\eta)\,\mathrm{d}\eta.$$

Since the left-hand side is equal to

$$\int_a^y \left\{ \int_\xi^y (y-\eta)^{-\alpha}\,(\eta-\xi)^{\alpha-1}\,\mathrm{d}\eta \right\} f(\xi)\,\mathrm{d}\xi$$

and

$$\int_{\xi}^{y} (y - \eta)^{-\alpha} (\eta - \xi)^{\alpha-1} \, d\eta = \int_{0}^{1} \lambda^{-\alpha}(1 - \lambda)^{\alpha-1} \, d\lambda = \frac{\pi}{\sin \pi\alpha}$$

holds, we get

$$\int_{a}^{y} f(\xi) \, d\xi = \frac{\sin \pi\alpha}{\pi} \int_{a}^{y} (y - \eta)^{-\alpha} g(\eta) \, d\eta . \tag{25}$$

If f is continuous from (25) it follows that

$$f(\xi) = \frac{\sin \pi\alpha}{\pi} \frac{d}{d\xi} \int_{a}^{\xi} (\xi - \eta)^{-\alpha} g(\eta) \, d\eta .$$

If a and $g(a)$ are finite, integration by parts gives

$$\int_{a}^{\xi} (\xi - \eta)^{-\alpha} g(\eta) \, d\eta = \frac{1}{1 - \alpha} \left[g(a) \, (\xi - a)^{1-\alpha} + \int_{a}^{\xi} (\xi - \eta)^{1-\alpha} g'(\eta) \, d\eta \right]$$

hence

$$f(\xi) = \frac{\sin \pi\alpha}{\pi} \left[g(a) \, (\xi - a)^{-\alpha} + \int_{a}^{\xi} (\xi - \eta)^{-\alpha} g'(\eta) \, d\eta \right].$$

(ii) If there exist two functions f_1 and f_2 satisfying (1) then $f = f_1 - f_2$ fulfills (1) with $g = 0$. From (25) it follows that $\int_{a}^{y} f(\xi) \, d\xi = 0$ for each $y \in \,]a, b[$. If f is continuous we get $f = 0$. In the other case the relation $f = 0$ holds in the sense of $\mathbf{L}^1([a, b])$.

(iii) These results can be found in Fenyö and Stolle [1], Tonelli [1]. ∎

Theorem 1.2 (Generalized Abel integral equation). *Let h be a strictly increasing function on $]a, b[$ and*

$$\int_{a}^{\eta} \frac{f(\xi) \, d\xi}{[h(\eta) - h(\xi)]^{1-\alpha}} = g(\eta), \quad a \leqq \eta \leqq b, \tag{26}$$

a generalized Abel integral equation. If there exists a continuous solution f then

$$f(\xi) = \frac{\sin \pi\alpha}{\pi} \frac{d}{d\xi} \int_{a}^{\xi} \frac{h'(\eta) \, g(\eta) \, d\eta}{[h(\xi) - h(\eta)]^{\alpha}}. \tag{27}$$

If a and $g(a)$ are finite and g' exists

$$f(\xi) = \frac{\sin \pi\alpha}{\pi} \left\{ \frac{h'(a)\, g(a)}{[h(\xi) - h(\eta)]^\alpha} + \int_a^\xi \frac{h'(\eta)\, g'(\eta)\, d\eta}{[h(\xi) - h(\eta)]^\alpha} \right\}. \tag{28}$$

Proof. Multiplying both sides of (26) by

$$\frac{h'(\eta)}{[h(y) - h(\eta)]^\alpha}, \qquad y \in \,]a, b[\,,$$

and integrating this relation we obtain for the left-hand side

$$\int_a^y f(\xi) \left\{ \int_\xi^y \frac{h'(\eta)\, d\eta}{[h(y) - h(\eta)]^\alpha\, [h(\eta) - h(\xi)]^\alpha} \right\} d\xi = \frac{\pi}{\sin \pi\alpha} \int_a^y f(\xi)\, d\xi. \tag{29}$$

The formulas (27) and (28) follow from equation (29). ∎

Theorem 1.3 (Instability of Abel's equation). *The solution* (23) *is instable, i.e., there exists a sequence $g_k \to g$ converging uniformly such that*

$$\int_a^\xi (\xi - \eta)^{-\alpha}\, g_k'(\eta)\, d\eta \to \infty \tag{30}$$

for fixed $\xi \in \,]a, b[$.

Proof. Let us formally write (1) in the form $Af = g$, where

$$Af(\eta) = \int_a^\eta (\eta - \xi)^{\alpha - 1} f(\xi)\, d\xi.$$

If $g(a) = 0$ it follows that

$$A^{-1}g(\xi) = \frac{\sin \pi\alpha}{\pi} \int_a^\xi (\xi - \eta)^{-\alpha}\, g'(\eta)\, d\eta. \tag{31}$$

If A^{-1} is continuous from $g_k \to g$ it follows that $A^{-1}g_k \to A^{-1}g$ in the sense of $L^1([a, b])$. For simplicity we take $a = 0$. Let

$$g_k'(\eta) := \begin{cases} (\xi - \eta)^{-1+\alpha} & \text{if } \; 0 \leqq \eta \leqq \xi - \dfrac{1}{k}, \\ \left(\dfrac{1}{k}\right)^{-1+\alpha} = k^{1-\alpha} & \text{if } \; \xi - \dfrac{1}{k} \leqq \eta \leqq \xi, \\ 0 & \text{if } \; \xi < \eta < b, \end{cases} \tag{32}$$

where $k = 1, 2, \ldots$ (see Fig. 1.5).

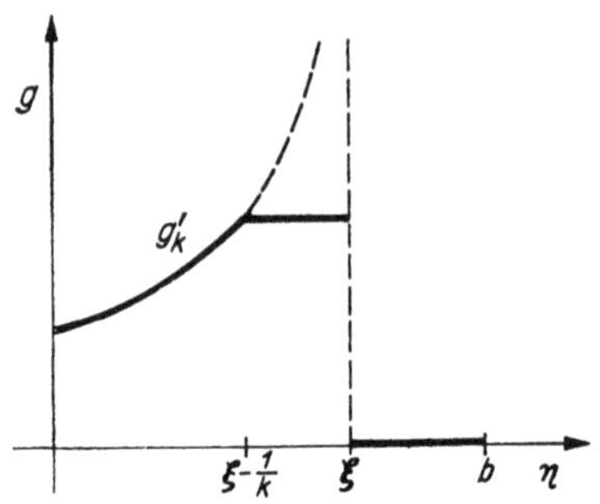

Fig. 1.5

The functions $g_k(\eta) = \int\limits_0^{\eta} g_k'(y)\,\mathrm{d}y$ have the form

$$g_k(\eta) = \begin{cases} \dfrac{1}{\alpha}\left[-(\xi-\eta)^\alpha + \xi^\alpha\right] & \text{if } 0 \leqq \eta \leqq \xi - \dfrac{1}{k}, \\[2ex] \dfrac{1}{\alpha}\left[-\dfrac{1}{k^\alpha} + \xi^\alpha\right] + k^{1-\alpha}\left(\eta - \xi + \dfrac{1}{k}\right) & \text{if } \xi - \dfrac{1}{k} \leqq \eta \leqq \xi, \\[2ex] \dfrac{1}{\alpha}\left[-\dfrac{1}{k^\alpha} + \xi^\alpha\right] + \dfrac{1}{k^\alpha} & \text{if } \xi \leqq \eta \leqq b. \end{cases} \tag{33}$$

The continuous functions g_k converge monotonically to a continuous function g. By Dini's theorem the convergence is uniform on $[a, b]$ to

$$g(\eta) = \frac{1}{\alpha}\left[-(\xi-\eta)^\alpha + \xi^\alpha\right] \quad \text{if } 0 \leqq \eta \leqq \xi,$$

$$g(\eta) = \frac{1}{\alpha}\,\xi^\alpha \qquad \text{if } \xi \leqq \eta \leqq b.$$

Further, the g_k' form a strictly increasing sequence, where the limit has the form

$$g_k(\eta) = \begin{cases} (\xi-\eta)^{-1+\alpha} & \text{if } 0 \leqq \eta \leqq \xi, \\ +\infty & \text{if } \eta = \xi, \\ 0 & \text{if } \xi < \eta \leqq b. \end{cases} \tag{34}$$

The function defined in (34) is integrable. Finally we get

$$\int\limits_0^{\xi} (\xi-\eta)^{-\alpha}\, g_k'(\eta)\,\mathrm{d}\eta = \lim_{k\to\infty}\left[\int\limits_0^{\xi-1/k} (\xi-\eta)^{-1}\,\mathrm{d}\eta + \int\limits_{\xi-1/k}^{\xi} (\xi-\eta)^{-\alpha}\, k^{1-\alpha}\,\mathrm{d}\eta\right]$$

The right-hand side tends to $+\infty$ for $k \to \infty$. ∎

In Keller [1] the first part of the following result is proved.

Theorem 1.4 (Non-uniqueness in Example 1.1). *If the function V in Example 1.1 is increasing in $0 \leqq s \leqq \sigma_1$, $V(\sigma_1)$ finite, decreasing in $\sigma_1 \leqq s \leqq \sigma_2$ and increasing again in $\sigma_2 \leqq s$, then the equation* (10) *does not have a uniquely deter-*

mined solution. In this case the inverse mapping $V \mapsto s(V)$ is multi-valued. The set of solutions of equation (10) *is non-denumerable.*

Proof. Let $s_1(V)$, $s_2(V)$, $s_3(V)$ be the inverse of $V(s)$ in the intervals considered (see Fig. 1.2). Then (10) can be written in the form

$$T(E) = (2m)^{1/2} \left[\int_0^{V(\sigma_1)} (E - V)^{-1/2} \frac{\mathrm{d}s_1}{\mathrm{d}V} \mathrm{d}V - \int_{V(\sigma_2)}^{V(\sigma_1)} (E - V)^{-1/2} \frac{\mathrm{d}s_2}{\mathrm{d}V} \mathrm{d}V + \int_{V(\sigma_2)}^{E} (E - V)^{-1/2} \frac{\mathrm{d}s_3}{\mathrm{d}V} \mathrm{d}V \right]. \tag{35}$$

We define $w(V)$, the width of depth V of the potential well, by

$$w(V) := s_3(V) - s_2(V), \qquad V(\sigma_2) \leqq V \leqq V(\sigma_1).$$

We assume $E \geqq V(\sigma_1)$. Then we can rewrite (35) as

$$T(E) = (2m)^{1/2} \left[\int_0^{V(\sigma_1)} (E - V)^{-1/2} \frac{\mathrm{d}s_1}{\mathrm{d}V} \mathrm{d}V + \int_{V(\sigma_1)}^{E} (E - V)^{-1/2} \frac{\mathrm{d}s_3}{\mathrm{d}V} \mathrm{d}V + \int_{V(\sigma_2)}^{V(\sigma_1)} (E - V)^{-1/2} \frac{\mathrm{d}w}{\mathrm{d}V} \mathrm{d}V \right]. \tag{36}$$

In formula (36) the travel time $T(E)$ involves only the width function $w(V)$ of the well, and does not depend upon its two sides $s_2(V)$ and $s_3(V)$ separately. Let us regard (36) as an integral equation for $\mathrm{d}s_3/\mathrm{d}V$ in the range $V \geqq V(\sigma_1)$. The solution is given by (9) for $V(\sigma_1) \leqq V$:

$$s_3(V) = s_3[V(\sigma_1)] + \frac{1}{\pi} \int_{V(\sigma_1)}^{V} (V - E)^{-1/2} \left[(2m)^{-1/2}\, T(E) - \int_0^{V(\sigma_1)} (E - z)^{-1/2} \frac{\mathrm{d}s_1}{\mathrm{d}z} \mathrm{d}z - \int_{V(\sigma_2)}^{V(\sigma_1)} (E - z)^{-1/2} \frac{\mathrm{d}w}{\mathrm{d}z} \mathrm{d}z \right] \mathrm{d}E. \tag{37}$$

This form of the solution can be simplified with the aid of the integral formula

$$\int_{V(\sigma_1)}^{V} (V - E)^{-1/2} (E - z)^{-1/2} \, \mathrm{d}E = 2 \sin^{-1} \left(\frac{E - z}{V - z} \right)^{1/2} \Bigg|_{V(\sigma_1)}^{V} = \pi - 2 \sin^{-1} \left(\frac{V(\sigma_1) - z}{V - z} \right)^{1/2}. \tag{38}$$

Here $\sin^{-1}$ denotes the inverse of sin. From (37) and (38) it follows that

$$s_3(V) = \frac{1}{(2m)^{1/2}\,\pi} \int\limits_{V(\sigma_1)}^{V} (V - E)^{-1/2}\, T(E)\, \mathrm{d}E$$
$$+ \frac{2}{\pi} \int\limits_{V(\sigma_2)}^{V(\sigma_1)} \frac{\mathrm{d}w}{\mathrm{d}z} \sin^{-1} \left(\frac{V(\sigma_1) - z}{V - z}\right)^{1/2} \mathrm{d}z + \cdots, \qquad V(\sigma_1) \leqq V. \tag{39}$$

Formula (39) shows that $s_3(V)$ is not unique, since $w(V)$ can be chosen arbitrarily. For each choice of $w(V)$, $s_3(V)$ is determined uniquely in the range $V(\sigma_1) \leqq V$, up to the next local maximum of $V(s)$. The function $\mathrm{d}s_1/\mathrm{d}V$ in (37) is given by (10) in terms of $T(E)$. Let φ be a continuously differentiable function such that $\varphi(V) = 0$ for all $V \notin [V(\sigma_2), V(\sigma_1)]$ and

$$s_3' - \varphi' \leqq 0 \quad \text{and} \quad s_3 - \varphi' \geqq 0\,.$$

Then $s - \varphi$ is decreasing and $s - \varphi$ increasing. We obtain

$$w(V) = \big(s_3(V) - \varphi(V)\big) - \big(s_2(V) - \varphi(V)\big) = s_3(V) - s_2(V)\,.$$

There exists a non-denumerable set of different arcs $s_2 - \varphi$ and $s_3 - \varphi$ satisfying (39) (see also Fig. 1.2). ▮

Remark 1.2. Theorem 1.4 shows that the information content of an inverse problem is basic in applications, i.e., which internal parameters of a system inaccessible to measurement can be uniquely determined by the external parameters of the system. In Theorem 1.4. the information content concerning V, which has maxima and minima, is too weak to uniquely determine $V(s)$. This result is a general principle in applications. Often not too much is known to uniquely determine the internal parameters (see also Sect. 1.5).

1.3. Some special differential equations of the first order

Using simple facts of calculus one can prove difficulties arising in inverse theory. The simplest ordinary differential equations defined on $\mathbb{R}$ have the full information content, especially can the basic theorems of measure theory and functional analysis be applied. In the plane $\mathbb{R}^2$ and the space $\mathbb{R}^3$ the same facts hold, but often the calculations are more difficult.

Let $f\colon \mathbb{R} \mapsto \mathbb{R}$ be a real-valued function satisfying $f(x) = 0$ for all $x \notin [a, b]$. We consider the simplest differential equation

$$u' = f\,, \qquad u(a) = 0\,. \tag{1}$$

Often one writes $u' = \partial u = \mathrm{d}u/\mathrm{d}x$. For $x \in\,]a, b[$ its solution is given by

$$u(x) = \int\limits_a^x f(y)\, \mathrm{d}y = \int H(x - y)\, f(y)\, \mathrm{d}y \tag{2}$$

where H is the *Heaviside function*, i.e.,

$$H(z) = 1 \quad \text{if} \quad z > 0, \qquad H(z) = 0 \quad \text{if} \quad z \leqq 0. \tag{3}$$

In (2) the second integral is extended over $\mathbb{R}$. That symbol is often used in mathematics. Now let f and u be defined on $[a, b]$, where f is continuous. Let us write (2) in the form $Af = u$. The inverse mapping A^{-1} can be determined by differentiating (2). Since f is continuous we get

$$f = A^{-1}u = u'. \tag{4}$$

In our considerations the linear space $\mathbf{C}([a, b])$ of all real-valued functions f defined and continuous on $[a, b]$ is applied. On $\mathbf{C}([a, b])$ the norm

$$\|f\|_{\mathbf{C}} := \max \{|f(x)|, x \in [a, b]\} \tag{5}$$

is introduced. From $|f(x)| \leqq \|f\|_{\mathbf{C}}$ it follows that $\|f_k\|_{\mathbf{C}} \to 0$ for $k \to \infty$ is equivalent to uniform convergence of the sequence $(f_k)_{k\in\mathbf{N}}$. Therefore every fundamental sequence $(f_k)_{k\in\mathbf{N}}$, i.e.,

$$\|f_{k'} - f_{k''}\|_{\mathbf{C}} < \varepsilon \quad \text{for all} \quad k', k'' > k_0(\varepsilon) \tag{6}$$

has a limit $f \in \mathbf{C}([a, b])$. The space $\mathbf{C}([a, b])$ is complete relative to the norm (5). A normed linear space which is complete is called a *Banach space*.

Further, on $\mathbf{C}([a, b])$ the following norm can be introduced:

$$\|f\|_{\mathbf{L}^p} := \left(\int_a^b |f(x)|^p \, \mathrm{d}x\right)^{1/p}, \qquad p \geqq 1. \tag{7}$$

The space $\mathbf{C}([a, b])$ is not complete relative to the norm (7). In integration theory it is proved that $\mathbf{C}([a, b])$ is dense in $\mathbf{L}^p([a, b])$ relative to (7) (Bauer [2], Bourbaki [1]). The spaces $\mathbf{L}^p([a, b])$ defined in Sect. 1.2 and Sect. 3.2 are also Banach spaces and fundamental in analysis.

Let $\mathbf{C}^1([a, b])$ be the linear space of all $f \in \mathbf{C}([a, b])$ having continuous derivatives of the first order in $]a, b[$ which can be continuously extended on the closure $[a, b]$. On $\mathbf{C}^1([a, b])$ the norm

$$\|f\|_{\mathbf{C}^1} := \max \{|f(x)|, x \in [a, b]\} + \max \{|f'(x)|, x \in [a, b]\} \tag{8}$$

is introduced. From $\|f_k\|_{\mathbf{C}^1} \to 0$ for $k \to \infty$ it follows that $f_k \to 0$ and $f'_k \to 0$ uniformly on $[a, b]$. The space $\mathbf{C}^1([a, b])$ is complete relative to the norm (8), also a Banach space. From (5) and (8) follows the inequality

$$\|f\|_{\mathbf{C}} \leqq \|f\|_{\mathbf{C}^1}. \tag{9}$$

Each sequence $(f_k)_{k\in\mathbf{N}}$ of functions $f_k \in \mathbf{C}^1([a, b])$, which converges relative to (8), also converges relative to (5). The convergence of (8) is finer than that of (5) but more difficult to realize in concrete applications. Often only the values of f can be measured like in computed tomography.

It is very easy to prove that the linear mapping

$$A: \mathbf{C}([a, b]) \mapsto \mathbf{C}([a, b])$$

defined in (2) satisfies the inequality

$$\|Af\|_{\mathbf{C}} \leqq M \|f\|_{\mathbf{C}} \tag{10}$$

which is equivalent to the continuity of A (DUNFORD and SCHWARTZ [1]). The same inequality holds with respect to (7), $p = 2$.

Using a counterexample one can prove that the inverse A^{-1} is not continuous. Let

$$u_k(x) = \frac{\sin kx}{k^{1/2}}, \qquad u_k'(x) = k^{1/2} \cos kx. \tag{11}$$

From

$$\|u_k\|_{\mathbf{C}} = k^{-1/2}, \qquad \|u_k'\|_{\mathbf{C}} = k^{1/2}, \qquad k \geqq k_0, \tag{12}$$

it follows that

$$\|u_k\|_{\mathbf{C}} \to 0 \quad \text{while} \quad \|A^{-1}u_k\|_{\mathbf{C}} = \|u_k'\|_{\mathbf{C}} \to \infty \quad \text{for} \quad k \to \infty.$$

If

$$A^{-1}: \mathbf{C}([a, b]) \mapsto \mathbf{C}([a, b])$$

is continuous relative to (5), the inequality

$$\|A^{-1}u_k\|_{\mathbf{C}} \leqq M_0 \|u_k\|_{\mathbf{C}}$$

must hold (see (10)). Therefore A^{-1} is discontinuous. Now let $a = 0$ and $b = 2$. Using the norm (7) we get for $p = 2$

$$\|u_k\|_{\mathbf{L}^2} = \pi k^{-1/2} \quad \text{while} \quad \|u_k'\|_{\mathbf{L}^2} = \pi k^{1/2}. \tag{13}$$

The mapping A^{-1} is also discontinuous with respect to (7). But with respect to the norm (8) the inverse A^{-1} becomes continuous since $\|u'\|_{\mathbf{C}} \leqq \|u\|_{\mathbf{C}^1}$. The same fact holds for the inverse mapping A^{-1} defined in (1.2.31). The integral operator contains derivatives of the first order. In applications often $\|u\|_{\mathbf{C}}$ or $\|u\|_{\mathbf{L}^2}$ can be measured only. Then A^{-1} is discontinuous, a difficult problem to be solved (BERTERO [1], GROETSCH [1], HOFMANN [1], IVANOV et al. [1], TIKHONOV and ARSENIN [1]).

Remark 1.3. In applications only the value u or certain functionals (integrals) of u can be measured (see computed tomography). That is one reason why in applications a lack of information exists in general. From this fact it follows that the inverse A^{-1} does not depend continuously on u. Additional information is necessary coming from the experiment.

If $b < x$ from (3) it follows that ($f(x) = 0$ for $b < x$)

$$u(x) = \int H(x - y) f(y) \, dy = \int_a^b f(y) \, dy = \text{const}. \tag{14}$$

From integration theory it is well known that different f can give the same value c, also in the case $f \geqq 0$. Therefore additional conditions on f are necessary. Let $f = hx$. Then

$$\int_a^b f(y)\,dy = \frac{h}{2}(b^2 - a^2) = c. \tag{15}$$

From this equation h can be uniquely determined.

If f is integrable the function u defined in (2) is continuous. The function u has more properties, namely u is absolutely continuous. In this case one has to apply the space $\mathbf{L}^1([a, b])$. The first who studied such properties was H. Lebesgue in 1904.

Let $\mathbf{C}_0^\infty(\mathbb{R})$ be the linear space of all continuous functions $\varphi\colon \mathbb{R} \mapsto \mathbb{R}$ with derivatives of arbitrary order and compact support $\operatorname{supp}\varphi$, where $\operatorname{supp}\varphi$ is the closure of

$$\{x\colon \varphi(x) \neq 0\}. \tag{16}$$

A set $M \subset \mathbb{R}$ is compact if and only if M is a bounded and closed set. A special φ is given by (see Fig. 1.6), $n = 1$,

$$\varphi_\varepsilon(x) := \begin{cases} \dfrac{c}{\varepsilon^n}\exp\left(-\dfrac{\varepsilon^2}{\varepsilon^2 - |x|^2}\right) & \text{if} \quad |x| < \varepsilon, \\ 0 & \text{if} \quad |x| \geqq \varepsilon. \end{cases} \tag{17}$$

Here c is to be determined such that

$$c \int_{|x|<1} \exp\left(-\frac{1}{1 - |x|^2}\right) dx = 1.$$

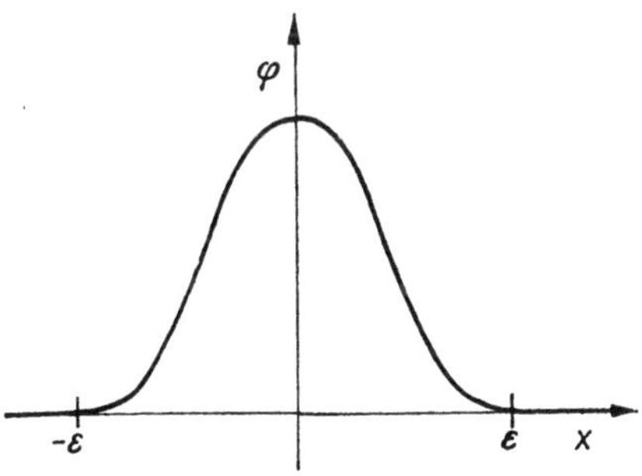

Fig. 1.6

Further $\varphi \in \mathbf{C}_0^\infty(\mathbb{R})$ can be constructed by considering the *convolution integral* $\varphi_\varepsilon * h$, where $h\colon \mathbb{R} \mapsto \mathbb{R}$ is a bounded integrable function vanishing outside a compact set. Here

$$(\varphi_\varepsilon * h)(x) := \int \varphi_\varepsilon(x - y)\,h(y)\,dy. \tag{18}$$

The integral is extended over $\mathbb{R}$. We remark that the $\varphi \in C_0^\infty(\mathbb{R})$, $\varphi \neq 0$, are not analytic on the whole $\mathbb{R}$. A function $\varphi: \mathbb{R} \mapsto \mathbb{R}$ is called *analytic* if it can be represented as a power series. The function $\varphi \in C_0^\infty(\mathbb{R})$ is zero outside a bounded set of $\mathbb{R}$. From this fact follows that an analytic φ must be identically zero.

For $\varphi \in C_0^\infty(\mathbb{R})$ we get

$$\int H(x-y)\,\varphi'(y)\,\mathrm{d}y = \int_{-\infty}^{x} \varphi'(y)\,\mathrm{d}y = \varphi(x). \tag{19}$$

Let $G(x, y) := H(x-y)$ and $P(\partial)\,\varphi := \varphi'$. Using these abbreviations we obtain (see Fig. 1.7)

$$\int G(x, y)\,P(\partial)\,\varphi(y)\,\mathrm{d}y = \varphi(x) \tag{20}$$

for all $\varphi \in C_0^\infty(\mathbb{R})$ and all $x \in \mathbb{R}$.

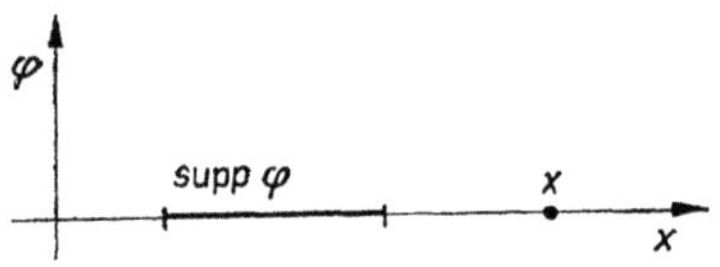

Fig. 1.7

The function $G: \mathbb{R} \times \mathbb{R} \mapsto \mathbb{R}$ is a so-called *fundamental solution* of the simplest differential operator

$$P(\partial)\,\varphi := \varphi'. \tag{21}$$

Relation (20) holds for each linear differential operator $P(x, \partial)\,u$ defined on the n-dimensional Euclidean space $\mathbb{R}^n$, $n \geqq 1$, having a locally integrable fundamental solution G. For $x \notin \operatorname{supp} \varphi$ the integral (20) is zero. From this fact follows the non-uniqueness of the inverse source problem, i.e., the determination of $f = P(x, \partial)\,\varphi$ by the values of the integral (20) outside $\operatorname{supp} \varphi$ is not possible. Further information about the density f must be known to uniquely determine f (see (15)).

We now consider the linear differential equation

$$u' - q(x)\,u = 0, \qquad u(a) = u_0. \tag{22}$$

The symbol $q(x)$ means that the function q depends on x. The solution u of (22) is given by

$$u(x) = u_0 \exp\left(\int_a^x q(y)\,\mathrm{d}y\right). \tag{23}$$

If $q(x) = q_0 = \text{const}$ we get

$$u(x) = u_0 \exp\big(q_0(x-a)\big). \tag{24}$$

In applications q is to be determined by measurement. If $q = q_0$ and $u(x^0) = g(x^0)$ is known, for fixed $x^0 \neq a$ we get

$$q_0 = \frac{1}{x^0 - a} \log \frac{g(x^0)}{u_0}. \tag{25}$$

From (24) and (25) it follows that the relationship between $u(x^0)$ and q_0 is non-linear and small changes of $u(x^0)$ can give arbitrary changes of q_0 if x^0 is near a.

It is well known that the initial-value problem (22) is uniquely determined by a and $u(a)$.

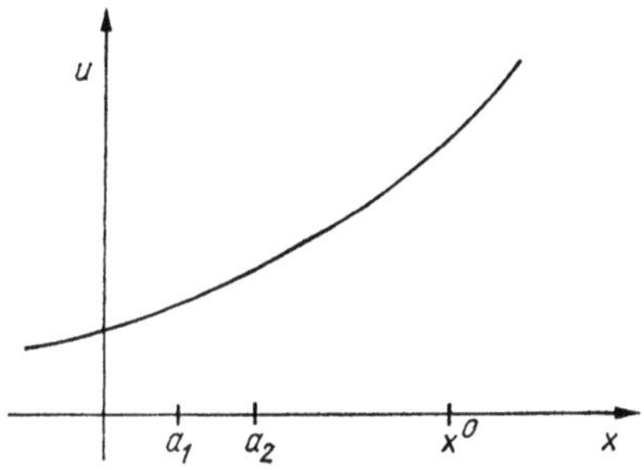

Fig. 1.8

Let the solution of (22) be known at x^0. A further inverse problem consists in determining a and $u(a)$ from $u(x^0)$. This special problem has infinitely many solutions (see Fig. 1.8). Often x is the time-coordinate. Additional information (coming from the experiment) is necessary to uniquely determine a and $u(a)$.

In the following we shall prove that the function $q(x)$ of (22) cannot be uniquely determined by the values of u outside an open set $]a_1, b_1[\subset [a, b]$. Let $\varphi\colon \mathbb{R} \mapsto \mathbb{R}$ be an integrable function satisfying

$$\operatorname{supp} \varphi \subset \,]a_1, b_1[\quad \text{and} \quad \int \varphi(x)\, \mathrm{d}x = 0.$$

The solution v of

$$v' - \big(q(x) + \varphi(x)\big)\, v = 0, \qquad v(a) = u_0, \tag{26}$$

is given by

$$v(x) = u_0 \exp\left(\int_a^x \big(q(y) + \varphi(y)\big)\, \mathrm{d}y\right). \tag{27}$$

If $a \leqq x \leqq a_1$ we get $u(x) = v(x)$. The same relation holds if $b_1 \leqq x$ because $\int \varphi(x)\, \mathrm{d}x = 0$. The function φ cannot be determined by measurements on the intervals $[a, a_1]$ and $[b_1, b]$. Let $q = 0$ and $u(0) = 1$. Then

$$v(x) = \exp\left(\int_0^x \varphi(y)\, \mathrm{d}y\right). \tag{28}$$

For $x \notin]a_1, b_1[$ we get $v(x) = u(x) = 1$. From (28) it follows that

$$\int_0^x \varphi(y)\, dy = \log v(x) \tag{29}$$

and

$$\varphi(x) = \frac{v'(x)}{v(x)}.$$

It is not possible to uniquely determine φ by finitely many measurements. May be that in inverse theory it is more useful to determine averages like

$$c_k := \int_{a_k}^{a_{k+1}} \varphi(y)\, dy, \qquad a_1 < a_2 < \cdots < a_N. \tag{30}$$

Let $\varrho : \mathbb{R} \mapsto \mathbb{R}$ be a function with first continuous derivative satisfying $\varrho(x) = 1$ outside $]a_1, b_1[$, $\varrho(x) > 0$ otherwise. Then

$$\big(\varrho(x)\, u\big)' = \varrho(x)\, u' + \varrho'(x)\, u =: P_\varrho(x, \partial)\, u. \tag{31}$$

Outside $]a_1, b_1[$ we get $P_\varrho(x, \partial)\, u = u'$, on $]a_1, b_1[$ the operator $P_\varrho(x, \partial)\, u$ depends on the special function ϱ. This construction is also possible in $\mathbb{R}^n$ and for nonlinear operators $P(x, \partial)\, u$. A special ϱ can be constructed in the following way:

$$\varrho(x) := 1 + \varphi_\varepsilon(x - x^0).$$

1.4. Differential equations of the second order

The one-dimensional Poisson equation has the form

$$-u'' = f, \tag{1}$$

where $f : \mathbb{R} \mapsto \mathbb{R}$ is a continuous or, in a more general way, an integrable function. A fundamental solution $G_L : \mathbb{R} \times \mathbb{R} \mapsto \overline{\mathbb{R}}$ of (1) is defined by

$$\int G_L(x, y)\, P(\partial)\, \varphi(y)\, dy = \varphi(x) \tag{2}$$

for all $\varphi \in C_0^\infty(\mathbb{R})$ and all $x \in \mathbb{R}$. Here $P(\partial)\, \varphi := -\varphi''$ and

$$G_L(x, y) = -(x - y)\, H(x - y) + c_1(x - y) + c_2. \tag{3}$$

Integrating by parts one can prove relation (2). For simplicity we use the Green function G_L^Ω relative to the open interval $\Omega =]0, 1[\subset \mathbb{R}$. The constants c_1 and c_2 are to be determined such that

$$G_L^\Omega(0, y) = G_L^\Omega(1, y) = 0 \quad \text{for all} \quad y \in]0, 1[.$$

We get

$$G_L^\Omega(x, y) = \begin{cases} x(1-y) & \text{if} \quad 0 \leqq x \leqq y, \\ (1-x)\, y & \text{if} \quad y \leqq x \leqq 1. \end{cases} \tag{4}$$

The boundary-value problem $-u'' = f$, $u(0) = u(1) = 0$, has the solution

$$u(x) = \int_0^1 G_L^\Omega(x, y)\, f(y)\, \mathrm{d}y.$$

Let $\varphi \in \mathbf{C}_0^\infty(\mathbb{R})$ with $\operatorname{supp} \varphi \subset\,]a, b[$, $f = P(\partial)\, \varphi$ and $x \notin\,]a, b[$. From (2) it follows that

$$\int_a^b G_L(x, y)\, f(y)\, \mathrm{d}y = 0 \quad \text{for all} \quad x \notin\,]a, b[\,. \tag{5}$$

Equation (5) proves the non-uniqueness of the inverse source problem. This relation also holds in $\mathbb{R}^n$, $n \geqq 2$. In this case G_L is the *logarithmic kernel* ($n = 2$) and the *Newtonian kernel* for $n \geqq 3$ (see (2.1.18) and (2.1.19)). Let

$$Af(x) := \int_a^b G_L(x, y)\, f(y)\, \mathrm{d}y, \tag{6}$$

where $y \in [a, b]$, $x \notin\,]a, b[$, $f \in \mathbf{C}([a, b])$. Then A is a mapping from $\mathbf{C}([a, b])$ into $\mathbf{C}(\mathbb{R} \setminus\,]a, b[)$. The kernel $\ker A$ of A is defined by all $f \in \mathbf{C}([a, b])$ satisfying $Af = 0$. From (5) it follows that special $f \in \ker A$ have the form $f = P(\partial)\varphi$, $\operatorname{supp} \varphi \subset [a, b]$. The operator defined in (6) has a too large kernel. This fact is typical for fundamental solutions of differential equations, where in the preceding case G_L is defined on the product space $(\mathbb{R} \setminus\,]a, b[) \times [a, b]$. In the case of the gravitational field the closed interval $[a, b]$ is to be replaced by the closure of a bounded open set $\Omega \subset \mathbb{R}^3$.

In the following we shall prove, that the coefficients of a linear differential operator of second order defined outside an open set $\Omega = \,]a, b[$ cannot be uniquely continued onto $[a, b]$. Such continuation problems often arise in applications in order to determine internal parameters of a physical system inaccessible to measurement (see formula (3.1.20), Remark 3.2 and Theorem 5.9). For simplicity let

$$P_0(\partial)\, u := -u''. \tag{7}$$

The fundamental system of $P_0(\partial)\, u = 0$ consists of

$$u_1 = 1 \quad \text{and} \quad u_2 = x. \tag{8}$$

Let $\varphi \in \mathbf{C}_0^\infty(\mathbb{R})$ with $\operatorname{supp} \varphi \subset\,]a, b[$. We consider the new fundamental system defined by

$$v_1 := 1 + \varphi, \qquad v_2 := x + \varphi. \tag{9}$$

Outside $]a, b[$ the relations $u_1 = v_1$ and $u_2 = v_2$ hold. Using the Wronskian

$$P_\varphi(x, \partial)\, u := \begin{vmatrix} u'' & u' & u \\ v_1'' & v_1' & v_1 \\ v_2'' & v_2' & v_2 \end{vmatrix}$$

we get

$$P_\varphi(x, \partial)\, u = \big(1 + g_2(x)\big)\, u'' + g_1(x)\, u' + g_0(x)\, u\,, \tag{10}$$

where $g_2(x) = g_1(x) = g_0(x) = 0$ outside $]a, b[$. One can prove that $\big(1 + g_2(x)\big)' = -g_1(x)$. Different φ give different $P_\varphi(x, \partial)\, u$ on $]a, b[$. Without additional conditions on g_2, g_1, g_0 the continuation onto $]a, b[$ is not unique. If one requires that the g_j are analytic functions then $\varphi = 0$. In this case the coefficients are uniquely determined, where $g_2 = g_1 = g_0 = 0$.

We have proved in the case $P(\partial)\, u = u'$ and $P(\partial)\, u = -u''$

Theorem 1.5. *Let $P(x, \partial)\, u$ be a linear differential operator, G a fundamental solution of $P(x, \partial)\, u$. Then the inverse source problem cannot be uniquely solved. Further, if the coefficients of $P(x, \partial)\, u$ are known only outside an open set $]a_1, b_1[\, \subset [a, b]$, then there exist non-denumerably many continuations of the coefficients of the $P(x, \partial)\, u$ defined on $[a, b] \setminus\,]a_1, b_1[$ onto $[a_1, b_1]$.*

Remark 1.4. In general the results of Theorem 1.5 also hold in $\mathbb{R}^n$ relative to linear partial differential equations and also relative to nonlinear partial differential equations if there are no additional conditions.

Remark 1.5. In Theorem 3.9 we shall prove the following: If the solution u is known on the whole space then the source term is uniquely determined. In Theorem 8.4 we shall prove that in this case the coefficients of $P(x, \partial)\, u = 0$ are determined up to a multiplicative function (Bliedtner and Hansen [1], Bukhgeim [1]).

Remark 1.6. In general the solution u of a boundary-value or initial-value problem of a fixed ordinary differential operator $P(x, \partial)\, u$ can be approximated, because the solution u is represented by a compact integral operator, by a polygon or by finite elements (see Fig. 1.9), while for inverse problems the

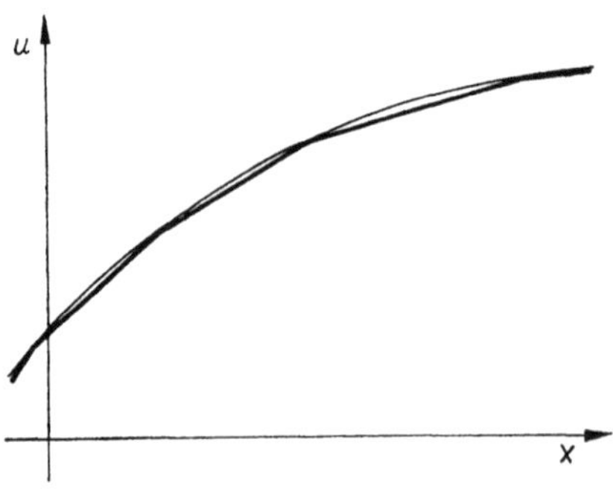

Fig. 1.9

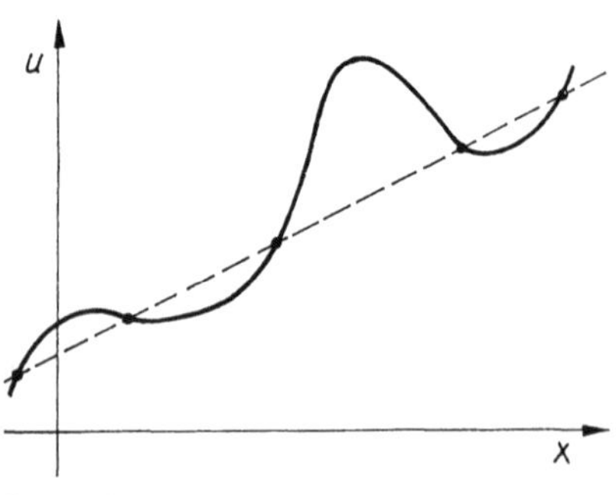

Fig. 1.10

information content often is too weak. In the last case the physical laws between the points, at which the solution is measured, must be known. In Fig. 1.10 one sees the result of Theorem 1.5 without any information between the m points. In geophysics and medicine it is very difficult to solve inverse problems. One must have a laboratory to try to get further information by experiments.

Remark 1.7. In applications the differential operator $P(x, \partial)\, u := \big(a(x)\, u'\big)' = 0$ is often important. From $\big(a(x)\, u'\big)' = 0$ it follows that $a(x) = c/u'(x)$. One immediately sees that a does not depend continuously on u, since $u \mapsto u'$ is discontinuous in the norm $\|\cdot\|_C$ or $\|\cdot\|_{L^2}$. If $a(x) = 1/\big(b(x) + \varphi(x)\big)$ one gets $u' = c\big(b(x) + \varphi(x)\big)$. Then the result of Sect. 1.3 holds.

Remark 1.8. If a depends on u we have to solve the equation $\big(a(u)\, u'\big)' = 0$. Then $u' = c/a(u)$ is to be solved. If $u = 1$ then $u' = 0$ and infinitely many $a(u)$ satisfy $\big(a(u)\, u'\big)' = 0$. In that case $a(u)$ cannot be uniquely determined by measurement of u at the boundary.

1.5. Inverse spectral problems

In 1929 V. Ambarzumian [1] proved the following uniqueness theorem. It is the first result in inverse spectral theory.

Theorem 1.6. *Let u be a solution of the differential equation*

$$P(x, \partial)\, u := \mu u'' - q(x)\, u + \lambda u = 0 \tag{1}$$

satisfying $u'(0) = u'(\pi) = 0$. Here $q(x)$ is a continuous function, μ a constant. Then the only differential equation

$$\varkappa u'' + \lambda u = 0, \qquad u'(0) = u'(\pi) = 0, \tag{2}$$

has eigenvalues $\lambda_k = \varkappa k^2$, $k = 1, 2, \ldots$.

In Barcilon and Turchetti [1], Barcilon [2, 3] the following result is proved.

Remark 1.9. If only finitely many eigenvalues are known then the results of Theorem 1.6 do not hold, i.e., there exist different operators having the same eigenvalues $\lambda_1, \ldots, \lambda_N$.

This book does not deal with inverse problems in spectral theory. Many books and papers have been written on this subject. See, for instance Bertero et al. [1], Chadan and Sabatier [1], Gladwell [1], Kravaris and Seinfeld [1], Levitan [1], Marchenko [1], Naats [1], Newton [1], Pao et al. [1], Pöschel and Trubowitz [1], Suzuki [1], Zakhariev and Suzko [1].

Especially nonlinear differential equations are studied by using results of inverse spectral problems (Ablowitz et al. [1], Dodd et al. [1], Zakharov et al. [1]).

1.6. Differential equations of higher order

Now let us consider some special ordinary differential equations of higher order. The functions u_j satisfying a system of differential equations of the second order can often be eleminated. Then they satisfy a differential equation of a higher order. In the preceding sections we have learned, that the information content of a single differential equation is often too weak. These differential equations correspond to mathematical models developed in the eighteenth and nineteenth centuries. Therefore the theory of differential equations of higher order is necessary (ANGER [4], SCHULZE and WILDENHAIN [1]).

The following differential equation is a characteristic representant of a differential equation of higher order. The solution of

$$u^{(k)} = \partial^k u := \frac{\mathrm{d}^k u}{\mathrm{d}x^k} = 0 \tag{1}$$

is the polynomial

$$u(x) = c_0 + c_1 x + \cdots + c_{k-1} x^{k-1}. \tag{2}$$

In applications often $k = 2m$. Then there are $2m$ constants $c_0, c_1, \ldots, c_{2m-1}$ to be determined. One immediately sees that the information content of (1) is greater than that of $u'' = 0$. We have the following possibilities to uniquely determine a solution:

$$u(a_1) = u_1, \qquad u(a_2) = u_2, \ldots, u(a_{2m}) = u_{2m} \tag{3}$$

where $a_1 < a_2 < \cdots < a_{2m}$, or

$$u(b_1) = u_1, u'(b_1) = u_2, \ldots, u(b_m) = u_{2m-1}, u'(b_m) = u_{2m}. \tag{4}$$

The conditions in (4) follow from (3) by considering

$$a_2 \to a_1, a_4 \to a_3, \ldots, a_{2m} \to a_{2m-1},$$

where

$$b_1 = a_1, b_2 = a_3, \ldots, b_m = a_{2m-1}.$$

The initial-value problem

$$u(a) = u_1, u'(a) = u_2, \ldots, u^{(2m-1)}(a) = u_{2m} \tag{5}$$

is a special case of the last consideration for $a_{2m} \to a_1 = a$.

The eigenvalue problem

$$u^{(4)} + \lambda u = 0, \qquad u(a) = u'(a) = u(b) = u'(b) = 0 \tag{6}$$

is of special interest for inverse spectral problems (BARCILON [1]).

1.7. Consequences for the theory of inverse problems

In studying inverse problems we often face a lack of information as we have seen in the preceding sections. Therefore the internal parameters of the system inaccessible to measurement cannot be uniquely determined. Additional a priori information is necessary which must be verified by experiments. In the following we shall try to give some ideas for interpreting special inverse problems arising in science and technology.
At first the definition of a stable solution f of an equation of the first kind $Af = g$ (HADAMARD [1], TIKHONOV and ARSENIN [1], BAUMEISTER [1]).

Definition 1.1 (Well-posed problem). Let X and Y be two Banach spaces, or in a more general way, two metric spaces. Further, let $A: X \mapsto Y$ be a mapping from X into Y. Following J. Hadamard (1923) the equation $Af = g$ describing the corresponding problem is called *well-posed*, if:

(i) For every $g \in Y$ there exists at least one f satisfying $Af = g$ (existence).
(ii) The element f satisfying $Af = g$ is uniquely determined (uniqueness). Then f is denoted by $f = A^{-1}g$.
(iii) The solution f depends continuously on g (stability).

If one of these three definitions is not fulfilled, the problem is called *not well-posed*, *ill-posed* or *improperly posed*.

Remark 1.10. One can find similar definitions of well-posed problems more adapted to practical problems in ENGL et al. [1], HOFMANN [1], IVANOV et al. [1], NATTERER [1], SABATIER [2], TIKHONOV and ARSENIN [1] and BECK et al. [1] including data error. A completely developed theory exists especially in the case of systems of linear equations $Af = g$. See, for instance, NASHED [1] and ZIELKE [1].

Remark 1.11. In Sect. 1.2—Sect. 1.5 we have seen, that many equations of the first kind are ill-posed. Often, the solution f is not uniquely determined. Further, some inverse mappings A^{-1}, like in computed tomography, contain derivatives. In practice, scientists often can measure only the function g and not its derivatives. From these facts it follows that the problem under consideration is ill-posed. We note that in most cases exists a lack of information, which cannot be remedied by any mathematical trickery (Lanczos, see COLTON and KRESS [1]). From the point of view of the job market, inverse problems are those, which students normally face when they leave the university (see BOLT [1]).

In the following we give some notations used in mathematical analysis. Let $\mathbb{R}^n$ be the n-dimensional Euclidean space, $x = (x_1, x_2, \ldots, x_n)$, $y = (y_1, y_2, \ldots, y_n)$ points of the $\mathbb{R}^n$,

$$|x - y| := \big((x_1 - y_1)^2 + \cdots + (x_n - y_n)^2\big)^{1/2} \tag{1}$$

the distance between x and y,

$$\partial_j u := \frac{\partial u}{\partial x_j}, \qquad \partial^\alpha u = \partial_1^{\alpha_1} u \ldots \partial_n^{\alpha_n} u := \frac{\partial^{\alpha_1 + \cdots \alpha_n} u}{\partial x_1^{\alpha_1} \ldots \partial x_n^{\alpha_n}} \tag{2}$$

partial derivatives, where $\alpha = (\alpha_1, \alpha_2, \ldots, \alpha_n)$ is a multiindex of non-negative integers, $|\alpha| = \alpha_1 + \cdots + \alpha_n$,

$$P(x, \partial)\, u := \sum_{|\alpha| \leqq m} a_\alpha(x)\, \partial^\alpha u = f(x) \tag{3}$$

a linear partial differential equation. The functions $a_\alpha(x)$ and $f(x)$ are given functions satisfying certain integrability conditions or other conditions. If the coefficients a_α are constant we write $P(\partial)\, u$. In applications a system of partial differential equations is often to be studied. Then $u = (u_1, \ldots, u_m)$ and $f = (f_1, \ldots, f_m)$ are vectors of functions. In science and technology the coefficients sometimes depend on u and certain derivatives of u. Then the problems considered become nonlinear.

The results of experience accumulated by scientists in their observation and interpretation of nature are often used in constructing mathematical models. These are frequently described by means of very complex systems of partial differential equations or other equations. One of the principal fields of activity of mathematicians and physicists has been the study of existence and uniqueness theorems for boundary-value and initial-value problems for individual differential equations (direct problems), as well as their stochastic treatment and numerical solution. The coefficients or the source term of the differential equations, which are internal parameters of the system described, are assumed given and the solution obtained is compared with measured values. When the relationship between the internal parameters and the measured values is one-to-one, this procedure yields correct results in general. In order to be able to make concrete statements about a system of partial differential equations one must often vastly simplify the system. This leads to a situation similar to that which arises when the system studied is modeled too coarsely, or the differential equations are discretized. In the case of the earth frequently only the gravitational field is taken into consideration. Sometimes one has too few or no measured values. The model simplification leads to intrinsic ambiguities, i.e., different values of the internal physical parameters can give the same external parameters (measured values). This situation is typical of complex questions in, for example, geophysics, solid-state physics, medicine, etc., and this renders the treatment of such problems considerably more difficult. For these reasons the interpretation of measured values must be undertaken as an inverse problem.

From the point of view of mathematics involved as well as that of information content, as the first step one can introduce and study the set of all solutions in so far as this is possible at the present state of mathematical development. For example, with the help of a scale the mass of a body, but not its physical composition or its geometric shape, can be determined. These steps are followed in every applied problem and are familiar from our everyday experience. By this systematic approach new solutions are revealed, which can be of significance in applications (see Sect. 5.4). Such systematic investigations constitute the great strength of mathematics. They often point toward new directions of technical development.

The answer to the following question is basic for applications. Let us notice that the integral over a function is a real number. Such quantities are sometimes called *functionals.*

A main problem in mathematical analysis. *Which functionals of the external parameters of a system are to be measured on which subset of* $\mathbb{R}^n$ *in order that certain coefficients or the source term or functionals of these coefficients or the source term inaccessible to measurement can be determined in a stable and unique manner?*

In the following chapters we shall see that this problem has been studied only in special cases (see also BALZER [1]).

2. Actual differential equations in applications

In this century science and technology have made great progress, with a highly developed measurement technique as one outcome. In geodesy the distance between earth and moon can be determined up to some centimetres, in optics special layers can be prepared having the thickness of atoms. For this reason mathematical modeling becomes more and more important to describe the new phenomena. As mentioned in Sect. 1.7 the new models are systems of partial differential equations or other equations. In many applications, several coefficients (*identification problems*) or the source term (*inverse source problem*) of a system of differential equations inaccessible to measurement are to be determined. In a formal way one can describe the relationship between the internal parameters f and the external parameters g of a system by equations of the first kind $Af = g$. The classical direct problem consists in determining Af, where the internal parameters (coefficients, source term) are assumed to be given. In order to determine $f = A^{-1}g$ the direct problem must be solved. In practice the solution is known only at a finite number of points, i.e., outside an open set. From this fact follows, as we have seen in Chapter 1, the existence of infinitely many solutions of the coefficients or the source term. To uniquely determine the coefficients or the source term additional a priori information is necessary which must come from the applied problem. One possibility is to study so-called one-dimensional inverse problems. In that case the coefficients or the source term depend only on one coordinate. It is very important to study inverse problems at first for special differential equations, especially the Laplace equation, the heat equation, and the wave equation. In this case solutions and potentials can often be calculated. These investigations give preliminary information about solutions of inverse problems and provide an intuitive feeling for other inverse problems.

2.1. The Laplace equation and elliptic differential equations of second order

In many applications such as gravimetry, electrostatics, hydrodynamics, and so on, the *Laplace equation*

$$\Delta u := \sum_{k=1}^{n} \partial_k^2 u = \sum_{k=1}^{n} \frac{\partial^2 u}{\partial x_k^2} = 0 \tag{1}$$

and the Poisson equation

$$-\Delta u = f \tag{2}$$

play an important role. One often writes

$$\Delta u = \operatorname{div} \operatorname{grad} u. \tag{3}$$

The solutions of (1) are called *harmonic functions.*

Following B. Riemann (1859), C. Neumann (1868) and A. Einstein (1916) a corrected time-depending gravitational potential u satisfies the *biharmonic equation* (Treder [1], v. Borzeskowski [1])

$$\Delta^2 u := \Delta(\Delta u) = 0. \tag{4}$$

Let $\Omega \subset \mathbb{R}^n$ be a bounded domain. The *Dirichlet problem* consists in determining a function $u \in \mathbf{C}^2(\Omega)$ satisfying

$$\Delta u = 0 \quad \text{in } \Omega, \qquad \lim_{x \to y} u(x) = h(y), \qquad y \in \partial\Omega, \tag{5}$$

where $h \in \mathbf{C}(\partial\Omega)$ is known. Often, the *Neumann problem*

$$\Delta u = 0 \quad \text{in } \Omega, \qquad \partial_n u(y) = h(y), \qquad y \in \partial\Omega, \tag{6}$$

is studied. Here $\partial_n u = \partial u/\partial n$ denotes the exterior normal derivative. The normal derivative can be defined only for smooth boundaries. Further, the following boundary value problems are studied:

$$\Delta u = 0 \quad \text{in } \Omega, \qquad \sum_{|\alpha| \leqq m} b_\alpha(y) \lim_{x \to y} \partial^\alpha u(x) = h(y), \qquad y \in \partial\Omega, \tag{7}$$

where $b_\alpha : \partial\Omega \mapsto \mathbb{R}$ and h are known functions. Problem (6) is a special case of (7).

Let $\Omega \subset \mathbb{R}^n$ be a bounded domain with smooth boundary $\partial\Omega$. The solution u of the Dirichlet problem is given by the *Poisson integral*

$$u(x) = \int_{\partial\Omega} h(y)\, P(x, y)\, \mathrm{d}S(y). \tag{8}$$

In the case of an open ball $B(x^0, R) \subset \mathbb{R}^n$ the *Poisson kernel* $P(x, y)$ has the form:

$$n = 2: P(x, y) = \frac{1}{2\pi R} \frac{R^2 - |x - x^0|^2}{|x - y|^2} \tag{9}$$

$$n = 3: P(x, y) = \frac{1}{4\pi R} \frac{R^2 - |x - x^0|^2}{|x - y|^3} \tag{10}$$

$$n \geqq 4: P(x, y) = \frac{1}{V_n R} \frac{R^2 - |x - x^0|^2}{|x - y|^n} \tag{11}$$

where V_n is the surface content of a ball with radius 1. Further, if $u : \Omega \mapsto \mathbb{R}$ is a harmonic function u is uniquely determined by its values on an arbitrary ball $B(x^0, R) \subset \Omega$. This theorem is proved by analytic continuation (O. G. KELLOGG [1], C. MIRANDA [1]). In an axiomatic approach of the Dirichlet problem one can find the continuation property as axioms (G_1) and (G_2) on page 48. In that approach the properties of harmonic functions find their counterpart in the axioms (BS_1), (BS_2), (BS_3).

The *Cauchy problem*

$$\Delta u = 0 \quad \text{in } \Omega, \qquad u(y) = h_0(y), \qquad \partial_n u(y) = h_1(y), \qquad y \in \partial\Omega, \tag{12}$$

sometimes studied for smooth boundaries is important for some problems in applications (HÖRMANDER [2], KNOPS [1], MIZOHATA [1], PAYNE [1], TRÈVES [1]). In general the solution u of (12) does not continuously depend on h_0, h_1. This special Cauchy problem is a not well-posed problem. J. Hadamard gave the following counterexample. Let

$$\Omega := \{x \in \mathbb{R}^2 : x_2 > 0\}, \qquad h_0(y) = \frac{\cos ky_1}{k}, \qquad h_1(y) = 0. \tag{13}$$

The solution u is given by

$$u(x) = \frac{(\cos kx_1) \cosh kx_2}{k}. \tag{14}$$

While h_0 and h_1 tend to zero for $k \to \infty$ the solution u tends to infinity. The Dirichlet problem, the Neumann problem, the Cauchy problem, and all the other initial- and boundary-value problems are so-called direct problems.

Now some important remarks concerning the *instability of the Dirichlet problem* by varying the boundary $\partial\Omega$. Let $\Omega \subset \mathbb{R}^n$ be a domain such that $\partial\Omega$ is compact. The solution u of the Dirichlet problem relative to Ω and boundary-values h is denoted by H_h (LANDKOF [1]). Suppose that $(\Omega_k)_{k\in\mathbb{N}}$ is an increasing sequence of domains $\Omega_k \subset \Omega$. Further, let $h_0 : \mathbb{R}^n \mapsto \mathbb{R}$ be a continuation of $h : \partial\Omega \mapsto \mathbb{R}$, $H_{h_0}^k$ the solution of the corresponding Dirichlet problem. If

$$\Omega = \bigcup_{k=1}^{\infty} \Omega_k \quad \text{we get} \quad H_h(x) = \lim_{k\to\infty} H_{h_0}^k(x). \tag{15}$$

The convergence is uniform on any compact set $K \subset \Omega$. Here variations of the boundary 'into' the domain will in general be essentially distinct from variations 'away' from the domain. In the last case we have

$$\Omega_k \supset \Omega \quad \text{and} \quad \bigcap_{k=1}^{\infty} \Omega_k = \Omega. \tag{16}$$

If a domain satisfies (16) and the relation (15) then the Dirichlet problem is called *stable* in Ω. In 1937 M. V. Keldych found a counterexample for the instability of the Dirichlet problem (LANDKOF [1]). These remarks are important for the numerical solution of the Dirichlet problem and other boundary-value problems of elliptic differential equations of second order.

The *fundamental solution* $G_L\colon \mathbb{R}^n \times \mathbb{R}^n \mapsto \overline{\mathbb{R}}$ of the Laplace equation is given in the form:

$$n = 1\colon G_L(x, y) := -(x-y)\,H(x-y) + c_1(x-y) + c_2, \tag{17}$$

$$n = 2\colon G_L(x, y) := \frac{1}{2\pi} \log \frac{1}{|x-y|}, \quad x \neq y, \quad G_L(x, x) = +\infty, \tag{18}$$

$$n \geqq 3\colon G_L(x, y) := \frac{1}{(n-2)\,V_n} \frac{1}{|x-y|^{n-2}}, \quad x \neq y, \quad G_L(x, x) = +\infty. \tag{19}$$

In the Euclidean space $\mathbb{R}^3$, where our physics holds, the fundamental solution G_L (*Newtonian kernel*) has the form

$$G_L(x, y) = \frac{1}{4\pi} \frac{1}{|x-y|}. \tag{20}$$

Using the formula

$$u(x) = \int_\Omega G_L(x, y)\,\big(-\Delta u(y)\big)\,dy + \int_{\partial\Omega} \left(u(y) \frac{\partial G_L(x, y)}{\partial n_y} - \frac{\partial u(y)}{\partial n_y} G_L(x, y)\right) dS(y) \tag{21}$$

one finds that G_L satisfies the equation

$$\varphi(x) = \int G_L(x, y)\,\big(-\Delta\varphi(y)\big)\,dy \tag{22}$$

for all $\varphi \in C_0^\infty(\mathbb{R}^n)$ and all $x \in \mathbb{R}^n$. Here $C_0^\infty(\mathbb{R}^n)$ denoted the linear space of functions $\varphi\colon \mathbb{R}^n \mapsto \mathbb{R}$ having compact support and derivatives of arbitrary order. In order to prove (22) we have to choose a domain $\Omega \supset \operatorname{supp}\varphi$. Then $\varphi(y) \equiv 0$ for all $y \notin \operatorname{supp}\varphi$.

Equation (22) can formally be written

$$-\Delta_y G_L(x, y) = \delta_x, \tag{23}$$

where δ_x is the Dirac measure at x and Δ_y means that one has to differentiate with respect to y. If u is a solution of $\Delta u = 0$ then $G_L + u$ is also a fundamental solution, which can be proved by using integration by parts. The *Dirac measure* δ_x is defined by

$$\delta_x(f) := f(x), \qquad f \in C_0^\infty(\mathbb{R}^n). \tag{24}$$

The *potential* $G_L\mu_\varrho$ *due to a volume distribution* ϱ is defined at $x \in \mathbb{R}^n$ by

$$G_L\mu_\varrho(x) := \int_\Omega G_L(x, y)\,\varrho(y)\,dy. \tag{25}$$

Example 2.1 (Unknown source term). The main inverse source problem relative to the Laplace equation is that of finding a (positive) mass distribution (measure) μ concentrated on the closure $\bar{\Omega} = \Omega \cup \partial\Omega$ of a given domain $\Omega \subset \mathbb{R}^n$ by the value g of its potential $G_L\mu$ on $\partial\Omega$, i.e.,

$$G_L\mu(x) = g(x) \quad \text{for all} \quad x \in \partial\Omega. \tag{26}$$

Example 2.2 (Inverse source problems in gravimetry). In gravimetry a mass distribution μ concentrated on $\bar{\Omega}$ is to be determined by $g = (g_1, g_2, g_3)$, the gradient of $G_L\mu$ on $\partial\Omega$, i.e.,

$$\operatorname{grad} G_L\mu(x) = g(x) = \big(g_1(x), g_2(x), g_3(x)\big) \quad \text{for all} \quad x \in \partial\Omega. \tag{27}$$

Further, along the orbit of a satellite only the quantity

$$|\operatorname{grad} G_L\mu(x)| = g(x) \tag{28}$$

can be measured (BACKUS [1]). In that case μ is to be determined by measurement (28).

Inverse source problems can also be formulated relative to the conditions (6), (7), (12). In geophysics magnetic fields (see BERDICHEVSKIJ and ZHDANOV [1], RÖSLER [1], ZHDANOV [1], ZIDAROV [1]), in electrocardiology electrostatic fields (see FRANZONE [1], PLONSEY [1]) and in electroencephalographic research (see NUNEZ [1], PLONSEY [1]) the *potential* W *of a double layer* is of special interest. It is defined at x by

$$W(x) := \int_{\partial\Omega} b(y) \frac{\partial G_L(x, y)}{\partial n_y} \, dS(y). \tag{29}$$

Example 2.3 (Unknown source term of W). The double layer (density) b is to be determined by measurement of W on the boundary $\partial\Omega$.

One can find further inverse problems relative to elliptic differential equations in Sect. 2.2. If these differential equations are time-independent one gets inverse problems for elliptic differential equations.

If a volume distribution $\mu = \mu_\varrho$ satisfies (26) or (27) or (28), and if ϱ is sufficiently smooth then the potential $u = G_L\mu_\varrho$ satisfies

$$-\Delta u = \varrho. \tag{30}$$

In applications it is often more useful to formulate problems relative to the potential $G_L\mu$, which is a potential energy. In that case one does not need derivatives. Energy integrals are the base to apply Hilbert space methods and Sobolev space methods to partial differential equations (HELMS [1], SCHULZE and WILDENHAIN [1]). Sometimes equation (30) is to strong. Considering special measures μ_ϱ having a density ϱ one cannot often find all solutions of a given problem. For instance, the measures (3.1.7) and (3.1.8) do not satisfy equation (30) in the classical sense (derivatives of a real function). The same considerations hold relative to the heat equation, the equation for carrier transport in semi-conductors, and so on. In the case of the equations for the carrier transport (see (2.4.7)) only measures having a density are considered.

Using our ideas to study the set of all measures whose potentials satisfy certain conditions on $\partial\Omega$ it is possible to find new solutions for applications (see Sect. 5.2–5.5).

Let us describe two further important problems relative to the Laplace equation to be solved. The determination of the earth's surface and its gravitational field leads to boundary-value problems (if the data are assumed to be continuously given on the earth's surface) and inverse problems (in the realistic case of discrete data). For this special inverse problem the points $y^k \in \bar{\Omega}_0$ and weights a_k are to be determined such that the potential $G_L\mu$, where $\mu = \sum a_k\delta_{yk}$, represents the earth's potential $G_L\mu_0$ sufficiently well (Barthelmes [1]). One can find some remarks concerning the special measures μ in Remark 4.9. *Molodensky's boundary-value problem* formulated in 1945 is the following (Moritz [1]): Assume that at each point of the topographic surface $S \subset \mathbb{R}^3$ we know both the gravity potential u and the gravity vector g. The surface S is unknown and is to be determined from the values u and g (a free boundary-value problem relative to the Laplace equation). The quantities are both measurable: Gravity measurements give the magnitude of the vector g and astronomical observations provide its direction. The potential u is given to an additive constant which is to be determined otherwise by a combination of spirit level and gravity measurements. In Prilepko [1] it is shown that Molodensky's problem is equivalent to an inverse source problem relative to the Laplace equation.

The principle of Molodensky's problem and its solution is simple. Let $\Omega = B(x^0, R) \subset \mathbb{R}^3$ be a ball and μ_{ϱ_0} the volume distribution with constant density $\varrho = \varrho_0$. Then $G_L\mu_{\varrho_0}$ on $\partial B(x, R)$ has the form

$$u = \frac{M}{R}, \qquad |g| = |\text{grad } u| = \frac{M}{R^2}.$$

Then we get

$$|g| = \frac{u}{R} \quad \text{and} \quad R = \frac{u}{|g|}.$$

This special problem gives relations between geometrical parameters and physical parameters. These considerations hold for an arbitrary domain. Assume first the surface S and the potential u on it to be given. Then the function u can be computed everywhere outside S (by a solution of the Dirichlet problem well known in potential theory) and the gravity vector g is obtained by

$$g = \text{grad } u \tag{31}$$

outside and on S. Symbolically we thus have on S

$$g = F(S, u). \tag{32}$$

Assume that it is possible to solve this equation for S,

$$S = F^{-1}(g, u). \tag{33}$$

Formally, this solves Molodensky's problem. The principle is simple, but the details are extremely intricate. The symbol, of course, denotes a complicated nonlinear operator, and the solution of (32) with respect to S involves a very 'hard' implicit function theorem of nonlinear analysis. Only recently, Molodensky's problem was attacked in this way by L. Hörmander [1], who succeeded in proving existence and uniqueness of the solution, but only for very smooth surfaces. For surfaces as rugged as the actual earth's topography the solution is still outstanding, and what is more, it is not even clear whether it can be attacked by means of existing mathematical tools (Grafarend and Rapp [1], Günther [1], Holota [1], W. Keller [1, 2], Moritz [2]).

The *Helmholtz equation*

$$\Delta u + V(x)\, u = 0 \tag{34}$$

is important in wave propagation (Bleistein [1], Colton and Kress [1], Newton [1], Ramm [2], Tikhonov and Samarski [1], Trèves [1], Zakharov and Suzko [1]) and in quantum theory. Often it is written in the form

$$\Delta u + V(x)\, u = \lambda u \tag{35}$$

to study spectral properties of this equation. Let $n = 1$ and $V(x) = \varkappa^2$, $\varkappa = 1, 2, \ldots$ Then the fundamental system of

$$u'' + \varkappa^2 u = 0 \tag{36}$$

is

$$u_1 = \sin \varkappa x\,, \qquad u_2 = \cos \varkappa x\,. \tag{37}$$

The solution $u_1 = \sin \varkappa x$ proves the existence of eigenfunctions, i.e., from $u(x) = 0$ for $x = 0$ and $x = \pi$ it follows that $u(x) \not\equiv 0$ on $[0, \pi]$. Further, from $u(x) \geqq 0$ on $\partial\Omega$ does not necessarily follow that $u(x) \geqq 0$ in Ω. But the maximum principle holds for sufficiently small Ω (Miranda [1]). In quantum physics equation (35) is studied on the whole $\mathbb{R}^n$ while in other areas of physics like mechanics the influence of the boundary $\partial\Omega$ of a given domain $\Omega \subset \mathbb{R}^n$ plays an important role for equations (34) and (35).

In the case $V(x) = \varkappa^2$ exists a concrete *fundamental solution* of (34). We give it in the following form:

$$n = 1\colon G_S(x, y) := \frac{\sin \varkappa(x - y)}{\varkappa}\, H(x - y)\,, \qquad \varkappa > 0\,, \tag{38}$$

$$n = 2\colon G_S(x, y) := \frac{1}{2\pi}\, N_0(\varkappa\, |x - y|)\,, \qquad x \neq y\,, \qquad G_S(x, x) = -\infty\,, \tag{39}$$

$$n = 3\colon G_S(x, y) := -\frac{1}{4\pi} \frac{\cos \varkappa\, |x - y|}{|x - y|}\,, \qquad x \neq y\,, \qquad G_S(x, x) = -\infty\,. \tag{40}$$

Here N_0 denotes the Neumann function of order zero. The direct problem consists in determining u satisfying a given boundary condition, where $V(x)$ is fixed. The inverse problem is that of finding $V(x)$ from all eigenvalues (see Examples 2.19 and 2.20). Further, inverse source problems can be formulated relative to the inhomogeneous Helmholtz equation.

Further, the *elliptic differential equation*

$$P(x, \partial)\, u := -\operatorname{div}\big(a(x) \operatorname{grad} u\big) + b(x)\, u = -\sum_{j,k=1}^{n} \partial_j\big(a_{jk}(x)\, \partial_k u\big) + b(x)\, u = f \tag{41}$$

is considered in inverse theory. Here the a_{jk} defined on $\Omega \subset \mathbb{R}^n$ are integrable functions satisfying $a_{jk} = a_{kj}$. Further $a(x) = \big(a_{jk}(x)\big)$ is a matrix satisfying

$$c_0 \sum_{j=1}^{n} \xi_j^2 \leqq \sum_{j,k=1}^{n} a_{jk}(x)\, \xi_j \xi_k \leqq c_1 \sum_{j=1}^{n} \xi_j^2, \qquad c_0 > 0. \tag{42}$$

In the case of an *isotropic medium*, i.e., with the coefficient not depending on the direction, we have

$$a_{11}(x) = \cdots = a_{nn}(x), \qquad a_{jk}(x) = 0 \quad \text{if} \quad j \neq k. \tag{43}$$

If $b(x) = 0$ then (41) reads

$$-\operatorname{div}\big(a(x) \operatorname{grad} u\big) = -\sum_{j=1}^{n} \partial_j\big(a(x)\, \partial_j u\big) = f. \tag{44}$$

In electricity u is the *potential*, $a(x)$ the *conductivity*, in heat conduction u is the *temperature*, $a(x)$ the *thermal conductivity* and in fluid through porous media u is the *pressure*, $a(x)$ the *permeability*. In applications the coefficients often depend on the solution u or its gradient. Then equation (41) is nonlinear (see also Examples 2.4–2.5).

Let $P(x, \partial)\, u = 0$ be the differential equation defined in (41) with $f = 0$. If $\Omega \subset \mathbb{R}^n$ is sufficiently small then a maximum principle holds for their solutions u. From this follows that the solution of the Dirichlet problem can be represented in the form (C. Miranda [1])

$$u(x) = H_h(x) = \int_{\partial\Omega} h(y)\, P(x, y)\, \mathrm{d}S(y). \tag{45}$$

Under certain conditions for the coefficients a_{jk} an analytic continuation property holds (C. Miranda [1], de la Pradelle [1]). The properties of the solution u of $P(x, \partial)\, u = 0$ find their counterpart in the axioms (G_1), (G_2), (BS_1), (BS_2) and (BS_3).

Let us formulate some basic facts concerning the solutions of an elliptic differential equation of second order, especially for the Laplace equation. Of cource, different methods give the same solutions. But there exists a uniform axiomatic approach for studying properties of the solutions of elliptic differential equations of second order. Further, one sees structural properties and relations to other fields like probability theory (Bauer [2], Bliedtner and Hansen [1], Doob [1]), Korovkin approximation (Bauer [1]), and so on.

Inverse problems relative to a given domain can be studied from a systematic point of view only by using structural properties of the corresponding mathematical model. These investigations concern the relations between the external parameters (measured values) and the internal parameters. Modern potential theory is that theory in partial differential equations having highest level. This theory holds relative to linear elliptic and parabolic differential equations of second order. Further, these investigations are possible in more abstract spaces.

In the following open sets are denoted by U and V. Now some properties of solutions of an elliptic differential equation of second order, for which the maximum principle holds.

Property I. Let $B(x^0, R) \subset \mathbb{R}^n$ be a ball. For every function $h \in C(\partial B(x^0, R))$ there exists a uniquely determined 'harmonic' function H_h which extends h continuously. Furthermore, $h \geqq 0$ implies $H_h \geqq 0$.

Property II (Harnack's second convergence theorem). Let $(h_k)_{k\in\mathbb{N}}$ be an increasing sequence of 'harmonic' functions h_k defined on a domain $\Omega \subset \mathbb{R}^n$. Then the upper envelope $\sup\limits_k h_k$, where $\left(\sup\limits_k h_k\right)(x) := \sup\limits_k h_k(x)$, is either harmonic or the constant $+\infty$ on Ω.

Further, the 'harmonic' functions have the following properties. To every open subset $U \neq \emptyset$ of $\mathbb{R}^n$ the set $\mathcal{H}(U)$ of all functions harmonic on U forms a linear subspace of $C(U)$ satisfying:

(G_1) Restriction of a harmonic function $h \in \mathcal{H}(U)$ to an open subset $V \neq \emptyset$ of U yields a function $h|_V \in \mathcal{H}(V)$.

(G_2) For every family $(U_k)_{k\in I}$ of open sets $U_k \neq \emptyset$ in $\mathbb{R}^n$ and every function h defined on $U = \bigcup\limits_{k\in I} U_k$ one has $h \in \mathcal{H}(U)$ provided that $h|_{U_k} \in \mathcal{H}(U_k)$ holds for all $k \in I$.

The harmonic functions form a so-called *sheaf* $\mathcal{H}$ of linear spaces $\mathcal{H}(U)$ of real-valued continuous functions (Bauer [2, 3], Bliedtner and Hansen [1]).

In axiomatic potential theory developed by M. Brelot (1958) and H. Bauer (1962) a sheaf $\mathcal{H}$ is given, i.e., to every open subset $U \neq \emptyset$ of the topological space X (for instance, the $\mathbb{R}^n$) is associated a linear subspace $\mathcal{H}(U)$ of $C(U)$, the linear space of real-valued continuous functions on U, in such a way that (G_1) and (G_2) are satisfied. The elements of $\mathcal{H}(U)$ will be called *harmonic functions* on U. The sheaf $\mathcal{H}$ is called *harmonic structure* on X if the following three axioms are satisfied:

(BS_1) $\mathcal{H}$ is *not degenerate*, i.e., for every point $x \in X$ there exists a harmonic function $h \in \mathcal{H}(U)$ defined in a neighbourhood U of x such that $h(y) > 0$ for all $y \in U$.

(BS_2) Base Axiom: the topology of X has a *base consisting of regular sets*, i.e., each open set $U \subset X$ can be represented as union of regular sets. Here a set $V \subset X$ is called *regular* (with respect to $\mathcal{H}$), if the Dirichlet problem is solvable without restrictions in the following sense: for every function $h \in C(\partial V)$ there exists a uniquely determined harmonic function $H_h \in \mathcal{H}(V)$ which extends h continuously. Furthermore, $h \geqq 0$ implies $H_h \geqq 0$.

(BS_3) Brelot's Convergence Axiom K_B: for every increasing sequence $(h_k)_{k\in\mathbb{N}}$ of harmonic function h_k defined on an open set U, one has $\sup_k h_k(x) < \infty$ for some $x \in \Omega \Rightarrow \sup_k h_k \in \mathcal{H}(U)$.

In the case of parabolic equations one has a weaker convergence axiom:

(BS_3') Doob's Convergence Axiom K_D: for every increasing sequence of harmonic functions h_k defined on an open set U, one has $\left\{x \in U : \sup_k h_k(x) < \infty\right\}$ dense in $U \Rightarrow \sup_k h_k \in H_U$.

In the case that $\mathcal{H}$ satisfies (BS_1), (BS_2), (BS_3) the sheaf $\mathcal{H}$ is called a *Brelot harmonic structure* and $(X, \mathcal{H})$ a *Brelot space*. If it satisfies (BS_1), (BS_2), (BS_3') the sheaf $\mathcal{H}$ is called a *Bauer harmonic structure* and $(X, \mathcal{H})$ a *Bauer space*. In the axiomatic approach no derivatives are used.

The solution u of the differential equation defined in (41) in the case $b < 0$, $f = 0$, form a Brelot space. If $b \geqq 0$, $f = 0$ the axioms locally hold. For a Brelot space or a Bauer space a potential theory can be developed (BLIEDTNER and HANSEN [1], HANSEN and HUEBER [1], STURM [1]).

2.2. Heat equation and parabolic differential equations of second order

The *heat equation* is defined by

$$P(\partial)\, u := \partial_t u - \Delta u = f. \tag{1}$$

For a cylinder $\Omega = \Omega_0 \times]0, T[$, where $\Omega_0 \subset \mathbb{R}^n$ is a bounded domain, initial-boundary value problems are studied. For $t = 0$ the initial condition

$$u(x, 0) = u(x, t)|_{t=0} = \lim_{t\to 0} u(x, t) = \Phi(x) \tag{2}$$

is prescribed, and on $\partial\Omega_0 \times [0, T[$ one can consider one of the boundary conditions (2.1.5), (2.1.6), (2.1.7). Such problems are called direct problems. It is possible to study a Dirichlet problem relative to an arbitrary domain $\Omega \subset \mathbb{R}^{n+1}$

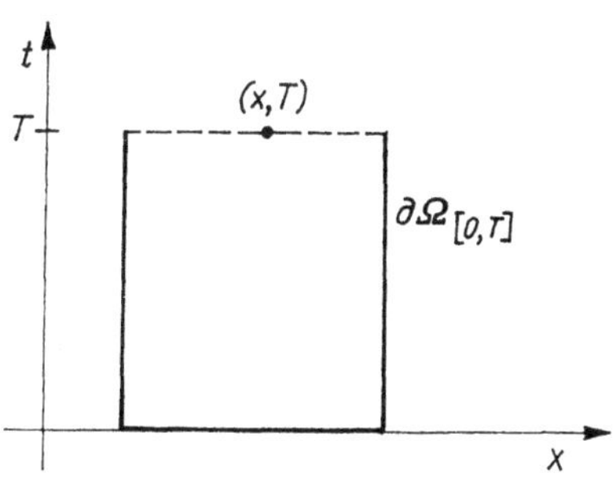

Fig. 2.1

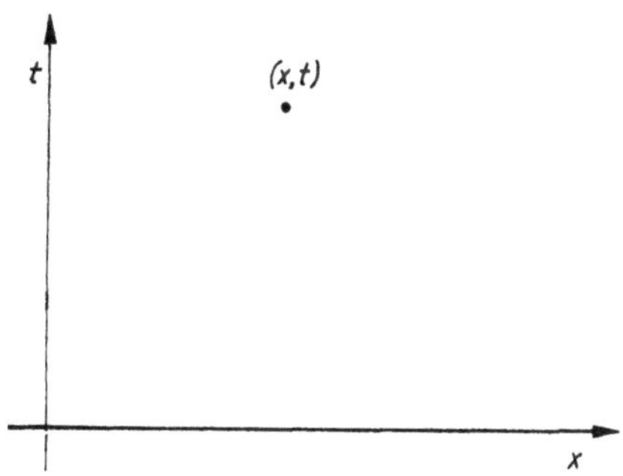

Fig. 2.2

(BLIEDTNER and HANSEN [1]). We have sketched first ideas for such a method on pages 48 and 49. One cannot describe boundary conditions on the whole Euclidean boundary. For instance, for the special cylinder $\Omega_0 \times]0, T[$ the values on $\Omega_0 \times \{T\}$ are determined by the values on

$$\partial\Omega_{[0,T]} := (\Omega_0 \times \{0\}) \cup (\partial\Omega_0 \times [0, T]). \tag{3}$$

If

$$\Omega_T := \{(x, t) \in \mathbb{R}^{n+1}: t > T = 0\}$$

then (see Fig. 2.2)

$$\partial\Omega_T := \{(x, t) \in \mathbb{R}^{n+1}: t = T = 0\}. \tag{4}$$

For the heat equation a *maximum principle* holds. Let the derivatives $\partial_t u$, $\partial_k^2 u$, $k = 1, \ldots, n$, be continuous on Ω and u continuous on $\bar{\Omega}$. Then

$$\min_{(y,\tau)\in\partial\Omega_{[0,t]}} u(y, \tau) \leqq u(x, t) \leqq \max_{(y,\tau)\in\partial\Omega_{[0,t]}} u(y, \tau). \tag{5}$$

From (5) the uniqueness of the initial-boundary value problem (Dirichlet problem) with boundary values on $\partial\Omega_0 \times]0, T[$ follows.

For equation (1) there exists a *fundamental solution* G_{H}, which is defined for $\tilde{x} = (x, t)$, $\tilde{y} = (y, \tau)$ by

$$G_{\mathrm{H}}(\tilde{x}, \tilde{y}) = \begin{cases} \dfrac{1}{\big(\pi(t-\tau)\big)^{n/2}} \exp\left(-\dfrac{\sum\limits_{k=1}^{n} (x_k - y_k)^2}{4(t-\tau)}\right) & \text{if } t > \tau, \\ 0 \quad \text{otherwise.} \end{cases} \tag{6}$$

For the domain Ω_T an initial-value problem (Cauchy problem) is studied. In that case a solution of (1) is to be determined such that

$$u(x, 0) = \lim_{t\to 0} u(x, t) = \Phi(x). \tag{7}$$

Its solution u is given by

$$u(x, t) = \int_{\mathbb{R}^n} \Phi(y)\, G_{\mathrm{H}}(x, t; y, 0)\, \mathrm{d}y, \qquad dy = \mathrm{d}y_1 \ldots \mathrm{d}y_n. \tag{8}$$

One can introduce and study potentials relative to the kernel (6). Formula (8) is a special potential.

The solution of the initial-boundary value problem

$$\partial_t u - \partial_x^2 u = 0, \qquad 0 < x < \infty, \qquad 0 < t, \tag{9}$$

$$u(x, 0) = \Phi(x), \qquad 0 < x < \infty, \tag{10}$$

$$u(0, t) = h(t), \qquad 0 < t, \tag{11}$$

is given by

$$u(x, t) = \int_0^\infty \Phi(y)\, G_{\mathrm{H}}^{\Omega}(x, t; y, 0)\, \mathrm{d}y + 2\int_0^t \partial_x G_{\mathrm{H}}(x, t; 0, t^*)\, h(t^*)\, \mathrm{d}t^*, \tag{12}$$

where G_H is the fundamental solution (6) and

$$G_H^\Omega(x, t; y, 0) = \frac{1}{2\sqrt{\pi t}}\left(\exp\left(-\frac{(x-y)^2}{4t}\right) - \exp\left(-\frac{(x+y)^2}{4t}\right)\right) \tag{13}$$

the Green function of the problem considered.

Further, the initial-boundary value problem

$$\partial_t u - \partial_x^2 u = 0, \qquad 0 < x < 1, \qquad 0 < t, \tag{14}$$

$$u(x, 0) = \Phi(x), \qquad 0 < x < 1, \tag{15}$$

$$u(0, t) = u(1, t) = 0, \qquad 0 < t, \tag{16}$$

has the form

$$u(x, t) = \int_0^1 G_H^\Omega(x, t; y, 0)\, \Phi(y)\, dy, \tag{17}$$

where

$$G_H^\Omega(x, t; y, 0) := 2 \sum_{k=1}^{\infty} (\sin k\pi x)(\sin k\pi y) \exp(-\pi^2 k^2 t) \tag{18}$$

is the Green function of this problem. In $\mathbb{R}^n$, $n \geqq 2$, similar initial-boundary value problems can be studied (Tikhonov and Samarski [1]).

A linear *parabolic equation* is defined by

$$P(x, \partial)\, u := c(x)\, \partial_t u - \operatorname{div}\big(a(x) \operatorname{grad} u\big) + b(x)\, u = f, \tag{19}$$

where $a(x)$ and $b(x)$ were defined in (2.1.41). One can find a theory for such equations in Alessandrini [1], Friedman [1], Trèves [1], Tikhonov and Samarski [1].

Let us formulate some inverse problems.

Example 2.4 (Unknown source term). The inverse problem amounts to finding the unknown source term F in the equation

$$\partial_t u - \Delta u = F(u), \qquad x \in \Omega_0, \qquad 0 < t, \tag{20}$$

from Φ, h and g in

$$u(x, 0) = \Phi(x), \qquad x \in \Omega_0, \tag{21}$$

$$u(x, t) = h(x, t), \qquad x \in \partial\Omega_0, \qquad 0 < t, \tag{22}$$

$$\partial_n u(x^0, t) = g(t), \qquad x^0 \in \partial\Omega_0, \qquad 0 < t. \tag{23}$$

We refer to the functions F, Φ and h as the proper data in this problem and the problem of finding u from $\{F, \Phi, h\}$ will be referred to as the direct problem (Pilant and Rundell [1], Lorenzi and Paparoni [1]).

Example 2.5 (Unknown coefficients in divergence form).

$$c(u)\,\partial_t u = \operatorname{div}\big(k(u)\operatorname{grad} u\big), \qquad x \in \Omega_0, \qquad 0 < t, \tag{24}$$

$$u(x, 0) = \Phi(x), \qquad x \in \Omega_0, \tag{25}$$

$$k(u)\,\partial_{\mathrm{n}} u = h(x, t), \qquad x \in \partial\Omega_0, \qquad 0 < t, \tag{26}$$

$$u(x^k, t) = g_k(t), \qquad x^k \in \partial\Omega_0, \qquad k = 1, \dots, n, \qquad 0 < t. \tag{27}$$

The functions Φ and h are the proper data in this problem, while the overspecification consists of the m functions $g_1, \dots, g_m$ corresponding to measurements of u at m distinct (and fixed) points on the boundary. The inverse problem consists then in finding the unknown coefficients $c(u)$, $k(u)$, given Φ, h and $g_1, \dots, g_m$ (DuChateau [1, 2], Dümmel and Handrock-Meyer [1], Handrock-Meyer [1], Hornung [1], Muzylev [1]).

Let $c(u) \equiv 0$. We define T by (Cannon and DuChateau [1])

$$v(x) = Tu(x) := \int\limits_{u(x^0)}^{u(x)} k(r)\,\mathrm{d}r. \tag{28}$$

Then

$$\partial_j v = (T'u)\,\partial_j u = k(u)\,\partial_j u$$

and

$$\partial_j^2 v = \partial_j\big(k(u)\,\partial_j u\big).$$

From (24) and (28) it follows that

$$\Delta v = \operatorname{div}\big(k(u)\operatorname{grad} u\big) \tag{29}$$

and

$$k(u)\,\partial_{\mathrm{n}} u = \partial_{\mathrm{n}} v. \tag{30}$$

If $k \geqq k_0 > 0$ the mapping $u \mapsto Tu$ is one-to-one. Its inverse is denoted by

$$u = S_k v. \tag{31}$$

The effect is thus to transform the undetermined coefficient problem into the determination of a known equation satisfying overspecified data depending linearly on an unknown function.

In Sect. 7.1, we shall prove uniqueness and existence for a special unknown coefficient problem. In that case the boundary is smooth. Let us prove two counterexamples (Pilant and Rundell [3]). In some cases it is possible to give the exact degree of non-uniqueness.

Example 2.6 (Counterexample). Let $B(0, 1)$ be the unit disk, $\partial\Omega_1$ and $\partial\Omega_2$ are the upper and lower semicircular boundaries. The coefficient $k(u)$ is to be determined by

$$\operatorname{div}\big(k(u)\operatorname{grad} u\big) = 0, \qquad x \in B(0, 1), \tag{32}$$

$$k(u)\,\partial_{\mathrm{n}_y} u(y) = 0, \qquad y \in \partial\Omega_2, \tag{33}$$

$$u(y) = 1, \qquad y \in \partial\Omega_1, \tag{34}$$

and the overspecified data

$$u(y) = g(y) = 1, \qquad y \in \partial\Omega_2. \tag{35}$$

For any $k(u)$, the function $u(x) \equiv 1$ is a solution of the above problem, and coincides with g on $\partial\Omega_2$.

Example 2.7 (Counterexample). Let $\Omega := \{(x_1, x_2): 0 < x_1 < 1, 0 < x_2 < 1\}$ with line $x_1 = 0$ corresponding to $\partial\Omega_2$, and let $u(x)$ satisfy

$$\operatorname{div}\big(k(u) \operatorname{grad} u\big) = 0, \qquad x \in \Omega, \tag{36}$$

$$\partial_{x_1} u(0, x_2) = \partial_{x_1} u(1, x_2) = 0, \tag{37}$$

$$u(x_1, 0) = 0, \qquad u(x_1, 1) = 1, \tag{38}$$

and the overspecified data

$$u(0, x_2) = g(x_2), \qquad g(0) = 0, \qquad g(1) = 1. \tag{39}$$

From (28), (29) and (30) it follows that

$$\Delta v(x) = 0 \quad \text{in} \quad \Omega, \tag{40}$$

$$\partial_{x_1} v(0, x_2) = 0, \qquad \partial_{x_1} v(1, x_2) = 0, \tag{41}$$

$$v(x_1, 0) = 0, \qquad v(x_1, 1) = \int_0^1 k(r)\, \mathrm{d}r, \tag{42}$$

$$v(0, x_2) = \int_0^{g(x_2)} k(r)\, \mathrm{d}r. \tag{43}$$

The solution of (40), (41) and (42) is given by

$$v(x) = x_2 \int_0^1 k(r)\, \mathrm{d}r. \tag{44}$$

From (43) and (44) we get

$$\int_0^{g(x_2)} k(r)\, \mathrm{d}r = x_2 \int_0^1 k(r)\, \mathrm{d}r, \tag{45}$$

which has to be solved for $k(u)$. If

$$c = \int_0^1 k(r)\, \mathrm{d}r \quad \text{and} \quad \eta = g(x_2)$$

from (45) it follows that

$$\int_0^{\eta} k(r)\, \mathrm{d}r = c g^{-1}(\eta) \tag{46}$$

showing that k is determined up to a constant multiple. Differentiating (46) and setting $x_2 = 0$ one can uniquely determine c if $g'(0) \neq 0$ and $k(0)$ is arbitrarily chosen,

$$g'(0)\, k(0) = c. \tag{47}$$

The solution of (46) depends on one parameter. If $k(0) \neq 0$ from $g'(0) = 0$ it follows that $c = 0$ and there is no solution $k(u)$ to the problem.

Example 2.8 (Unknown coefficient).

$$\partial_t u - a(u)\, \Delta u = f(x, t), \qquad x \in \Omega_0, \qquad 0 < t, \tag{48}$$

$$u(x, 0) = \Phi(x), \qquad x \in \Omega_0, \tag{49}$$

$$u(x, t) = h(x, t), \qquad x \in \partial\Omega_0, \qquad 0 < t, \tag{50}$$

$$a\big(h(x^0, t)\big)\, \partial_{n_x} u(x^0, t) = g(t), \qquad x^0 \in \partial\Omega_0, \qquad 0 < t. \tag{51}$$

The functions f, Φ and h are the proper data in this problem, while the overspecification consists in the function g at x^0 (Beznoshtshenko [1]).

Example 2.9 (Counterexample). Let u satisfy

$$\partial_t u - a\partial_x^2 u = -3x^2, \qquad -\infty < x < +\infty, \qquad 0 < t, \tag{52}$$

$$u(x, 0) = \frac{x^4}{12}, \qquad -\infty < x < +\infty, \tag{53}$$

$$u(0, t) = g(t) = -2t^2, \qquad 0 < t. \tag{54}$$

If a is a constant then

$$u(x, t) = (a^2 - 3a)\, t^2 + ax^2 t + \frac{x^4}{12} - 3tx^2 \tag{55}$$

is the uniquely determined solution of (52), (53). Substituting (55) into (54) we obtain

$$(a^2 - 3a)\, t^2 = -2t^2$$

and $a_1 = 1$, $a_2 = 2$. At $x = 0$ and $t = 0$ we have grad $u(0, 0) = 0$ and $g'(0) = 0$. One can find further uniqueness and nonuniqueness results in Kravaris and Seinfeld [2]. See also Richter [1], [2].

Example 2.10 (Unknown boundary coefficient).

$$\partial_t u - \Delta u = 0, \qquad x \in \Omega_0, \qquad t > 0, \tag{56}$$

$$u(x, 0) = \Phi(x), \qquad x \in \Omega_0, \tag{57}$$

$$\partial_{n_x} u - p(u) = h(x, t), \qquad x \in \partial\Omega_0, \qquad t > 0, \tag{58}$$

$$u(x^0, t) = g(t), \qquad x^0 \in \partial\Omega_0, \qquad t > 0. \tag{59}$$

Here the proper data consists of Φ and h while g is the overspecified data amounting to a measurement of u at a fixed point x^0 on the boundary. The inverse problem then amounts to finding the unknown coefficient p which appears in the boundary condition from Φ, g and h.

Remark 2.1. In Example 2.3 through Example 2.10 additional conditions are necessary, for instance, F and k are analytic functions. Without such conditions an inverse problem often cannot be uniquely solved.

In order to study an inverse source problem for the heat equation like in Example 2.1 solutions of the *adjoint heat equation* play an important role. It is defined by

$$P^*(\partial)\, u := -\partial_t u - \Delta u = f(x, t). \tag{60}$$

In this case the boundary

$$\partial\Omega^*_{[T,0]} := (\Omega_0 \times \{T\}) \cup (\partial\Omega_0 \times [0, T]) \tag{61}$$

is of special interest (see Fig. 2.3). Especially, we shall consider the half-space

$$\Omega^*_T := \{(x, t) \in \mathbb{R}^{n+1} : t < T\} \tag{62}$$

with the boundary

$$\partial\Omega^*_T := \{(x, t) \in \mathbb{R}^{n+1} : t = T\}. \tag{63}$$

Now some inverse problems, where the diffusion coefficient $a(t)$ is changing with time. A physical example of such a problem arises from heat conduction in a material that is undergoing radioactive decay or damage. The thermal conductivity varies with the degree of decay (Cannon [2], Rossi Casta [1]).

Example 2.11 (Interior measurement for a finite cylinder). The inverse problem amounts to finding the thermal conductivity $a(t) > 0$ from

$$\partial_t u = a(t)\, \partial_x^2 u, \qquad 0 < x < 1, \qquad 0 < t, \tag{64}$$

$$u(x, 0) = \sin \pi x, \qquad 0 < x < 1, \tag{65}$$

$$u(0, t) = u(1, t) = 0, \qquad 0 < t, \tag{66}$$

$$u\left(\frac{1}{2}, t\right) = g(t), \qquad 0 < t. \tag{67}$$

The overspecified data is g from which $a(t)$ is to be determined.

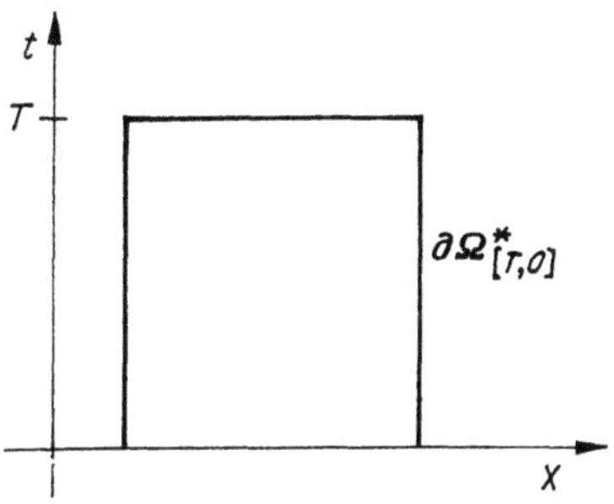

Fig. 2.3

Example 2.12 (Boundary-flux measurement for a finite cylinder). We consider the determination of $a(t) > 0$ and u such that the pair (a, u) satisfies

$$\partial_t u = a(t)\, \partial_x^2 u, \qquad 0 < x < 1, \qquad 0 < t, \tag{68}$$

$$u(x, 0) = \sin \pi x, \qquad 0 < x < 1, \tag{69}$$

$$u(0, t) = u(1, t) = 0, \qquad 0 < t, \tag{70}$$

$$-a(t)\, \partial_x u(0, t) = g(t), \qquad 0 < t. \tag{71}$$

Example 2.13 (Boundary-flux measurement for a semi-infinite conductor). The pair $a(t) > 0$ and u is to be determined such that (a, u) satisfies

$$\partial_t u = a(t)\, \partial_x^2 u, \qquad 0 < x < \infty, \qquad 0 < t, \tag{72}$$

$$u(x, 0) = 0, \qquad 0 < x < \infty, \tag{73}$$

$$u(0, t) = h_0, \qquad h_0 = \text{const}, \qquad 0 < t, \tag{74}$$

$$-a(t)\, \partial_x u(0, t) = g(t), \qquad 0 < t. \tag{75}$$

The overspecified data is g from which $a(t)$ is to be determined.

One further example is the estimation of the heating history experienced by a shuttle reentering the earth's atmosphere from space (see Fig. 2.4). Similar problems are to be solved in different areas of physics (Beck et al. [1]).

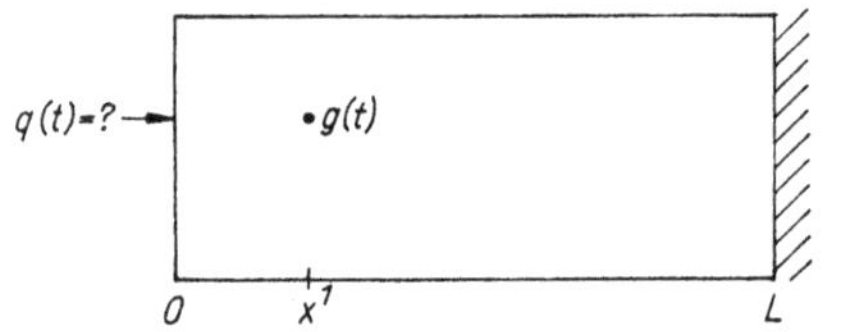

Fig. 2.4

Example 2.14 (Estimation of the surface heat flux). Let

$$\varrho(u)\, c(u)\, \partial_t u = \partial_x\big(k(u)\, \partial_x u\big) \tag{76}$$

be the heat equation. The coeffients k, the thermal conductivity, ϱ, the density, and c, the specific heat, are known functions of the temperature u. The temperature u satisfies

$$u(x, 0) = \Phi(x), \qquad 0 < x < L, \tag{77}$$

$$\partial_x u(L, t) = 0, \qquad 0 < t. \tag{78}$$

From

$$u(x^1, t) = g(t), \qquad 0 < t, \tag{79}$$

where $0 < x^1 < L$, the surface heat flux

$$q(t) = -k \partial_x u(0, t) \tag{80}$$

is to be determined. In $\mathbb{R}^n$ similar problems are to be solved.

Example 2.15 (Unknown source term). Let μ be a positive mass distribution concentrated on $\bar{\Omega} = \bar{\Omega}_0 \times [0, T]$,

$$G_{\mathrm{H}}\mu(\tilde{x}) := \int G_{\mathrm{H}}(\tilde{x}, \tilde{y}) \, \mathrm{d}\mu(\tilde{y}), \qquad \tilde{x} = (x, t), \qquad \tilde{y} = (y, t), \tag{81}$$

its potential. The inverse problem consists in determining the mass distribution μ by the values of $G_{\mathrm{H}}\mu$ on $\partial\Omega^*_{[T,0]}$, i.e.,

$$G_{\mathrm{H}}\mu(\tilde{x}) = h(\tilde{x}), \qquad \tilde{x} \in \partial\Omega^*_{[T,0]}. \tag{82}$$

Since G_{H} satisfies

$$\int G_{\mathrm{H}}(\tilde{x}, \tilde{y}) \, P(\partial) \, \varphi(\tilde{y}) \, \mathrm{d}\tilde{y} = \varphi(\tilde{x}), \qquad \mathrm{d}y = \mathrm{d}y_1 \ldots \mathrm{d}y_n \, \mathrm{d}y_{n+1}, \tag{83}$$

for all $\varphi \in \mathbf{C}_0^\infty(\mathbb{R}^{n+1})$ and all $\tilde{x} \in \mathbb{R}^{n+1}$, the inverse source problem (Example 2.15) cannot be uniquely solved without additional information. The right-hand side of (83) is zero if $x \notin \operatorname{supp} \varphi$.

Example 2.16 (Control problem in heat conduction). Let u be a solution of

$$\partial_t u = \Delta u \quad \text{in} \quad \Omega = \Omega_0 \times]0, T[\tag{84}$$

with

$$u(x, 0) = \Phi(x), \qquad x \in \Omega_0, \tag{85}$$

$$u(x, t) = 0 \quad \text{if} \quad (x, t) \in \partial\Omega_0 \times]0, T[. \tag{86}$$

Further, let h be a given function defined on $\Omega_0 \times \{T\}$. The initial-values $\Phi(x)$ are to be determined such that

$$h(x) = u(x, T) \tag{87}$$

(Lions and Duvaut [1], Zeidler [1]).

The paper of Aifantis [1] rationalizes diffusion in solids on the basis of a differential equation of balance expressing conservation of momentum for the diffusing species. The balance equation contains a tensor, modelling the stress supported by the diffusing species, and a diffusive force vector, modelling the exchange of momentum between the diffusing species and the species of the solid matrix. These two quantities, which are not identified in the classical diffusion interpretations, are the basic ingredients of the theory. The applicability of the theory to a variety of problems, ranging from metallurgy to polymer physics and geophysics, is illustrated in the paper of Aifantis. Relative to these equations inverse problems can be formulated and studied. But the solution of such inverse problems is more difficult. See also Michel [2], Shitzer and Eberhart [1] and system (2.4.6).

2.3. Wave equation and hyperbolic differential equations of second order

The simplest form of the wave equation is

$$\partial_t^2 u - c^2 \Delta u = f, \tag{1}$$

where c is a constant. We study the initial-boundary value problem for a cylinder $\Omega = \Omega_0 \times]0, T[\subset \mathbb{R}^{n+1}$. The solution u is to be determined by the initial-values

$$\begin{aligned} u(x, 0) &= \lim_{t\to 0} u(x, t) = \Phi_0(x), \\ u_t(x, 0) &= \lim_{t\to 0} \partial_t u(x, t) = \Phi_1(x), \qquad x \in \Omega_0, \end{aligned} \tag{2}$$

and the boundary-values

$$u(x, t) = h(x, t), \qquad x \in \partial\Omega_0 \times]0, T[, \qquad t > 0. \tag{3}$$

The more general form of (1) is the *linear hyperbolic equation* of second order (TIKHONOV and SAMARSKI [1], TRÈVES [1])

$$n^2(x)\, \partial_t^2 u - \operatorname{div}\big(a(x) \operatorname{grad} u\big) + b(x)\, u = f, \tag{4}$$

where $a(x)$ and $b(x)$ were defined in (2.1.41). For waves of the form $u(x, t) = u(x) \exp(-i\omega t)$ with frequency $\omega > 0$ we see that for (1) the space-dependent part u satisfies the Helmholtz equation (with $f = 0$)

$$\Delta u + k^2 u = 0, \qquad k^2 = \omega^2/c^2. \tag{5}$$

Similar results hold for equation (4).

Example 2.17 (Unknown coefficient)

$$n^2(x)\, \partial_t^2 u - \Delta u = f, \tag{6}$$

$$u(x, 0) = \partial_t u(x, 0) = 0, \qquad x \in \Omega_0, \tag{7}$$

$$\partial_n u(x, t) = h(x, t), \qquad x \in \partial\Omega_0, \qquad 0 < t, \tag{8}$$

$$u(x, t) = g(x, t), \qquad x \in \partial\Omega_0, \qquad 0 < t. \tag{9}$$

The conditions (7) and (8) are the proper data, while the overspecification consists in the function g defined on $\partial\Omega_0 \times]0, \infty[$. The inverse problem consists in finding the coefficient $n(x)$ for given f. If $n(x)$ is fixed one can attempt to determine the source term f (CASSINIS [1], FITZGIBBON [1], LAVRENT'EV et al. [1], ROMANOV [1], SABATIER [1]).

Example 2.18 (One-dimensional inverse problems). Let

$$\mu(x_3)\, \partial_t^2 u = \Delta u - q(x_3)\, u + f(t)\, \delta(x_1)\, \delta(x_2)\, \delta(x_3) \tag{10}$$

be a wave equation with

$$u(x, 0) = 0, \qquad u_t(x, 0) = 0, \qquad \partial_{x_3} u(x_1, x_2, 0) = 0, \tag{11}$$

where the $\delta(x_j)$ are Dirac measures at $x_j = 0$. The inverse problem consists in determining the coefficients μ and q depending only on the variable x_3 by measurement of u at the two points $(x_1^1, x_2^1, 0)$ and $(x_1^2, x_2^2, 0)$, i.e.,

$$u(x_1^1, x_2^1, 0, t) = g_1(t), \qquad u(x_1^2, x_2^2, 0, t) = g_2(t) \tag{12}$$

(LAVRENT'EV et al. [1], ROMANOV et al. [1]).

Remark 2.2. Instead of (7), (8) and (9) other conditions can be introduced. The main problem is to have sufficient information to uniquely solve the inverse problem. In practice (9) can be measured only at a finite number of points. In this case more information concerning the special problem considered is necessary. In general such additional information can be obtained only in laboratories using different experiments.

Problems for acoustic waves are governed by the Helmholtz equation (5). Let $\Omega \subset \mathbb{R}^3$ be a bounded domain with smooth boundary $\partial\Omega$ and u a solution of (5) satisfying the Sommerfeld radiation condition

$$\left\langle \frac{x}{|x|}, \operatorname{grad} u(x) \right\rangle - iku(x) = o\left(\frac{1}{|x|}\right) \quad \text{for} \quad x \to \infty, \tag{13}$$

where $\langle x, z \rangle := \sum_{j=1}^{n} x_j z_j$.

Then u has the asymptotic behaviour

$$u(x) = \frac{e^{ik|x|}}{|x|} F_0(\vartheta_1, \vartheta_2) + o\left(\frac{1}{|x|}\right). \tag{14}$$

The function F_0 is called the *far-field pattern* or *radiation pattern* of u. We denote by u^i a fixed plane wave solution of the Helmholtz equation, i.e.,

$$u^i(x) = e^{ik\langle x, a \rangle}.$$

Example 2.19 (Inverse problem in scattering theory). Let $u = u^i + u^s$ be a solution of the Helmholtz equation

$$\Delta u + k^2 u = 0$$

in $\mathbb{R}^3 \setminus \Omega$ such that u^s satisfies (13) and $u(x) = 0$ on $\partial\Omega$. From the knowledge of the far-field pattern of u^s determine Ω (COLTON and KRESS [1], RAMM [1]).

One can find further inverse problems in scattering theory in ALEKSEEV and CHEBOTARËV [1], BALTES [1, 2, 3], CHANDAN and SABATIER [1], FADDEEV [1], NEWTON [1], NOVIKOV and KHENKIN [1], PÖSCHEL and TRUBOWITZ [1], RAMM [2], ZAKHARIEV and SUSKO [1].

The propagation of seismic waves in an elastic anisotropic continuum in absence of body forces is governed by the elastodynamic equations

$$\varrho(x)\, \partial_{tt}^2 u_i = \sum_{j,k,l=1}^{3} \partial_j (c_{ijkl} \partial_k u_l), \qquad i = 1, 2, 3, \tag{15}$$

where $\varrho(x)$ is the density, u_i the (small) displacement from equilibrium, t the time, (c_{ijkl}) the elastic tensor satisfying the symmetry relationships

$$c_{ijkl} = c_{jikl} = c_{ijlk} = c_{klij}.$$

If the medium is isotropic the elastic tensor satisfies

$$c_{ijkl} = \lambda\delta_{ij}\delta_{kl} + \mu(\delta_{ik}\delta_{jl} + \delta_{il}\delta_{jk}), \tag{16}$$

where λ and μ are the Lamé parameters and

$$\delta_{ij} = 1 \quad \text{if} \quad i = j, \qquad \delta_{ij} = 0 \quad \text{if} \quad i \neq j.$$

In classical seismology, one usually assumes that the energy of the waves with high frequency travels as 'rays' according to the laws of geometrical optics. This is a rather drastic approximation to the theory since the effects of diffraction, the continuous reflection from velocity gradient, and the continuous conversion of waves are all neglected (see Example 1.2, CASSINIS [1], CAROLL and SANTOSA [1]). For the system (15) one-dimensional inverse problems have been studied (YAKHNO [1]).

Now we should like to write down some fundamental solutions of hyperbolic differential equations. By the transformation $\xi = x - ct$, $\eta = x + ct$ from (1) it follows that $v(\xi, \eta) := u(x_1(\xi, \eta), x_2(\xi, \eta))$ satisfies for $f = 0$

$$\frac{\partial^2 v}{\partial \xi\, \partial \eta} = 0. \tag{17}$$

More generally we consider

$$P(\partial)\, u := \frac{\partial^n u}{\partial x_1 \dots \partial x_n} = 0. \tag{18}$$

A fundamental solution of (18) has the form

$$G_{\mathrm{D}}(x, y) := H(x_1 - y_1)\, H(x_2 - y_2) \dots H(x_n - y_n). \tag{19}$$

In $\mathbb{R}^2$, i.e., $\tilde{x} = (x, t)$, $\tilde{y} = (y, \tau) \in \mathbb{R}^3$, we have

$$G_{\mathrm{W}}(x, y) = \begin{cases} \dfrac{1}{2\pi\sqrt{(t-\tau)^2 - |x - y|^2}} & \text{if} \quad t - \tau > |x - y|, \\ 0 \quad \text{otherwise}. \end{cases} \tag{20}$$

For $\tilde{x} = (x, t) \in \mathbb{R}^4$ see HÖRMANDER [2] and TRÈVES [1]. While for $x \in \mathbb{R}^2$ the fundamental solution is locally integrable this fact does not hold in $\mathbb{R}^n$ for $n \geqq 3$. In that case one has to apply the theory of distributions developed by S. L. Sobolev and L. Schwartz in the period 1937–1952 (see HÖRMANDER [2], TRÈVES [1]).

The *electromagnetic field* is governed by the Maxwell equations. Let $E = (E_1, E_2, E_3)$ be the electric field vector and $H = (H_1, H_2, H_3)$ the magnetic field vector. The Maxwell system reads

$$\begin{aligned} \operatorname{curl} H &= \varepsilon \partial_t E + \sigma E + j, \\ \operatorname{curl} E &= -\mu \partial_t H, \end{aligned} \tag{21}$$

where $j = (j_1, j_2, j_3)$ is the current density, $\varepsilon(x) = (\varepsilon_{jk}(x))$ a symmetric matrix of electric permeabilities, $\mu(x) = (\mu_{jk}(x))$ a symmetric matrix of magnetic permeabilities and $\sigma(x) = (\sigma_{jk}(x))$ a symmetric matrix of electric conductivities. Initial and boundary conditions are to be added to uniquely determine a solution. Special solutions of (21) satisfy the wave equation or the Helmholtz equation. It is of special interest to note that the bioelectric phenomena of an organism are also governed by the Maxwell equations (FRANZONE [1], NUNEZ [1], PLONSEY [1]). Further, in optics a lot of inverse problems are studied (BALTES [1, 2, 3], BERTERO et al. [1], COLTON [1], KAISER and KAISER [1]).

Since the coefficients of a differential equation cannot be uniquely determined at each point $x \in \Omega_0 \subset \mathbb{R}^3$ from measurements on the boundary $\partial\Omega_0$, one studies simpler problems in which the coefficients depend only on one variable. For instance, one requires in $\mathbb{R}^3$ that

$$E|_{t<0} \equiv 0, \qquad H|_{t<0} \equiv 0,$$

i.e., the initial-values of the components of E and H be identically zero, the coefficients depend only on x_3, the vectors E and H satisfy certain boundary conditions on $x_3 = 0$ and at a finite number of boundary points $(x_1^k, x_2^k, 0)$, $k = 1, 2, \ldots, m$, the vectors E and H or certain functionals of these vectors can be measured (ROMANOV [1], ROMANOV et al. [1]). If the problem is studied for an arbitrary domain $\Omega_0 \subset \mathbb{R}^3$ the inverse problem is very difficult to be solved, further physical fields are to be taken into consideration.

Another important differential equation in mathematical physics is the Schrödinger equation. In quantum mechanics the wave function $u(x, t)$ of a particle of mass m in a potential $V(x)$ satisfies the Schrödinger equation (CHADAN and SABATIER [1], NEWTON [1], NOVIKOV and KHENKIN [1], ZAKHARIEV and SUSKO [1])

$$i\hbar\partial_t u + \left(\frac{\hbar^2}{2m} \Delta u - V(x)\, u\right) = 0, \qquad x \in \mathbb{R}^n. \tag{22}$$

If $u(x, t) = T(t)\, \psi(x)$ we get

$$\Delta\psi + \frac{2m}{\hbar^2} \left(E - V(x)\right) \psi = 0. \tag{23}$$

Here $\hbar$ is the Planck constant divided by 2π, and E is the energy of the particle. Equation (23) has quadratically integrable solutions, not identical to zero, if and only if E is equal to one of the discrete sets of eigenvalues E_k, $k = 0, 1, 2, \ldots$, of equation (23).

Example 2.20 (Inverse spectral problem). The direct problem of equation (23) is that of finding the eigenvalues E_k for given $V(x)$. The inverse problem is that of finding the potential $V(x)$ given the eigenvalues E_k, $k = 0, 1, 2, \ldots$

In the special case

$$P(\partial)\,u := \frac{1}{\mathrm{i}}\,\partial_t u - \Delta u \tag{24}$$

a fundamental solution G is (TRÈVES [1])

$$G(\tilde{x}, \tilde{y}) = H(t-\tau)\,\frac{\mathrm{e}^{-\mathrm{i}(n-2)\pi/4}}{4(t-\tau)^{n/2}}\,\mathrm{e}^{-\frac{|x-y|^2}{4\mathrm{i}(t-\tau)}}. \tag{25}$$

One can find fundamental solutions for the more general Schrödinger equation in HÖRMANDER [2].

Example 2.21 (Linearized inverse problem). Let the solution u of the Helmholtz equation

$$\Delta u + V(x)\,u = 0 \tag{26}$$

satisfy the boundary condition $u|_{\partial\Omega} = h$. By means of the overspecification $\partial_{\mathrm{n}} u|_{\partial\Omega} = g$ the coefficient $V(x)$ is to be determined. We have seen in relation (1.3.24) that the relationship between the measured values g and the coefficient of a differential equation is strongly nonlinear in general. In order to apply linear theories we decompose the problem approximatively into two linear problems. Let $V(x) = V_0(x) + V_1(x)$ and $u = u_0 + u_1$. If $V_1(x)\,u_1$ is sufficiently small one considers the linearized problem

$$\Delta u_0 + \Delta u_1 + V_0(x)\,u_0 + V_0(x)\,u_1 + V_1(x)\,u_0 = 0,$$

where

$$\Delta u_0 + V_0(x)\,u_0 = 0, \qquad u_0|_{\partial\Omega} = h, \tag{27}$$

$$\begin{aligned} &\Delta u_1 + V_0(x)\,u_1 = -V_1(x)\,u_0, \qquad u_1|_{\partial\Omega} = 0,\\ &\partial_{\mathrm{n}} u_0|_{\partial\Omega} + \partial_{\mathrm{n}} u_1|_{\partial\Omega} = g. \end{aligned} \tag{28}$$

Similar investigations are possible for each identification problem (ANGER and KLEINE [1], ROMANOV [1, 2]).

Remark 2.3. If $V_0(x) = 0$ then the solution of (27) is given by the Poisson integral

$$u_0(x) = \int_{\partial\Omega} h(y)\,P(x, y)\,\mathrm{d}S(y), \tag{29}$$

the solution of (28) by

$$u_1(x) = \int_{\Omega} G_{\mathrm{L}}^{\Omega}(y, x)\,V_1(y)\,u_0(y)\,\mathrm{d}y, \tag{30}$$

where G_{L}^{Ω} is the Green function of $-\Delta u$. Then for $z \in \partial\Omega$ we obtain

$$\begin{aligned} g(z) - \partial_{\mathrm{n}} u_0(z) = \partial_{\mathrm{n}} u_1(z) &= \int_{\Omega} \partial_{\mathrm{n}_z} G_{\mathrm{L}}^{\Omega}(y, z)\,V_1(y)\,u_0(y)\,\mathrm{d}y\\ &= \int_{\Omega} P(y, z)\,V_1(y)\,u_0(y)\,\mathrm{d}y, \end{aligned} \tag{31}$$

which is an integral equation of the first kind for V_1.

2.4. Systems of partial differential equations

As we have seen most inverse problems in differential equations do not have a uniquely determined solution. The coefficients and the source term of the differential equation, which are internal parameters inaccessible to measurement, are to be determined by measurements in the boundary $\partial\Omega$ of a given domain $\Omega_0 \subset \mathbb{R}^3$. In order to get more information for the problem under consideration it is necessary to study systems of different physical fields.

One possibility is to consider simultaneously the *seismic field* and the *gravitational field* (ALEKSEEV and BUBNOV [1, 2]),

$$\varrho(x)\, \partial_t^2 u - \operatorname{div}\big(\lambda(x) \operatorname{grad} u\big) = f(x, t), \tag{1}$$

$$\Delta v = -4\pi\varrho(x), \tag{2}$$

where

$$\begin{aligned} &u(x, 0) = 0, \qquad u_t(x, 0) = 0, \qquad \partial_n u(x, t) = h_1(x, t), \\ &x \in \partial\Omega_0, \qquad 0 < t, \end{aligned} \tag{3}$$

$$v(x) = v_0(x), \qquad x \in \partial\Omega_0. \tag{4}$$

For the inverse problem the overspecification consists of

$$u(x) = v_0(x), \qquad x \in \partial\Omega_0. \tag{5}$$

Here $\Omega_0 \subset \mathbb{R}^3$ is a bounded domain with smooth boundary $\partial\Omega_0$, $\varrho(x)$ the mass density, $\lambda(x)$ the first Lamé coefficient in elasticity.

In the following we write down some further systems of differential equations. For such systems inverse problems are to be studied in a similar way as for the foregoing ones.

A homogeneous and isotropic *thermoelastic slab* is an example for a further coupled field. Let $0 \leqq x \leqq l$, u the displacement, θ the temperature, λ and μ the Lamé coefficients, ϱ the mass density, c the specific heat per unit mass, α the coefficient of thermal expansion, k the thermal conductivity, θ_0 a fixed temperature. The system reads (DAY [1], JENTZSCH [1], LEIS [1])

$$\begin{aligned} &k\theta_{xx} = \varrho c \theta_t + (3\lambda + 2\mu)\, \alpha\theta_0 u_{xt}, \\ &(\lambda + 2\mu)\, u_{xx} = (3\lambda + 2\mu)\, \alpha\theta_x + \varrho u_{tt}. \end{aligned} \tag{6}$$

One has to add initial and boundary conditions.

The basic equations for carrier transport in semiconductors – the *van Roosbroeck system* – in $\mathbb{R}^3$ have the form (GAJEWSKI [1], JEROME [1], WANG [1])

$$\begin{aligned} &-\Delta u = \frac{q}{\varepsilon}\,(f + p - n), \\ &q\, \partial_t n = \operatorname{grad} j_n - qR, \qquad j_n = q(D_n \operatorname{grad} n - \mu_n\, n \operatorname{grad} u), \\ &q\, \partial_t p = -\operatorname{grad} j_p - qR, \qquad j_p = -q(D_p \operatorname{grad} p + \mu_p\, p \operatorname{grad} u). \end{aligned} \tag{7}$$

Here u is the electrostatic potential, n the mobile electron density and p the mobile hole density. These functions are defined on the cylinder $\Omega_0 \times [0, \infty[$ where $\Omega_0 \subset \mathbb{R}^3$ is a bounded domain. Further, R depending on u, n, and p is the net recombination rate, j_n and j_p are the electron and hole current density, f the density, D_n and D_p the diffusion coefficients, ε the dielectric constant, μ_n and μ_p the electron and hole mobilities, q the electron charge. Initial and boundary values are to be added. In KLUGE [1] some inverse problems for this system are studied. A model exists for the carrier transport in semiconductors consisting of a system of eight partial differential equations (WANG [1]). It may be that in $\mathbb{R}^3$ this system cannot be completely solved by mathematics and computers. Each model simplification must be replaced by experiences of the relevant physical processes.

One can find further systems of differential equations in LEIS [1].

2.5. Equations of higher order

The solutions u and v of the Cauchy-Riemann system

$$\begin{aligned} \partial_{x_1} u &= \partial_{x_2} v, \\ \partial_{x_2} u &= -\partial_{x_1} v \end{aligned} \tag{1}$$

separately satisfy

$$\Delta u = 0, \qquad \Delta v = 0. \tag{2}$$

Often, the solutions u_j, $j = 1, \ldots, m$, of a system of partial differential equations can be eliminated. Then the u_j satisfy equations of higher order. We sketch some results for such equations using as example the biharmonic equation

$$\Delta^2 u := \Delta(\Delta u) = 0. \tag{3}$$

Considering the one-dimensional case $u^{(4)} = 0$ one sees that the values of u and u' can be described at a and b (see Sect. 1.6). Especially the Dirichlet problem for a bounded domain $\Omega \subset \mathbb{R}^n$ with smooth boundary $\partial\Omega$ consists in determining a solution $u \in \mathbf{C}^4(\Omega) \cap \mathbf{C}^1(\bar{\Omega})$ satisfying (SCHULZE and WILDENHAIN [1], TIKHONOV and SAMARSKI [1])

$$\Delta^2 u = 0 \quad \text{in } \Omega, \qquad u(y) = h_0(y) \quad \text{and} \quad \partial_n u(y) = h_1(y) \quad \text{on } \partial\Omega. \tag{4}$$

Here h_0 and h_1 are given functions. In the case of a ball $B(0, R) \subset \mathbb{R}^2$ one can prove that u has the form

$$u(x) = (|x|^2 - R^2)\, u_1 + u_2, \tag{5}$$

where u_1 and u_2 are harmonic functions. Then for $y \in \partial B(0, R)$

$$u_2(y) = h_0(y) \tag{6}$$

and

$$2Ru_1(y) + \frac{\partial u_2(y)}{\partial r} = h_1(y). \tag{7}$$

From (6) it follows that

$$u_2(x) = \frac{1}{2\pi} \int_0^{2\pi} h_0(R, \beta) \frac{R^2 - |x|^2}{|x - y|^2} \, \mathrm{d}\beta \tag{8}$$

where $y = (R, \beta)$. The function

$$x \to 2Ru(x) - \frac{|x|}{R} \frac{\partial u_2(x)}{\partial r} \tag{9}$$

with boundary values (7) is harmonic in $B(0, R)$. Then we get for $x \in B(0, R)$, $x = (|x|, \alpha)$,

$$2Ru_1(x) + \frac{|x|}{R} \frac{\partial u_2(x)}{\partial r} = \frac{1}{2\pi} \int_0^{2\pi} h_1(R, \beta) \frac{R^2 - |x|^2}{|x - y|^2} \, \mathrm{d}\beta . \tag{10}$$

From this equation one can calculate u_1. The final formula is

$$u(x) = \frac{(|x|^2 - R^2)}{2\pi R} \frac{1}{2} \int_0^{2\pi} \frac{-h_1(R, \beta) \, \mathrm{d}\beta}{R^2 + |x|^2 - 2R\,|x| \cos(\alpha - \beta)}$$
$$+ \int_0^{2\pi} \frac{h_0(R, \varphi) \big(R - |x| \cos(\alpha - \beta)\big) \, \mathrm{d}\beta}{\big(R^2 + |x|^2 - 2R\,|x| \cos(\alpha - \beta)\big)^2} . \tag{11}$$

A potential theory for equations of higher order was developed in ANGER [4], ATAKHODZHAEV [1], SCHULZE and WILDENHAIN [1]. It is possible to study an inverse source problem relative to $\Delta^2 u$ by considering the formulas (4) and (11) (KLEINE [2, 3], ŠKORVANEK [1]). One can find applications to problems in geophysics in KOSAREV et al. [1], ŠKORVANEK [1] and unsolved inverse problems for systems of partial differential equations in ANGER [16].

3. Potentials

3.1. Classical potentials

In classical potential theory the Riemann integral is used to define potentials relative to a kernel G. For most classical problems the Riemann integral is sufficient. In order to apply the results of functional analysis from a systematic point of view, it is necessary to know a little about measure theory. This aspect will be sketched in Sect. 3.3.

In $\mathbb{R}^n$, $n \geqq 2$, the kernels are often continuous for $x \neq y$ and infinite for $x = y$. Let $x, y \in \mathbb{R}^3$ and G_L the Newtonian kernel. Then one defines the *potential* $G_\mathrm{L}\mu_\varrho$ *of a volume distribution* μ_ϱ at x by

$$G_\mathrm{L}\mu_\varrho(x) := \int_\Omega G_\mathrm{L}(x, y)\, \varrho(y)\, \mathrm{d}y\,, \tag{1}$$

the *potential* $G_\mathrm{L}\mu_\sigma$ *of a surface distribution* μ_σ (single layer) by

$$G_\mathrm{L}\mu_\sigma(x) := \int_S G_\mathrm{L}(x, y)\, \sigma(y)\, \mathrm{d}S(y) \tag{2}$$

and the *potential* $G_\mathrm{L}\mu_\tau$ *of a line distribution* μ_τ by

$$G_\mathrm{L}\mu_\tau(x) := \int_\mathscr{C} G_\mathrm{L}(x, y)\, \tau(y)\, \mathrm{d}s(y)\,. \tag{3}$$

Here $\Omega \subset \mathbb{R}^3$ is a bounded domain, S a smooth and bounded (closed) surface, for instance, $S = \partial\Omega$, and $\mathscr{C}$ a smooth and bounded (closed) curve. Similar potentials can be defined in $\mathbb{R}^n$. In $\mathbb{R}^1$ the boundary $\partial\Omega$ of an open interval $]a, b[$ consists of the two points a and b.

In functional analysis it is more convenient to use linear functionals (often called linear forms) than point functions. Let $\mathrm{C}_0(\mathbb{R}^n)$ be the space of all continuous functions $\varphi\colon \mathbb{R}^n \to \mathbb{R}$ with compact support $\operatorname{supp}\varphi$. For $f \in \mathrm{C}_0(\mathbb{R}^n)$ we define a *volume distribution* μ_ϱ by

$$\mu_\varrho(f) := \int_\Omega f(y)\, \varrho(y)\, \mathrm{d}y\,, \qquad \mathrm{d}y = \mathrm{d}y_1 \ldots \mathrm{d}y_n \quad \text{volume element}\,, \tag{4}$$

a *surface distribution* μ_σ by

$$\mu_\sigma(f) := \int_S f(y)\, \sigma(y)\, \mathrm{d}S(y)\,, \qquad \mathrm{d}S(y) \quad \text{surface element}\,, \tag{5}$$

and a *line distribution* μ_τ by

$$\mu_\tau(f) := \int_\mathscr{C} f(y)\, \tau(y)\, \mathrm{d}s(y)\,, \qquad \mathrm{d}s(y) \quad \text{arc length element}\,. \tag{6}$$

The densities are Riemann integrable functions. More correctly one has to write the restriction $f|_\Omega = \text{rest}_\Omega f$ under the integral (4), and so on. But in integration theory the symbol f is used for all integrals.

In our theory on inverse problems the following measures $\mu_{\delta,\varrho}$ and $\mu_{\delta,\sigma}$ play an important role. Using only the Riemann integral one cannot define these mass distributions which contain point masses. Let

$$\mu_{\delta,\varrho}(f) := \sum_{k=1}^{m_1} a_k f(y^k) + \int_\Omega f(y)\,\varrho(y)\,\mathrm{d}y, \tag{7}$$

$$\mu_{\delta,\sigma}(f) := \sum_{k=1}^{m_2} b_k f(z^k) + \int_{\partial\Omega} f(y)\,\sigma(y)\,\mathrm{d}S(y), \tag{8}$$

where $f \in \mathbf{C}_0(\mathbb{R}^n)$, $y^k \in \bar{\Omega}$, $k = 1, \ldots, m_1$, and $z^k \in \bar{\Omega}$, $k = 1, \ldots, m_2$. In Sect. 3.3 we shall define the potential $G_{\mathrm{L}}\mu$ relative to a general mass distribution (measure) μ. This definition also holds for the special $\mu_{\delta,\varrho}$ and $\mu_{\delta,\sigma}$. Sometimes we define our integrals for functions $f \in \mathbf{C}(K)$. Here $K = \bar{\Omega}$ or $K = S$ or $K = \partial\Omega$ or another compact set $K \subset \mathbb{R}^n$. The space $\mathbf{C}(K)$ is a Banach space relative to the norm $\|\cdot\|_{\mathbf{C}}$ and more convenient to handle than $\mathbf{C}_0(\mathbb{R}^n)$. The last space is a so-called locally compact space (Dunford and Schwartz [1]).

Let $x \notin \bar{\Omega}$. Then $y \mapsto f(y) = G_{\mathrm{L}}(x, y)$ defines a continuous function on $K = \bar{\Omega}$, and the potential at x is calculated by (4). If $x \in \bar{\Omega}$ we have to extend the integral (4) to unbounded functions, which is sketched in Sect. 3.2. A large part of integration theory deals with such extensions to integrable functions. This step is necessary in order to apply the results of functional analysis.

In classical integration theory the following properties of integrals are proved. Here $\mu(f)$ can be one of the integrals (4), (5), (6) or (7), (8). Let f_1, f_2, f be elements of $\mathbf{C}_0(\mathbb{R}^n)$ (or $\mathbf{C}(K)$), $c_1, c_2 \in \mathbb{R}$. Then

$$\mu(c_1 f_1 + c_2 f_2) = c_1\mu(f_1) + c_2\mu(f_2), \tag{9}$$

$$|\mu(f)| \leqq M\,\|f\|_{\mathbf{C}}. \tag{10}$$

If ϱ, σ, τ are non-negative, then

$$\mu(f) \geqq 0 \quad \text{for all} \quad f \geqq 0. \tag{11}$$

From inequality (11) and (9) inequality (10) follows (see Proposition 3.1). A mapping $\mu\colon \mathbf{C}_0(\mathbb{R}^n) \to \mathbb{R}$ satisfying (9) is called a linear form or linear functional. If additionally μ satisfies (10) then μ is continuous. If μ satisfies (9) and (11) it is called a positive linear form. As we have mentioned each positive linear form is continuous.

Often, the potential W of a double layer is used. It is defined at $x \in \mathbb{R}^3$ by

$$W(x) := \int_S \left(\frac{\partial}{\partial_{n_y}} G_{\mathrm{L}}(x, y)\right) b(y)\,\mathrm{d}S(y). \tag{12}$$

While the integrals (4)–(6) define continuous linear forms on $\mathbf{C}_0(\mathbb{R}^n)$ (resp.

on $\mathbf{C}(K)$), the potential W is generated by a linear form L, where

$$L(f) := \int_S \frac{\partial}{\partial \mathfrak{n}_y} f(y)\, \mathrm{d}S(y) = \sum_{j=1}^{n} \int_S \frac{\partial f(y)}{\partial y_j} \cos(y_j, \mathfrak{n})\, \mathrm{d}S(y) \tag{13}$$

is continuous on $\mathbf{C}_0^1(\mathbb{R}^n)$ (resp. $\mathbf{C}^1(K)$) with respect to $\|\cdot\|_{\mathbf{C}^1}$. Such continuous linear forms L are called distributions (generalized functions) of the first order. The theory of distributions is developed in Hörmander [2]. In order to solve boundary-value problems from the numerical point of view the potentials $G_L\mu_\sigma$ and W are often used (see Gründemann [1]).

Potentials can be defined relative to more general kernels $G\colon \mathbb{R}^n \times \mathbb{R}^n \to \overline{\mathbb{R}}$, which are locally integrable. An example is the kernel G_H of the heat equation or the kernel G_D defined in (2.3.19). Let us calculate some special potentials relative to G_L and G_H.

Example 3.1 (Potential of a homogeneous ball). Let $\Omega = B(x^0, R) \subset \mathbb{R}^3$ be an open ball and $\sigma(y) = 1/4\pi R^2$. We define

$$\delta_{x^0,R}(f) := \frac{1}{4\pi R^2} \int_{\partial B(x^0,R)} f(x)\, \mathrm{d}S(x) \tag{14}$$

and

$$m_{x^0,R}(f) := \frac{3}{4\pi R^3} \int_{B(x^0,R)} f(x)\, \mathrm{d}x\,. \tag{15}$$

The integrals (14) and (15) are averages often used in potential theory. Using spherical coordinates

$$\begin{aligned} &y_1 = r \sin\vartheta_1 \cos\vartheta_2, \qquad y_2 = r\sin\vartheta_1 \sin\vartheta_2, \qquad y_3 = r\cos\vartheta_1, \\ &0 < \vartheta_1 < \pi, \qquad 0 \leqq \vartheta_2 < 2\pi, \end{aligned} \tag{16}$$

one can explicitly calculate $G_L m_{x^0,R}$, where in consequence of the symmetry of this potential the value can be studied at $x = (0, 0, x_3 - x_3^0)$. We obtain

$$\begin{aligned} G_L\delta_{x^0,R}(x) &= \frac{1}{4\pi R^2} \int_{\partial B(x^0,R)} \frac{\mathrm{d}S(y)}{4\pi\,|x - y|} \\ &= \frac{1}{4\pi R^2} \int_{\vartheta_2=0}^{2\pi} \left(\int_{\vartheta_1=0}^{\pi} \frac{R^2 \sin\vartheta_1\, \mathrm{d}\vartheta_1\, \mathrm{d}\vartheta_2}{4\pi(|x - x^0|^2 + R^2 - 2R\,|x - x^0| \cos\vartheta_1)^{1/2}} \right) \\ &= \frac{1}{2\,|x - x^0|\, 4\pi R} \left[(|x - x^0|^2 + R^2 - 2R\,|x - x^0|\cos\vartheta_1)^{1/2}\right]_0^\pi \\ &= \frac{1}{8\pi\,|x - x^0|\, R} \left[\big||x - x^0| + R\big| - \big||x - x^0| - R\big|\right]. \end{aligned} \tag{17}$$

Therefore we get

$$G_L\delta_{x^0,R}(x) = \frac{1}{4\pi\,|x - x^0|} \quad \text{if} \quad |x - x^0| \geqq R,$$
$$G_L\delta_{x^0,R}(x) = \frac{1}{4\pi R} \qquad \text{if} \quad |x - x^0| \leqq R. \tag{18}$$

The following formulas hold

$$m_{x^0,R}(f) = \frac{3}{4\pi R^3} \int\limits_{B(x^0,R)} f(y)\,\mathrm{d}y_1\,\mathrm{d}y_2\,\mathrm{d}y_3$$
$$= \int\limits_{r=0}^{R} \left(\frac{1}{4\pi r^2} \int\limits_{\partial B(x^0,r)} f(y)\,\mathrm{d}S(y) \right) \frac{3r^2}{R^3}\,\mathrm{d}r = \int\limits_{r=0}^{R} \delta_{x^0,r}(f)\,\mathrm{d}m(r), \tag{19}$$

where $\mathrm{d}m(r) = 3r^2\,\mathrm{d}r/R^3$. If $|x - x^0| \geqq R$ from (15) it follows that

$$G_L m_{x^0,R}(x) = \int\limits_{r=0}^{R} G_L\delta_{x^0,r}(x)\,\mathrm{d}m(r) = \frac{1}{4\pi\,|x - x^0|} \int\limits_0^R \frac{3r^2}{R^3}\,\mathrm{d}r$$
$$= \frac{1}{4\pi\,|x - x^0|} = G_L\delta_{x^0}(x) \tag{20}$$

and if $|x - x^0| \leqq R$

$$G_L m_{x^0,R}(x) = \frac{1}{4\pi\,|x - x^0|} \int\limits_0^{|x-x^0|} \frac{3r^2}{R^3}\,\mathrm{d}r + \int\limits_{|x-x^0|}^{R} \frac{1}{4\pi r}\,\frac{3r^2}{R^3}\,\mathrm{d}r$$
$$= \frac{1}{4\pi R^3} \left(\frac{3}{2} R^2 - \frac{1}{2}\,|x - x^0|^2 \right). \tag{21}$$

If $R \searrow 0$ (decreasing sequence) we obtain

$$G_L m_{x^0,R}(x) \nearrow G_L(x, x^0) \tag{22}$$

(increasing sequence) and

$$G_L\mu_\varrho(x^0) = \lim_{R\to 0} \int\limits_\Omega G_L m_{x^0,R}(y)\,\varrho(y)\,\mathrm{d}y = \int\limits_\Omega G_L(x^0, y)\,\varrho(y)\,\mathrm{d}y, \tag{23}$$

which can be considered as definition of the potential of a volume distribution at $x^0 \in \bar{\Omega}$.

Example 3.2 (Lebesgue spine). In the development of potential theory the counterexample of H. Lebesgue (1913) played an important role (Helms [1]). It showed that for a special domain $\Omega_0 \subset \mathbb{R}^3$ whose boundary $\partial\Omega_0$ is smooth except at the origin $(0, 0, 0) \in \partial\Omega_0$, the Dirichlet problem cannot be solved in the classical sense at (0, 0, 0). The domain Ω_0 is called *Lebesgue spine*. Let us consider the line segment $\mathscr{E} := \{(\xi, 0, 0) : 0 \leqq \xi \leqq 1\} \subset \mathbb{R}^3$. The Newtonian potential $G_L\mu_\tau$ of this line segment is defined by

$$G_L\mu_\tau(x) := \int_0^1 \frac{\xi \, d\xi}{\left((\xi - x_1)^2 + x_2^2 + x_3^2\right)^{1/2}}. \tag{24}$$

The integral considered has the value

$$\begin{aligned} G_L\mu_\tau(x) &= \left((1 - x_1)^2 + x_2^2 + x_3^2\right)^{1/2} - (x_1^2 + x_2^2 + x_3^2)^{1/2} \\ &\quad + x_1 \log \left|1 - x_1 + \left((1 - x_1)^2 + x_2^2 + x_3^2\right)^{1/2}\right| \\ &\quad + x_1 \log |x_1 + (x_1^2 + x_2^2 + x_3^2)^{1/2}| - x_1 \log (x_2^2 + x_3^2). \end{aligned} \tag{25}$$

The first four terms tend to 1 if $(x_1, x_2, x_3) \to (0, 0, 0)$. Now let $x_1 > 0$. Then

$$\lim_{|x| \to 0} G_L\mu_\tau(x) = 1 - \lim_{|x| \to 0} x_1 \log r^2, \tag{26}$$

where $r^2 = x_2^2 + r_3^2$ (see Fig. 3.1). If $x_1^k = r^2$ then $G_L\mu_\tau(x) \to 1$.

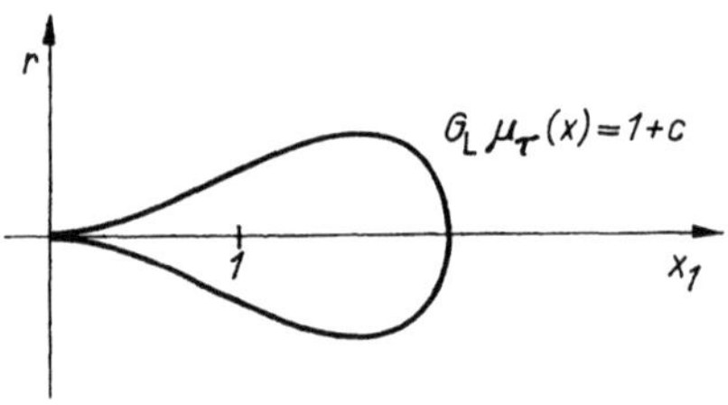

Fig. 3.1

If, however, this approach is along a path for which $x_1 > 0$ and $r = \exp(-c/2x_1)$, $c > 0$, then

$$G_L\mu_\tau(x) \to 1 + c.$$

Outside $\mathscr{E}$ the Newtonian potential $G_L\mu_\tau$ is harmonic, but along different paths defined by

$$G_L\mu_\tau(x) = 1 + c_j, \qquad j = 1, 2, \tag{27}$$

the harmonic function has different values. The point (0, 0, 0) is called an irregular boundary point. If Ω is not smooth we have to introduce the notion of an irregular boundary point, at which the Dirichlet problem is not solvable in the classical sense. Then (2.1.5) holds on $\partial\Omega$, except a set of capacity zero (Helms [1]).

Example 3.3 (Potential of a domain bounded by an ellipse). In the following we want to compute the logarithmic potential of a domain $\Omega \subset \mathbb{R}^2$ bounded by an ellipse using results of complex analysis. Similar investigations are possible relative to a domain $\Omega \subset \mathbb{R}^2$ bounded by algebraic curves (Herglotz

[2]) and in $\mathbb{R}^3$ to an ellipsoid (KELLOGG [1], HERGLOTZ [2]). *Elliptic coordinates* in $\mathbb{R}^2$ have the form (NEUMANN [1])

$$
\begin{aligned}
y_1 &= k\left(\frac{e^{\lambda} + e^{-\lambda}}{2}\right)\cos\vartheta = k\cosh\lambda\cos\vartheta,\\
y_2 &= k\left(\frac{e^{\lambda} - e^{-\lambda}}{2}\right)\sin\vartheta = k\sinh\lambda\sin\vartheta,
\end{aligned}
\tag{28}
$$

where $0 \leqq \lambda < \infty$, $0 \leqq \vartheta < 2\pi$. For $\lambda > 0$ fixed one gets an ellipse with semi-axis

$$
a = k\cosh\lambda, \qquad b = k\sinh\lambda, \qquad a + b = k\,e^{\lambda}. \tag{29}
$$

The ellipses are confocal. Its foci are

$$
\xi_1 = (-k, 0), \qquad \xi_2 = (k, 0).
$$

Let

$$
r_1^2 := (y_1 + k)^2 + y_2^2, \qquad r_2^2 := (y_1 - k)^2 + y_2^2.
$$

From (28) it follows that

$$
\begin{aligned}
r_1 &= k(\cosh\lambda + \cos\vartheta), \qquad & r_2 &= k(\cosh\lambda - \cos\vartheta),\\
r_1 + r_2 &= 2k\cosh\lambda, & r_1 - r_2 &= 2k\cos\vartheta.
\end{aligned}
\tag{30}
$$

For fixed ϑ we get a hyperbola. Now let us consider two arbitrary points

$$
\begin{aligned}
x_1 &= k\cosh\lambda\cos\vartheta, \qquad & x_2 &= k\sinh\lambda\sin\vartheta,\\
y_1 &= k\cosh\lambda'\cos\vartheta', & y_2 &= k\sinh\lambda'\sin\vartheta'.
\end{aligned}
\tag{31}
$$

The distance between x and y is given by the formula

$$
\begin{aligned}
r^2 = |x - y|^2 &= k^2[\cosh(\lambda + \lambda') - \cos(\vartheta + \vartheta')]\\
&\quad\times[\cosh(\lambda - \lambda') - \cos(\vartheta - \vartheta')]\\
&= \frac{k^2}{2}\,e^{2\lambda}\,|1 - \exp\{-(\lambda + \lambda') + i(\vartheta + \vartheta')\}|^2\\
&\quad\times|1 - \exp\{-(\lambda - \lambda') + i(\vartheta - \vartheta')\}|^2.
\end{aligned}
\tag{32}
$$

Since $\log(1 - w) = \log|1 - w| + i\arg(1 - w)$ and

$$
\log\frac{1}{1 - w} = \sum_{k=1}^{\infty}\frac{w^k}{k}
$$

we obtain for $\lambda' < \lambda$ with $K = \log(2/k)$ and

$$
w = e^{\{-(\lambda+\lambda') + i(\vartheta+\vartheta')\}} \text{ resp. } w = e^{\{-(\lambda-\lambda') + i(\vartheta-\vartheta')\}}
$$

the development

$$\log \frac{1}{r} = \log \frac{2}{k} + \log \mathrm{e}^{-\lambda} + \log 1/|1 - \exp\{-(\lambda + \lambda') + \mathrm{i}(\vartheta + \vartheta')\}|$$

$$+ \log 1/|1 - \exp\{-(\lambda - \lambda') + \mathrm{i}(\vartheta - \vartheta')\}|$$

$$= K - \lambda + 2 \sum_{k=1}^{\infty} \frac{1}{k \exp(k\lambda)}$$

$$\times \{\cosh k\lambda' \cos k\vartheta \cos k\vartheta' + \sinh k\lambda' \sin k\vartheta \sin k\vartheta'\}, \tag{33}$$

and for $\lambda < \lambda'$

$$\log \frac{1}{r} = K - \lambda' + 2 \sum_{k=1}^{\infty} \frac{1}{k \exp(k\lambda')}$$

$$\times \{\cosh k\lambda \cos k\vartheta \cos k\vartheta' + \sinh k\lambda \sin k\vartheta \sin k\vartheta'\}. \tag{34}$$

For the arc length of an ellipse $\lambda = \lambda'$ we get if $y = y(\lambda', \vartheta')$

$$\mathrm{d}s_1(y) = \sqrt{\psi_0}\, \mathrm{d}\vartheta',$$

and for the arc length of a hyperbola $\vartheta = \vartheta'$

$$\mathrm{d}s_2(y) = \sqrt{\psi_0}\, \mathrm{d}\lambda',$$

where

$$\psi_0(\lambda', \vartheta') := \frac{k^2}{2} (\cosh 2\lambda' - \cos 2\vartheta') = k^2(\cosh^2 \lambda' - \cos^2 \vartheta'). \tag{35}$$

From (30) it follows that

$$\psi_0 = r_1 r_2 = \left((y_1 + k)^2 + y_2^2\right)^{1/2} \left((y_1 - k)^2 + y_2^2\right)^{1/2} \tag{36}$$

and

$$\mathrm{d}y = \mathrm{d}y_1\, \mathrm{d}y_2 = \psi_0\, \mathrm{d}\lambda'\, \mathrm{d}\vartheta'. \tag{37}$$

Let Ω_{λ_0} be the interior of the ellipse defined by $\lambda \leqq \lambda_0$.

Theorem 3.1. *The potential $G_L\mu_{\sigma_0}$ of an ellipse $\lambda = \lambda_0$ with density $\sigma = \sigma_0\sqrt{\psi_0}$ at $x = x(\lambda, \vartheta)$ has the value*

$$G_L\mu_\sigma(x) = M\left[K - \lambda - \frac{1}{2} \mathrm{e}^{-2\lambda} \cos 2\vartheta\right] \quad \text{if} \quad x \notin \Omega_{\lambda_0} \tag{38}$$

and

$$G_L\mu_\sigma(x) = M[A_0 - 2B_0 \cosh \lambda \cos 2\vartheta] \quad \text{if} \quad x \in \Omega_{\lambda_0}. \tag{39}$$

Here $M = \sigma_0 \int\limits_0^{2\pi} \sqrt{\psi_0(\lambda_0, \vartheta')}\, \mathrm{d}s_1 = \sigma_0 \int\limits_0^{2\pi} \psi(\lambda_0, \vartheta')\, \mathrm{d}\vartheta' = \sigma_0 \pi k^2 \cosh 2\lambda_0$ *is the total mass and* $A_0 = K - \lambda_0$, $B_0 = (\exp -2\lambda_0)/4 \cosh 2\lambda_0$. *Further, the potential* $G_L \mu_{\varrho_0}$ *of a domain* Ω_{λ_0} *with constant density* ϱ_0 *at* $x \notin \Omega_{\lambda_0}$ *has the value*

$$G_L \mu_{\varrho_0}(x) = M_0 \left[K - \lambda - \frac{1}{2} \mathrm{e}^{-2\lambda} \cos 2\vartheta \right] \quad \text{if} \quad x \notin \Omega_{\lambda_0}, \tag{40}$$

where M_0 *is the total mass,* $M_0 = \varrho_0 \pi ab = \frac{\varrho_0}{2} k^2 \sinh 2\lambda_0$.

Proof. For the purpose of our book it is sufficient to calculate the potential $G_L \mu_{\varrho_0}$ at $x \notin \Omega_{\lambda_0}$. Following (35) and (37) we get for $x = x(\lambda, \vartheta)$, $y = y(\lambda', \vartheta')$

$$\begin{aligned} \int\limits_{\Omega_{\lambda_0}} G_L(x, y)\, \varrho_0\, \mathrm{d}y &= \varrho_0 \int\limits_{\lambda'=0}^{\lambda_0} \left(\int\limits_{\vartheta'=0}^{2\pi} G_L(x, y)\, \psi_0(\lambda', \vartheta')\, \mathrm{d}\vartheta' \right) \mathrm{d}\lambda' \\ &= \varrho_0 \int\limits_0^{\lambda_0} \left(\int\limits_0^{2\pi} \psi_0(\lambda', \vartheta') \log \frac{1}{r}\, \mathrm{d}\vartheta' \right) \mathrm{d}\lambda'. \end{aligned} \tag{41}$$

The total mass M of Ω_{λ_0} has the form

$$M = \varrho_0 \pi ab = \varrho_0 \pi k^2 \cosh \lambda_0 \sinh \lambda_0 = \frac{\varrho_0}{2} \pi k^2 \sinh 2\lambda_0 .$$

At first we have to calculate the interior integral. Using formulas (33) and (35) we obtain

$$\int\limits_0^{2\pi} \psi_0(\lambda', \vartheta') \log \frac{1}{r}\, \mathrm{d}\vartheta' = \pi k^2 \cosh 2\lambda' \left[K - \lambda - \frac{\mathrm{e}^{-2\lambda}}{2} \cos 2\vartheta \right]. \tag{42}$$

From (41) it follows that

$$\int\limits_{\Omega_{\lambda_0}} G_L(x, y)\, \varrho_0\, \mathrm{d}y = \frac{\varrho_0}{2} \pi k^2 \sinh 2\lambda_0 \left[K - \lambda - \frac{\mathrm{e}^{-2\lambda}}{2} \cos 2\vartheta \right], \tag{43}$$

which proves formula (40). In order to prove (38) we apply relation (42),

$$\begin{aligned} G_L \mu_{\sigma_0}(x) &= \sigma_0 \int \sqrt{\psi_0} \log \frac{1}{r}\, \mathrm{d}s_1 = \sigma_0 \int\limits_{\vartheta'=0}^{2\pi} \psi_0(\lambda_0, \vartheta') \log \frac{1}{r}\, \mathrm{d}\vartheta' \\ &= \sigma_0 \pi k^2 \cosh 2\lambda_0 \left[K - \lambda - \frac{\mathrm{e}^{-2\lambda}}{2} \cos 2\vartheta \right]. \end{aligned} \tag{44}$$

Formula (39) follows by using (34). ∎

Remark 3.1. The following considerations are important for inverse source problems. Let ϱ_0 be determined such that $M = 1$. We obtain

$$\int_{\Omega_{\lambda_0}} G_L(x, y)\, \varrho_0 \, dy = \int_{\lambda'=0}^{\lambda_0} \left(\frac{1}{\pi k^2 \cosh 2\lambda'} \int_{\vartheta'=0}^{2\pi} \psi_0(\lambda', \vartheta') \log \frac{1}{r} \, d\vartheta' \right) \frac{2 \cosh 2\lambda'}{\sinh 2\lambda_0} \, d\lambda' .$$

Let

$$dm(\lambda') := \frac{2 \cosh 2\lambda'}{\sinh 2\lambda_0} \, d\lambda' .$$

The mass distribution

$$d\gamma(\vartheta') := \frac{\psi_0(\lambda', \vartheta')}{\pi k^2 \cosh 2\lambda'} \, d\vartheta' = \frac{\sqrt{\psi_0(\lambda', \vartheta')}}{k^2 \cosh 2\lambda'} \, ds_1$$

gives the potential (see formula (42))

$$G_L\gamma(x) = \left[K - \lambda - \frac{1}{2} e^{-2\lambda} \cos 2\vartheta \right] \quad \text{if} \quad x \notin \Omega_{\lambda_0} .$$

Then

$$\int_{\Omega_{\lambda_0}} G_L(x, y)\, \varrho_0 \, dy = \int_{\lambda'=0}^{\lambda_0} \big(G_L\gamma(x)\big) \, dm(\lambda') . \tag{45}$$

Formula (45) is similar to formula (19).

Remark 3.2. If $\lambda = 0$ we obtain a (double) straight line segment joining the two foci. In this case

$$\psi_0(0, \vartheta') = \frac{k^2}{2} (1 - \cos 2\vartheta') , \qquad 0 \leqq \vartheta' < 2\pi ,$$

or

$$\psi_0(y_1) := 2(k + y_1)(k - y_1) , \qquad -k \leqq y_1 \leqq k . \tag{46}$$

The potential $G_L\mu_{\varrho_0}$ (see formula (43)) can be analytically continued up to this straight line segment and replaced by the potential $G_L\mu_\sigma$ with $\sigma = \sqrt{\psi_0}$ (see formula (44)). This result is similar to (20).

Remark 3.3. Similar considerations are possible for domains $\Omega_0 \subset \mathbb{R}^2$ bounded by algebraic curves (Herglotz [2]). Further, the potential of an ellipsoid in $\mathbb{R}^3$ can be calculated (Herglotz [2], Kellogg [1]).

Remark 3.4. Concrete formulas for potentials $G_L\mu_\varrho$, where $\Omega \subset \mathbb{R}^2$ is bounded by polygons, and its analytic continuation can be found in Strakhov and Brodsky [1]. Further, in this paper one can find remarks concerning the uniqueness of the inverse source problem for $G_L\mu_{\varrho_0}$, where $\Omega \subset \mathbb{R}^2$ is bounded by lemniscates and ϱ_0 is constant. In $\mathbb{R}^3$ the results are different (STRAKHOV and BRODSKY [2]).

In most books on partial differential equations the following theorem is proved. It is used in our considerations on inverse source problems of the heat equation.

Theorem 3.2. *Let G_H be the fundamental solution of the heat equation defined in (2.2.6), $I_n := \{(x, t) \in \mathbb{R}^{n+1} : t = t_0\}$, $f: I_n \mapsto \mathbb{R}$ a continuous bounded function. Then*

$$G_H\mu_f(x, t) := \int_{I_n} G_H\big((x, t), (y, t_0)\big) f(y) \, dy_1 \dots dy_n, \; t > t_0, \tag{47}$$

is a solution of the heat equation satisfying $G_H\mu_f(x, t) \to f(y)$ for $(x, t) \to (y, t_0)$. Further, if $f(y) = 1$ then

$$G_H\mu_f(x, t) = 1 \quad \text{for all} \quad t > t_0. \tag{48}$$

3.2. Some uniqueness theorems for classical potentials

In Sect. 3.1 we have proved that different mass distributions, for instance, $m_{x^0,R}$, $\delta_{x^0,R}$, δ_{x^0}, can give the same potential outside a bounded domain $\Omega \subset \mathbb{R}^n$. In this special case Ω is a ball $B(x^0, R)$. If we require that the mass distribution μ_ϱ has a constant density $\varrho = \varrho_0$ and $\Omega = B(x^0, R)$, then ϱ_0 is uniquely determined by the values of the potential $G_L\mu_{\varrho_0}$ outside Ω. This result holds for all convex domains $\Omega \subset \mathbb{R}^n$ and a constant density ϱ_0. Analyzing the proof of this theorem one sees that the condition $\partial_{y_k}\varrho(y) = 0$, k fixed, is sufficient to prove uniqueness for the special μ_ϱ. The proof of this theorem uses the divergence theorem. If it is applied to the fundamental solution G_L one immediatly sees that the left-hand side of this identity is a dipole distribution, while the surface integral is the potential of a single layer having positive and negative values. Such results are important for magnetic fields.

We notice that without restrictions the mass distribution (measure) μ cannot be uniquely determined by its potential outside the support of this distribution, which will be discussed from a general point of view in Chap. 5. In 1914 a first uniqueness theorem for bodies of rotation with constant density was proved by HERGLOTZ [2]. In 1938 NOVIKOV [1] proved a uniqueness theorem for star-like bodies with constant densities. One can find generalizations of this theorem and further counterexamples in SMITH [1], PRILEPKO [2, 3], STRAKHOV and BRODSKY [1], ISAKOV [1], BRODSKY [1] and other authors. The stability of such inverse source problems was studied by LORENZI and PAGANI [1], PAGANI [1]. In this direction a first paper was published by RAPOPORT [1] in 1941.

The following theorem was proved by G. Lauricella and P. Pizetti in 1909 to 1911 (see OGANESYAN and STAROSTENKO [1], MARUSSI [1]). In mathematical analysis relation (2) is more useful than relation (1). Using (2) results of functional analysis can be systematically applied to problems in potential theory. This relation is the basis for our investigations in Chap. 5.

Theorem 3.3. *Let* μ_{ϱ_1} *and* μ_{ϱ_2} *be two volume distributions defined in* (3.1.4). *If*

$$G_L\mu_{\varrho_1}(x) = G_L\mu_{\varrho_2}(x) \quad \text{for all} \quad x \notin \Omega, \tag{1}$$

then

$$\int_\Omega u(y)\,\varrho_1(y)\,\mathrm{d}y = \int_\Omega u(y)\,\varrho_2(y)\,\mathrm{d}y \tag{2}$$

holds for all $u \in \mathbf{C}^2(\Omega) \cap \mathbf{C}(\bar{\Omega})$ *satisfying* $\Delta u = 0$ *in* Ω.

Proof. a) Let $x \notin \Omega$ and $u(y) = G_L(x, y)$. Then $u \in \mathbf{C}(\bar{\Omega})$ and $\Delta u = 0$ in Ω. From (2) it follows that

$$G_L\mu_{\varrho_1}(x) = G_L\mu_{\varrho_2}(x).$$

The potentials are equal outside Ω.

b) Let μ_σ be a surface distribution defined in (3.1.5). Integrating (1) with respect to $\sigma(x)\,\mathrm{d}S(x)$ one gets

$$\int_{\partial\Omega} G_L\mu_{\varrho_1}(x)\,\sigma(x)\,\mathrm{d}S(x) = \int_{\partial\Omega} G_L\mu_{\varrho_2}(x)\,\sigma(x)\,\mathrm{d}S(x). \tag{3}$$

The following relations hold:

$$\begin{aligned}\int_{\partial\Omega} G_L\mu_{\varrho_k}(x)\,\sigma(x)\,\mathrm{d}S(x) &= \int_{\partial\Omega}\Big(\int_\Omega G_L(x, y)\,\varrho_k(y)\,\mathrm{d}y\Big)\,\sigma(x)\,\mathrm{d}S(x)\\ &= \int_\Omega\Big(\int_{\partial\Omega} G_L(x, y)\,\sigma(x)\,\mathrm{d}S(x)\Big)\,\varrho_k(y)\,\mathrm{d}y\\ &= \int_\Omega G_L\mu_\sigma(y)\,\varrho_k(y)\,\mathrm{d}y = \int_\Omega u(y)\,\varrho_k(y)\,\mathrm{d}y,\end{aligned} \tag{4}$$

where $u(y) = G_L\mu_\sigma(y)$ is harmonic in Ω. From (3) and (4) it follows that

$$\int_\Omega u(y)\,\varrho_1(y)\,\mathrm{d}y = \int_\Omega u(y)\,\varrho_2(y)\,\mathrm{d}y. \tag{5}$$

In the following let $n = 3$. In Proposition 4.2 we shall prove that each harmonic polynomial h_k defined on $\bar{\Omega}$ can be represented in the form

$$h_k(y) = G_L\mu_{\sigma_k}(y),$$

where σ_k is defined on $\partial\Omega$. By Proposition 4.3 the set of all restrictions $\mathrm{rest}_{\partial\Omega}\, G_L\mu_{\sigma_k}$ is dense in $\mathbf{C}(\partial\Omega)$. From the first Harnack convergence theorem (see Doob [1], Kellogg [1]) it follows that (5) holds for all $u \in \mathbf{C}^2(\Omega) \cap \mathbf{C}(\bar{\Omega})$ satisfying $\Delta u = 0$ in Ω. One can proof a similar result in $\mathbb{R}^n$, $n \geqq 2$, if the Green function G_L^Ω is zero on $\partial\Omega$ (Theorem 4.5). ∎

A set $M \subset \mathbb{R}^n$ is *convex*, if for arbitrary $x^1, x^2 \in M$ and all t satisfying $0 \leqq t \leqq 1$

$$tx^1 + (1 - t)\,x^2 \in M. \tag{6}$$

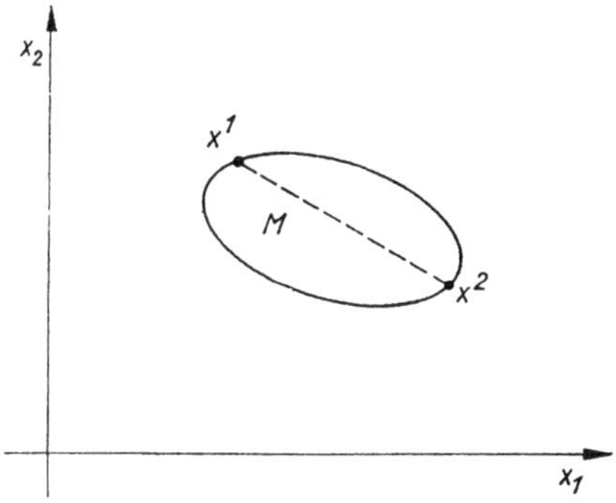

Fig. 3.2

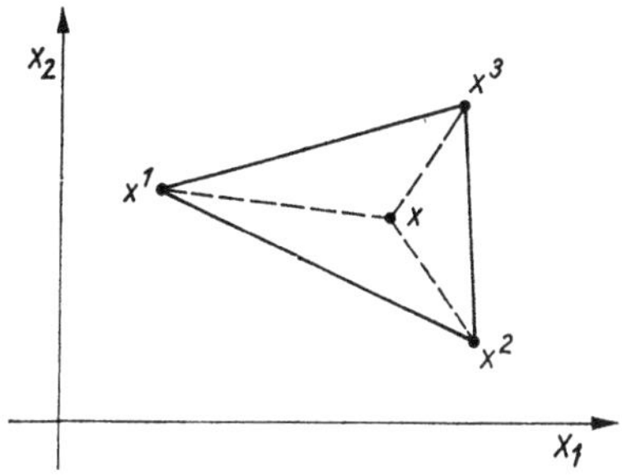

Fig. 3.3

This definition also holds for linear infinite-dimensional spaces X like the space $\mathbf{L}^p(K)$ and $\mathbf{C}(K)$, $K \subset \mathbb{R}^n$ compact set. In Fig. 3.2 and Fig. 3.3 one can find convex sets in $\mathbb{R}^2$. In Fig. 3.3 $x \in M$ is uniquely determined by its extremal points $x^1, x^2, x^3 \in M$. These extremal points are uniquely determined. In infinite-dimensional spaces the extremal points of a convex set also play a fundamental role (CHOQUET [1]). We apply this theory to inverse source problems in Chap. 5.

For the following theorem the smoothness of the boundary can be weakened. It is only important that the divergence theorem holds. We prove the Lemma of Novikov for convex domains (see PRILEPKO [2]).

Theorem 3.4 (Lemma of P. S. Novikov). *Let $\Omega \subset \mathbb{R}^n$, $n \geqq 2$, be a bounded, convex domain with sufficiently smooth boundary $\partial\Omega$. Further, let $\varrho_j\colon \bar{\Omega} \mapsto \mathbb{R}$, $j = 1, 2$, be continuous functions such that $\partial_{y_k}\varrho_j(y) = 0$ for all $y \in \bar{\Omega}$, k fixed. Then from $G_{\mathrm{L}}\mu_{\varrho_1}(x) = G_{\mathrm{L}}\mu_{\varrho_2}(x)$ for all $x \notin \Omega$ it follows that $\varrho_1 = \varrho_2$.*

Proof. a) If $\varrho(y) := \varrho_1(y) - \varrho_2(y) \not\equiv 0$ then ϱ has positive and negative values on $\partial\Omega$. From $\varrho(y) \geqq 0$ on $\partial\Omega$ and $\partial_{y_k}\varrho(y) = 0$ it follows that $\varrho(y)$ is constant on the straight line segment joining the two boundary points $y^1, y^2 \in \partial\Omega$ (see Fig. 3.2). From $\varrho(y) \geqq 0$ we obtain $\varrho(y) \geqq 0$ on $\bar{\Omega}$. For non-negative functions u we get

$$\int_\Omega u(y)\,\varrho_1(y)\,\mathrm{d}y > \int_\Omega u(y)\,\varrho_2(y)\,\mathrm{d}y,$$

which contradicts (2).

b) From Theorem 3.3 we obtain

$$\int_\Omega u(y)\,\varrho_1(y)\,\mathrm{d}y = \int_\Omega u(y)\,\varrho_2(y)\,\mathrm{d}y \tag{7}$$

for all $u \in \mathbf{C}^2(\Omega) \cap \mathbf{C}(\bar{\Omega})$ satisfying $\Delta u = 0$ in Ω. Then

$$\int_\Omega u(y)\,\varrho(y)\,\mathrm{d}y = 0 \tag{8}$$

for all harmonic functions u. Especially we consider the derivative

$$\partial_{y_k}u \in \mathbf{C}^2(\Omega) \cap \mathbf{C}(\bar{\Omega}),$$

which is also harmonic in Ω. Then

$$\partial_{y_k}\big(u(y)\,\varrho(y)\big) = \big(\partial_{y_k}u(y)\big)\,\varrho(y) + u(y)\,\partial_{y_k}\varrho(y) = \big(\partial_{y_k}u(y)\big)\,\varrho(y).$$

From the divergence theorem (see KELLOGG [1]) we get

$$\int_\Omega \varrho(y)\,\partial_{y_k}u(y)\,\mathrm{d}y = \int_\Omega \partial_{y_k}\big(u(y)\,\varrho(y)\big)\,\mathrm{d}y = \int_{\partial\Omega} u(y)\,\varrho(y)\cos(y_k,\mathfrak{n})\,\mathrm{d}S(y), \tag{9}$$

where $\mathfrak{n}$ is the outward normal at $y \in \partial\Omega$. By (8) the left-hand side is zero. Therefore

$$\int_{\partial\Omega} u(y)\,\varrho(y)\cos(y_k,\mathfrak{n})\,\mathrm{d}S(y) = 0. \tag{10}$$

Let $y^0 \in \partial\Omega$ be a point where the product $\varrho(y^0)\cos(y_k,\mathfrak{n})$ is positive or negative. We consider the case where it is positive. Because $\partial\Omega$ is sufficiently smooth the function $y \mapsto \cos(y_k,\mathfrak{n})$ is continuous. From the continuity of ϱ on $\partial\Omega$ the existence of a neighbourhood $U(y^0)$ of y^0 follows, such that

$$\varrho(y)\cos(y_k,\mathfrak{n}) > 0 \quad \text{for all} \quad y \in U(y^0).$$

If there exists a positive harmonic function u_0 which is zero on $\partial\Omega \setminus U(y^0)$ and strictly positive on $U(y^0)$ from (10) it follows that

$$\int_{\partial\Omega} u_0(y)\,\varrho(y)\cos(y_k,\mathfrak{n})\,\mathrm{d}S(y) > 0, \tag{11}$$

which is a contradiction to (10). Therefore $\varrho(y) = \varrho_1(y) - \varrho_2(y) = 0$.

c) We have to prove the existence of a harmonic function u_0 satisfying

$$\partial_{y_k}u_0 \in \mathbf{C}^2(\Omega) \cap \mathbf{C}(\bar\Omega) \quad \text{and} \quad u_0(x) \to u_0(y) = 0 \quad \text{for} \quad x \to y \in \partial\Omega \setminus U(y^0)$$

and $u_0(x) \to u_0(y) > 0$ on $U(y^0)$. It is sufficient to prove the existence of a u_0 such that $\partial_{y_k}u_0 \in \mathbf{C}^2(\Omega) \cap \mathbf{C}(\bar\Omega)$ and the divergence theorem can be applied. Let Ω_0 be a further domain such that $\Omega \subset \Omega_0$ and $\partial\Omega \cap \partial\Omega_0 = \partial\Omega \setminus U(y^0)$ (see Fig. 3.4).

The Green function $G_{\mathrm{L}}^{\Omega_0}$ of Ω_0 exists and has the property that

$$x \mapsto \partial_{x_k}G_{\mathrm{L}}^{\Omega_0}(x,y), \qquad k = 1, 2, \ldots, n,$$

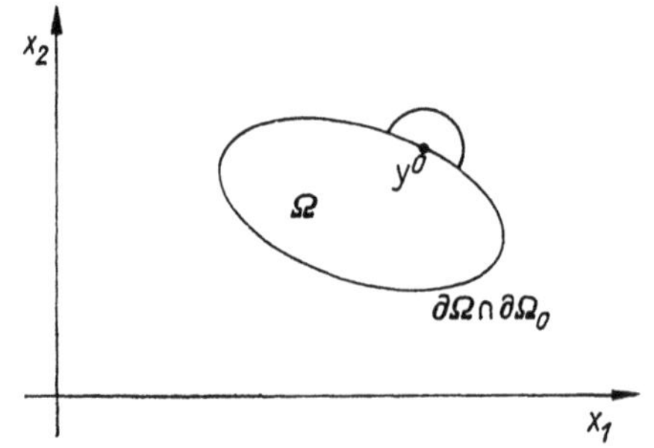

Fig. 3.4

is continuous on $\partial\Omega_0$, $y \in \Omega$ fixed (see LICHTENSTEIN [1]). On $U(y^0) \subset \partial\Omega$ we consider a sufficiently smooth function σ satisfying $\operatorname{supp} \sigma \subset U(y^0)$. Then the function u_0 defined by

$$u_0(x) = G_L^{\Omega_0}\mu_\sigma(x) = \int_{U(y^0)} G_L^{\Omega_0}(x, y)\, \sigma(y)\, dS(y) \tag{12}$$

has the properties to prove (11). ∎

Remark 3.5. If we put in (9) $u(y) = G_L(x, y)$, $x \notin \Omega$, and $\varrho\colon \bar{\Omega} \mapsto R$ satisfies $\partial_{y_k}\varrho(y) = 0$, then the divergence theorem reads

$$\int_\Omega \partial_{y_k} G_L(x, y)\, \varrho(y)\, dy = \int_{\partial\Omega} G_L(x, y)\, \varrho(y) \cos(y_k, \mathfrak{n})\, dS(y). \tag{13}$$

The left-hand side is the potential of a special dipole distribution (see (3.1.13)) called by L. Schwartz a distribution of the first order. From 1945–1952 L. Schwartz developed the theory of distributions, which is a powerful tool in mathematical analysis (see HÖRMANDER [2]). The theory of distributions has a long history. Studying the wave equation by means of its fundamental solution certain divergent integrals appear. In that direction there are important contributions by V. Volterra (1895), J. Hadamard (1923), S. L. Sobolev (1936), M. Riesz (1949) and other scientists. From (13) it follows that the special dipole distribution can be replaced by a potential of a single layer μ_σ, where

$$\sigma(y) := \varrho(y) \cos(y_k, \mathfrak{n}).$$

If ϱ is positive then σ has positive and negative values, which depend on the sign of $\cos(y_k, \mathfrak{n})$. The formula (13) has consequences for the inverse problem of magnetic fields, especially in geophysics (RÖSLER [1]) and neurosciences (NUNEZ [1], PLONSEY [1]).

Remark 3.6. In the papers of STRAKHOV and BRODSKY [1] and BRODSKY [1] one can find in $\mathbb{R}^2$ examples for the unique determination of ϱ by the exterior potential $G_L\mu_\varrho$ and the non-uniqueness of ϱ for non-convex polygons. Let $\Omega \subset \mathbb{R}^3$ be a polyhedron homeomorphic to a sphere. If $\varrho = \varrho_0 = \text{const}$ then the density ϱ_0 is uniquely determined by its exterior potential $G_L\mu_{\varrho_0}$ (STRAKHOV and BRODSKY [2]). That result is different to corresponding results in $\mathbb{R}^2$. One can find investigations about the stability of ϱ in LORENZI and PAGANI [1] and PAGANI [1].

Remark 3.7. If the potential $G_L\mu_\varrho$ is known on the surface $\partial\Omega$ it can be computed everywhere outside $\bar{\Omega}$ by a solution of the Dirichlet problem. The boundary $\partial\Omega$ has dimension $n - 1$. From Theorem 3.4 it follows that the special density depends on $(n - 1)$ variables and is therefore uniquely determined by its potential outside Ω. It seems that this is a general principle in potential theory. If measurements of the potential are possible on a set having dimension $n - 1$ then the mass distribution ϱ can be uniquely determined on a set having dimension $n - 1$. See also the results in Chap. 4 and Chap. 5, especially the case of a harmonic density ϱ, i.e., $\Delta\varrho = 0$, and the case of extremal elements of the set $\mathfrak{B}(\mu)$.

It is not difficult to give further counterexamples for the non-uniqueness of the inverse problem for dipole distributions. Let $\Omega \subset \mathbb{R}^n$ be a domain with smooth boundary $\partial\Omega$,

$$u(x) = \int_{\partial\Omega} u(y)\, P(x, y)\, \mathrm{d}S(y)$$

the solution of the Dirichlet problem, where in the case of a ball $\Omega = B(x^0, R)$ the Possion kernel P was defined in (2.1.8). If $u(y) = G_{\mathrm{L}}(y, z)$, $z \notin \Omega$, we obtain (see (3.4.44))

$$G_{\mathrm{L}}(x, z) = \int_{\partial\Omega} G_{\mathrm{L}}(y, z)\, P(x, y)\, \mathrm{d}S(y).$$

Differentiating this relation with respect to x we get

$$\partial_x^\alpha G_{\mathrm{L}}(x, z) = \int_{\partial\Omega} G_{\mathrm{L}}(y, z)\, \partial_x^\alpha P(x, y)\, \mathrm{d}S(y) = \int_{\partial\Omega} G_{\mathrm{L}}(y, z)\, \sigma_x(y)\, \mathrm{d}S(y), \quad (14)$$

where

$$\sigma_x(y) := \partial_x^\alpha P(x, y).$$

If $z \notin \Omega$, the multipole $\partial_x^\alpha G_{\mathrm{L}}(x, z)$ concentrated at the point $x \in \Omega$ – a special distribution of order $|\alpha|$ – can be replaced by a potential of a single layer μ_{σ_x}. Integrating the left-hand side of (14) with respect to a volume distribution $\varrho(x)\,\mathrm{d}x$ one gets the potential of multipoles concentrated on $\bar{\Omega}$. The potential has then the form

$$\int_{\Omega} \left(\partial_x^\alpha G_{\mathrm{L}}(x, z)\right) \varrho(x)\, \mathrm{d}x = \int_{\partial\Omega} G_{\mathrm{L}}(y, z)\, \sigma_x(y)\, \mathrm{d}S(y), \quad (15)$$

also the potential of a special distribution of order $|\alpha|$. In that direction a potential theory has to be further developed. One can find a potential theory for distributions of finite order with finite energy integrals in Deny [1]. The remarks concerning the non-uniqueness of the inverse dipole problem may have consequences in brain research (Nunez [1], Plonsey [1]) and geophysics (Rösler [1], Zidarov [1]).

3.3. Some remarks on integration theory

In mathematics it is often more suitable to consider the energy integral

$$\int G_{\mathrm{L}}\mu\, \mathrm{d}\lambda = \int G_{\mathrm{L}}\lambda\, \mathrm{d}\mu \quad (1)$$

than the potential $G_{\mathrm{L}}\mu$, where $G_{\mathrm{L}}\lambda$ is continuous. The potential $G_{\mathrm{L}}\mu$ can be discontinuous (see Example 3.2) while the mapping

$$G_{\mathrm{L}}\lambda \mapsto \int G_{\mathrm{L}}\lambda\, \mathrm{d}\mu \quad (2)$$

is a continuous linear form on the space of continuous functions. In classical potential theory λ can be a measure of the special form μ_ϱ or μ_σ, which produce continuous potentials $G_L\mu_\varrho$ resp. $G_L\mu_\sigma$. In the case of an arbitrary bounded domain $\Omega \subset \mathbb{R}^n$ the measure λ can be concentrated on the boundary $\partial\Omega$. In applications the given measure μ is concentrated on $\bar{\Omega}$ but the potential $G_L\mu$ is only known on $\partial\Omega$ and by solving the Dirichlet problem, outside $\bar{\Omega}$. In geodesy, geophysics, solid-state physics, and so on, the boundaries $\partial\Omega$ are not smooth. But these domains have the property that the set of all restrictions $\operatorname{rest}_{\partial\Omega} G_L\lambda$, where λ is concentrated on $\partial\Omega$ and $G_L\lambda$ continuous, is dense in $C(\partial\Omega)$. Using relation (1) it is possible to apply the results of linear analysis and on convex sets on infinite-dimensional functional spaces to inverse problems. Later we introduce a measure μ as a linear form on the space of continuous functions. Let μ be concentrated on $\bar{\Omega}$. Then $y \mapsto f(y) = G_L(x, y)$, $x \notin \bar{\Omega}$, is continuous on $\bar{\Omega}$. In this way the potential $G_L\mu(x)$ can be defined as the value of the linear form at f. If $x \in \bar{\Omega}$ then we consider an increasing sequence $G_1, G_2, \ldots$ of functions continuous on $\bar{\Omega} \times \bar{\Omega}$ (see Example 3.1) and define for a positive measure μ the potential $G_L\mu$ as the limit of the increasing sequence $G_1\mu, G_2\mu, \ldots$, where the $G_k\mu$ are continuous.

The reader not familiar with the foundations of measure theory should consider relation (1) only for the classical measures $\mu = \mu_\varrho$ and $\lambda = \mu_\sigma$ and notice that the set of all functions $\operatorname{rest}_{\partial\Omega} G_L\mu_\sigma$ is dense in $C(\partial\Omega)$. The first time relation (1) was applied to inverse problems by G. Lauricella in 1910–1912 using the special measures μ_ϱ and μ_σ. Similar relations like (1) hold for each locally integrable fundamental solution G. In the case of a non-symmetric kernel the adjoint potential $G^*\mu$ is to be considered. In the case of the wave equation and equations of higher order potentials relative to a distribution of finite order are to be defined. Such potentials consist of finitely many sums (ANGER [4], DENY [1], SCHULZE and WILDENHAIN [1]). *We notice that in relation* (3.3.1) *we do not need derivatives nor any assumptions on the smoothness of* $\partial\Omega$. *There is a one-to-one relation between problems in physics and corresponding problems on the space of continuous functions* (*see Sect.* 4.1).

At first we prove some properties for classical measures. In Definition 3.1 we introduce the notion of a measure having the properties of Theorem 3.1. Using a limit like in Example 3.1 it is possible to define the potential $G_L\mu$ of a measure μ. We often write $\mu(f)$ instead of $\int f(x)\, d\mu(x)$, and so on.

Theorem 3.5. *Let μ be one of the classical measures defined in* (3.1.4), (3.1.5) *or* (3.1.6). *To each compact set $K \subset \mathbb{R}^3$ there exists a constant $M_K \geqq 0$, such that for all $f \in C_0(\mathbb{R}^n)$ satisfying* $\operatorname{supp} f \subset K$ *the inequality*

$$|\mu(f)| \leqq M_K \|f\|_C \tag{3}$$

holds.

Proof. We prove the inequality for

$$\mu_\varrho(f) = \int_\Omega f(x)\, \varrho(x)\, dx. \tag{4}$$

If $f \in \mathrm{C}_0(\mathbb{R}^n)$ the integral can be extended over an unbounded open set $\Omega \subset \mathbb{R}^n$. Let $\operatorname{supp} f \subset K$. Then

$$\int_{\Omega} f(x)\,\varrho(x)\,\mathrm{d}x = \int_{\Omega \cap K} f(x)\,\varrho(x)\,\mathrm{d}x. \tag{5}$$

Using a fundamental property of classical integration theory we obtain

$$\begin{aligned}\left|\int_{\Omega\cap K} f(x)\,\varrho(x)\,\mathrm{d}x\right| &\leqq \int_{\Omega\cap K} |f(x)|\,|\varrho(x)|\,\mathrm{d}x \leqq \left(\max_{x\in\Omega\cap K}|f(x)| \int_{\Omega\cap K}|\varrho(x)|\,\mathrm{d}x\right)\\ &\leqq M_K \max_{x\in\mathbb{R}^n}|f(x)| = M_K\,\|f\|_{\mathrm{C}},\end{aligned} \tag{6}$$

where

$$M_K = \int_{\Omega\cap K} |\varrho(x)|\,\mathrm{d}x. \;\blacksquare \tag{7}$$

The Dirac measure δ_x and the measures (3.1.7) and (3.1.8) also satisfy (3).

Remark 3.8. Let $\Omega = \mathbb{R}^n$ and $\varrho = 1$. Then

$$\mu_\varrho(f) = \int_{\mathbb{R}^n} f(x)\,\mathrm{d}x \tag{8}$$

exists for each $f \in \mathrm{C}_0(\mathbb{R}^n)$ but not for $f = 1$. Therefore the space $\mathrm{C}_0(\mathbb{R}^n)$ consisting of all continuous functions having a compact support is necessary to develop a measure theory on the basis of linear form (integrals). If $\Omega \subset \mathbb{R}^n$ is a bounded domain and $K = \bar{\Omega}$ then the continuous linear forms on $\mathrm{C}(K)$ can be applied. These special linear forms can be considered as linear forms on $\mathrm{C}_0(\mathbb{R}^n)$.

We now sketch some results of integration theory on the real line $\mathbb{R} = \mathbb{R}^1$. Let $\varrho\colon [a, b] \mapsto \mathbb{R}$ be an integrable function. We define

$$\chi_1(x) := \int H(x-y)\,\varrho(y)\,\mathrm{d}y = \int H(x-y)\,\mathrm{d}\mu_1(y), \tag{9}$$

where H is the Heaviside function defined in (1.3.3) and

$$\mu_1(f) := \int_a^b f(y)\,\varrho(y)\,\mathrm{d}y. \tag{10}$$

Further, we introduce

$$\chi_2(x) := \sum_{k=1}^{\infty} c_k H(x-y^k) = \int H(x-y)\,\mathrm{d}\mu_2(y), \tag{11}$$

where, formally written,

$$\mu_2 = \sum_{k=1}^{\infty} c_k \delta_{y^k}. \tag{12}$$

The functions χ_1 and χ_2 are potentials of equation (1.3.1). Let $\varrho \geqq 0$, then χ_1 is an increasing function (see Fig. 3.5), χ_2 is a *step function* (see Fig. 3.6).

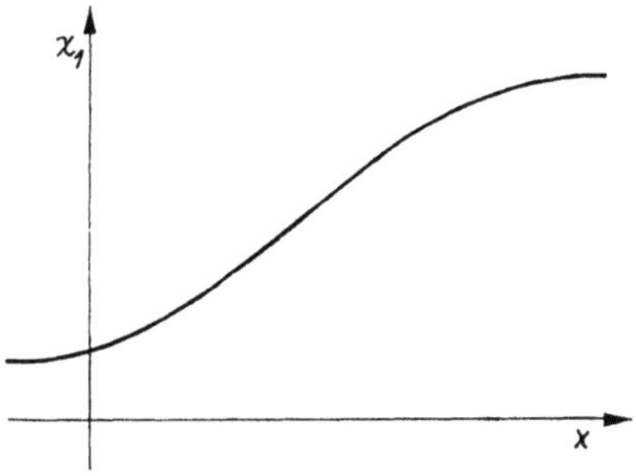

Fig. 3.5

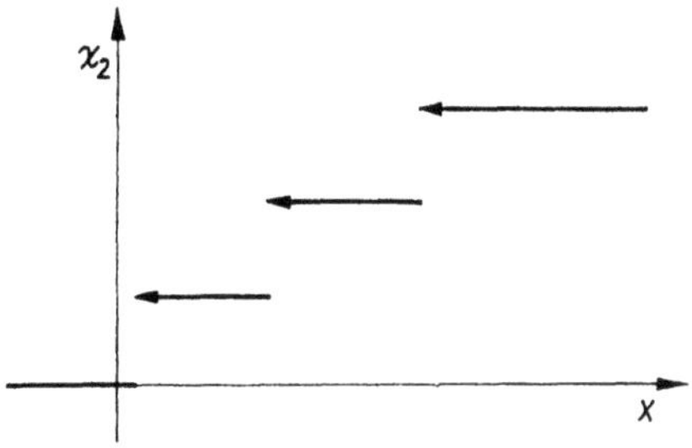

Fig. 3.6

Let μ be a positive measure with $\int \mathrm{d}\mu = M$ (see Definition 3.1). Then there exists a one-to-one relation (see Theorem 3.14) between μ and a point function $\chi\colon \mathbb{R} \mapsto \mathbb{R}$ defined by

$$\chi(x) := \int H(x - y)\, \mathrm{d}\mu(y). \tag{13}$$

This definition also holds in $\mathbb{R}^n$ (see (2.3.19)). We have already studied χ_1 and χ_2 which are parts of χ. Further, there exists a third part (Halmos [1]) $\chi_3 := \chi - \chi_1 - \chi_2$, the so-called *singular part*, having the property that χ_3 is continuous and $\chi_3'(x) = 0$ almost everywhere (i.e., $\chi_3'(x) = 0$ except a set of Lebesgue measure zero). Let us construct such a function χ_3.

In applications often only χ_1 and χ_2 are studied. In order to develop a rich theory using results of functional analysis the third part χ_3 of χ plays sometimes an important role. Such special functions have found applications in physics (Peitgen and Richter [1]). Let K_c be the *Cantor set* defined as in Fig. 3.7.

Fig. 3.7

Let $I_{0,1}$ be the open interval $\left]\frac{1}{3}, \frac{2}{3}\right[$. Then $K_1 := [0, 1] \setminus \left]\frac{1}{3}, \frac{2}{3}\right[$ consists of the two closed intervals $\left[0, \frac{1}{3}\right]$ and $\left[\frac{2}{3}, 1\right]$. Let $K_2 := K_1 \setminus \left]\frac{1}{9}, \frac{2}{9}\right[\cup \left]\frac{7}{9}, \frac{8}{9}\right[$. The set K_2 consists of four closed intervals, namely $\left[0, \frac{1}{9}\right]$, $\left[\frac{2}{9}, \frac{3}{9}\right]$, $\left[\frac{6}{9}, \frac{7}{9}\right]$, $\left[\frac{8}{9}, 1\right]$. In the next step we take away the middle third of each of these four intervals and get a set K_3. The Cantor set

$$K_c := \bigcap_{j=1}^{\infty} K_j \tag{14}$$

is that which is left after removing thirds in this way ad infinitum. After j steps, the total length of the union of the remaining closed intervals is $(2/3)^j$.

Since $K_c \subset K_j$ the measure of K_c is $< \varepsilon$, i.e. K_c is a non-denumerable set having the Lebesgue measure zero. The last property is a possibility to introduce sets of Lebesgue measure zero. We now define the function χ_3 as follows:

$$\chi_3(x) = 0 \quad \text{if} \quad x \leqq 0, \qquad \chi_3(x) = 1 \quad \text{if} \quad x \geqq 1, \qquad \chi_3(x) = \frac{1}{2} \quad \text{on} \quad \left]\frac{1}{3}, \frac{2}{3}\right[,$$

$$\chi_3(x) = \frac{1}{4} \quad \text{on} \quad \left]\frac{1}{9}, \frac{2}{9}\right[, \qquad \chi_3(x) = \frac{3}{4} \quad \text{on} \quad \left]\frac{7}{9}, \frac{8}{9}\right[, \tag{15}$$

$$\chi_3(x) = \frac{1}{8}, \frac{3}{8}, \frac{5}{8}, \frac{7}{8} \quad \text{(from the left to the right)}$$

on the open middle third parts of the four still remaining closed intervals, and so on. In this way the increasing function χ_3 is defined on $[0, 1] \setminus K_c$. This set is dense on $[0, 1]$. We define for $x \in K_c$

$$\chi_3(x) := \lim_{t \to x} \chi_3(t), \qquad t \in [0, 1] \setminus K_c. \tag{16}$$

Now χ_3 is defined on $[0, 1]$ and continuous. We get $\chi_3'(x) = 0$ for $x \notin K_c$, i.e., $\chi_3'(x) = 0$ holds on $\mathbb{R}$ except a set of Lebesgue measure zero.

A further singular function is *Salem's function* defined in the following way (Salem [1]). Let us consider in $\mathbb{R}^2$ the straight line $\overline{xy}$ joining the point $x = (x_1, x_2)$ to the point $y = (y_1, y_2)$ of the Cartesian coordinates $y_1 = x_1 + \Delta x_1$, $y_2 = x_2 + \Delta x_2$, $\Delta x_1 > 0$, $\Delta x_2 > 0$. Let λ_0, λ_1 be two numbers, essentially positive, such that $\lambda_0 + \lambda_1 = 1$, $\lambda_0 \neq \lambda_1$. Let us now consider the point z whose coordinates are $x_1 + (\Delta x_1)/2$, $x_2 + \lambda_0 \Delta x_2$. The vertical distance between x and z is $\lambda_0 \Delta x_2$, and the vertical distance between z and y is $\lambda_1 \Delta x_2$. If we replace the straight line $\overline{xy}$ by the broken line x, z, y, we say that we perform the transformation $T(\lambda_0, \lambda_1)$ on $\overline{xy}$. Let now f_0 be for $0 \leqq x \leqq 1$ the function equal to x, joining the origin $O = (0, 0)$ to the point $A(1, 1)$. Let us perform the transformation $T(\lambda_0, \lambda_1)$ on $\overline{OA}$. We get a broken line f_1. Let us perform the transformation $T(\lambda_0, \lambda_1)$ on each those two straight lines. Proceeding in the same way we get after p operations a function $f_p(x)$ strictly increasing $\big(f_p(0) = 0, f_p(1) = 1\big)$. The f_p converges uniformly to a continuous function f satisfying $f(0) = 0$, $f(1) = 1$, for which $f'(x) = 0$ holds almost everywhere. Putting $\chi_3(x) = 0$ if $x \leqq 0$, $\chi_3(x) = f(x)$ if $0 \leqq x \leqq 1$, $\chi_3(x) = 1$ if $x > 1$, we get a singular function strictly increasing on $[0, 1]$.

The function χ defined in (13) is left-side continuous, i.e.,

$$\lim_{t \to x^-} \chi(t) = \chi(x), \tag{17}$$

and continuous at x^0 if and only if $\mu(\{x^0\}) = 0$. For more details see Halmos [1]. Let $\chi = \chi_1 + \chi_2 + \chi_3$ be given, where χ_1 has the form (9), χ_2 the form (11) and χ_3 is continuous satisfying $\chi_3'(x) = 0$ almost everywhere. Beginning the integration theory with χ it is possible to construct the measure μ defined on subsets of $\mathbb{R}$ in the following way: Let $[a', b'[\subset \mathbb{R}$ be a semi-closed interval. We define the measure of $[a', b'[$ by

$$\mu([a', b'[) := \chi(b') - \chi(a'). \tag{18}$$

Using limits it is possible to introduce μ on certain subsets (the Borel sets) of $\mathbb{R}$ (see HALMOS [1]). If the function χ is given the *Riemann-Stieltjes integral* is defined similar to the Riemann integral in the following way. Let

$$a = x_0 < x_1 < x_2 < \cdots < x_{n-1} < x_n = b$$

be a decomposition, such that $x_k - x_{k-1}$ tends to zero for $n \to \infty$. Then

$$\int f(x)\,\mathrm{d}\chi(x) := \lim_{n\to\infty} \sum_{k=1}^{n} f(\xi_k)\left(\chi(x_k) - \chi(x_{k-1})\right), \qquad \xi_k \in [x_k, x_{k-1}]. \tag{19}$$

The Riemann-Stieltjes integral (19) exists if f is continuous. In the case of

$$\chi_0(x) = H(x - x^0) \tag{20}$$

we get

$$\int f(x)\,\mathrm{d}\chi_0(x) = f(x^0). \tag{21}$$

The measure which corresponds to χ_0 is the Dirac measure δ_{x^0}. Let $K = [a, b]$, $-\infty < a < b < +\infty$. In 1909 F. Riesz proved the following fundamental Theorem 3.6. In functional analysis a linear form L on $\mathbf{C}(K)$, i.e.,

$$L(c_1 f_1 + c_2 f_2) = c_1 L(f_1) + c_2 L(f_2)$$

for all $f_1, f_2 \in \mathbf{C}(K)$ and all $c_1, c_2 \in \mathbb{R}$, is continuous if and only if the following inequality holds (DUNFORD and SCHWARTZ [1]):

$$|L(f)| \leqq M\|f\|_{\mathbf{C}}, \quad \text{where} \quad \|f\|_{\mathbf{C}} := \max\{|f(x)|, x \in K\}. \tag{22}$$

Theorem 3.6. *To each continuous linear form $L\colon \mathbf{C}(K) \to \mathbb{R}$ there exists a uniquely determined χ of bounded variation such that L can be represented as Riemann-Stieltjes integral*

$$L(f) = \int f(x)\,\mathrm{d}\chi(x), \qquad f \in \mathbf{C}(K). \tag{23}$$

Remark 3.9. In 1918 J. Radon proved that a similar representation holds for each compact set $K \subset \mathbb{R}^n$. In $\mathbb{R}^n$ there exists a uniquely determined measure μ concentrated on K such that

$$L(f) = \int f\,\mathrm{d}\mu, \qquad f \in \mathbf{C}(K), \tag{24}$$

holds. Therefore one often writes μ instead of L.

Remark 3.10. Writing $f(\xi_k) = \delta_{\xi_k}(f)$ and $c_k = \chi(x_k) - \chi(x_{k-1})$ in (19) one sees that

$$\int f(x)\,\mathrm{d}\chi(x) = \lim_{n\to\infty} \sum_{k=1}^{n} c_k \delta_{\xi_k}(f) \quad \text{for each continuous } f, \tag{25}$$

i.e., the Riemann-Stieltjes integral is the so-called *weak limit* of finite sums of Dirac measures. In physics the right-hand side of (25) is often used to introduce the notion of a mass distribution. This definition gives possibilities to approximate a given field like the earth's gravitational field. The points ξ_k and the weights c_k are to be determined by measurements (see Sect. 2.1 and BARTHELMES [1]).

A measure μ is defined as a continuous linear form on $\mathbf{C}(K)$ resp. on $\mathbf{C}_0(\mathbb{R}^n)$, where in the case of a measure concentrated on K the space $\mathbf{C}(K)$ is easier to handle. Further, the measure μ can be introduced as the weak limit of sums of a finite number of Dirac measures (BOURBAKI [1]). Then this linear form must be extended by limits to discontinuous functions. Let 1_B be the *characteristic* or *indicator function* of the set $B \subset \mathbb{R}^n$, i.e.,

$$1_B(x) = 1 \quad \text{if} \quad x \in B, \qquad 1_B(x) = 0 \quad \text{if} \quad x \notin B. \tag{26}$$

Then the measure $\mu(B)$ of B is defined by $\mu(1_B) =: \mu(B)$, and so on.

Definition 3.1. A *measure* is a linear form $\mu\colon \mathbf{C}_0(\mathbb{R}^n) \mapsto \mathbb{R}$ such that for each compact set $K \subset \mathbb{R}^n$ there exists a constant $M_K \geqq 0$ satisfying

$$|\mu(\varphi)| \leqq M_K \|\varphi\|_{\mathbf{C}} \tag{27}$$

for each $\varphi \in \mathbf{C}_0(\mathbb{R}^n)$ with $\operatorname{supp} \varphi \subset K$. One often writes $\int \varphi \, d\mu$ for $\mu(\varphi)$.

Definition 3.2. A linear form $\mu\colon \mathbf{C}_0(\mathbb{R}^n) \mapsto \mathbb{R}$ is called *positive,* if $\mu(\varphi) \geqq 0$ for all $\varphi \geqq 0$.

Special measures are μ_ϱ, μ_σ, μ_τ, $\mu_{\delta,\varrho}$, $\mu_{\delta,\sigma}$ defined in (3.1.4), (3.1.5), (3.1.6), (3.1.7), and (3.1.8). If the densities ϱ, σ, τ and the a_k and b_k are positive then these measures are positive.

Proposition 3.1. *If* $\mu(\varphi) \geqq 0$ *for all* $\varphi \geqq 0$ *satisfying* $\operatorname{supp} \varphi \subset K$ *then* (27) *holds for all* $\varphi \in \mathbf{C}_0(\mathbb{R}^n)$ *with* $\operatorname{supp} \varphi \subset K$.

Proof. Let $\varphi_0 \in \mathbf{C}_0(\mathbb{R}^n)$ be a function with $\varphi_0 \geqq 0$ and $\varphi_0(x) = 1$ on K. At first we consider functions $\varphi \in \mathbf{C}_0(\mathbb{R}^n)$ satisfying $\|\varphi\|_{\mathbf{C}} \leqq 1$. Then

$$\varphi_0 + \varphi \geqq 0 \quad \text{and} \quad \varphi_0 - \varphi \geqq 0 \text{ on } K.$$

From $\mu(\psi) \geqq 0$ for all $\psi \geqq 0$ it follows that

$$\mu(\varphi_0 + \varphi) \geqq 0 \quad \text{and hence} \quad -\mu(\varphi_0) \leqq \mu(\varphi).$$

In a similar way we obtain from

$$\mu(\varphi_0 - \varphi) \geqq 0 \quad \text{always} \quad \mu(\varphi_0) \geqq \mu(\varphi).$$

Using both inequalities we get

$$-\mu(\varphi_0) \leqq \mu(\varphi) \leqq \mu(\varphi_0)$$

which is equivalent to

$$|\mu(\varphi)| \leqq \mu(\varphi_0) =: M_K.$$

If $\|\varphi\|_{\mathbf{C}} \geqq 1$ we take $\psi = \varphi/\|\varphi\|_{\mathbf{C}}$. Then

$$|\mu(\varphi/\|\varphi\|_{\mathbf{C}})| \leqq M_K \quad \text{or} \quad |\mu(\varphi)| \leqq M_K\|\varphi\|_{\mathbf{C}}. \blacksquare$$

In BOURBAKI [1] it is proved that each measure μ can be represented as difference of two positive measures μ^+ and μ^-, where

$$\mu^+(\varphi) := \sup \left\{\int \psi \,\mathrm{d}\mu,\ \psi \in \mathbf{C}_0(\mathbb{R}^n),\ 0 \leqq \psi \leqq \varphi\right\}, \tag{28}$$

$$\mu^-(\varphi) := \sup \left\{-\int \psi \,\mathrm{d}\mu,\ \psi \in \mathbf{C}_0(\mathbb{R}^n),\ 0 \leqq \psi \leqq \varphi\right\}. \tag{29}$$

$$|\mu| := \mu^+ + \mu^-. \tag{30}$$

Then $\mu = \mu^+ - \mu^-$. Further, we define

$$\|\mu\| := \sup \{\mu(\varphi),\ \varphi \in \mathbf{C}_0(\mathbb{R}^n),\ \|\varphi\| \leqq 1\}. \tag{31}$$

Definition 3.3. The support $\operatorname{supp} \mu$ of a measure $\mu: \mathbf{C}_0(\mathbb{R}^n) \mapsto \mathbb{R}$ is the *complement of the greatest open set* Ω_0 such that $\mu(\varphi) = 0$ for all $\varphi \in \mathbf{C}_0(\mathbb{R}^n)$ satisfying $\operatorname{supp} \varphi \subset \Omega_0$.

Remark 3.11. The support of the Dirac measure δ_{x^0} is the set $\{x^0\}$, the support of μ_ϱ is contained in Ω, the support of μ_σ in $\partial\Omega$.

Let $\mu: \mathbf{C}_0(\mathbb{R}^n) \mapsto \mathbb{R}$ be a measure with $\operatorname{supp} \mu \subset K$, $\nu: \mathbf{C}(K) \mapsto \mathbb{R}$ a continuous linear form. Let $f := \operatorname{rest}_K \varphi$, $\varphi \in \mathbf{C}_0(\mathbb{R}^n)$. The function f is called the *restriction* of φ on K, φ a *continuation* of f. To each f there exist different continuations (Tietze's theorem, see DUNFORD and SCHWARTZ [1]). Let $K = \bar{\Omega}$, $\Omega \subset \mathbb{R}^n$ be an open bounded set and

$$J(K) := \{\varphi \in \mathbf{C}_0(\mathbb{R}^n) : \varphi(x) = 0 \text{ on } K\} \tag{32}$$

a so-called *ideal* of $\mathbf{C}_0(\mathbb{R}^n)$, i.e., $\psi \in J(K)$ and $\varphi \in \mathbf{C}_0(\mathbb{R}^n)$ yield $\varphi\psi \in J(K)$. The difference of two continuations φ_1, φ_2 of f is an element of $J(K)$,

$$\varphi_1 - \varphi_2 \in J(K).$$

Let $f = \operatorname{rest}_K \varphi$, $\varphi \in \mathbf{C}_0(\mathbb{R}^n)$. We define

$$\nu_1(f) := \int \varphi \,\mathrm{d}\mu, \tag{33}$$

which is a continuous linear form on $\mathbf{C}(K)$ since

$$|\nu_1(f)| = \left|\int \varphi \,\mathrm{d}\mu\right| \leqq M_K \sup \{|\varphi(x)|,\ x \in K\} = M\|f\|_{\mathbf{C}}.$$

If $\nu: \mathbf{C}(K) \mapsto \mathbb{R}$ is a continuous linear form on $\mathbf{C}(K)$, then we define

$$\mu(\varphi) := \int \operatorname{rest}_K \varphi \,\mathrm{d}\nu, \tag{34}$$

where φ is a continuation of f. The support of μ is on K and

$$|\mu(\varphi)| \leqq M \sup \{|f(x)|,\ x \in K\} = M\|\varphi\|_{\mathbf{C}} \tag{35}$$

holds. There is a one-to-one relation between ν and μ with $\operatorname{supp}\mu \subset K$. In measure theory one does not distinguish between ν and μ.

We have to extend a measure $\mu: \mathbb{C}_0(\mathbb{R}^n) \mapsto \mathbb{R}$ to more general functions f. This extension is made by limits, especially by sequences of increasing resp. decreasing functions. It is well known that a real number $a \in \mathbb{R}$ can be introduced as the limit of an increasing (decreasing) sequence of rational numbers. The system of real numbers is closed in the following sense: each increasing (decreasing) sequence of real numbers has a limit $b \in \overline{\mathbb{R}}$. If we consider the limit of an increasing (decreasing) sequence of continuous functions, this limit is a lower (upper) semicontinuous function. The following fundamental property holds: the limit of an increasing (decreasing) sequence of lower (upper) semicontinuous functions is lower (upper) semicontinuous. In applications many functions are lower semicontinuous. Examples are the fundamental solutions (2.1.19), (2.2.6), (2.3.19) and the indicator (characteristic) function 1_U of an open set $U \subset \mathbb{R}^3$. The indicator function 1_K of a compact set $K \subset \mathbb{R}^3$ is upper semicontinuous. Therefore the class of semicontinuous functions is important. In relation (40) we compare the classical definition of the potential of a volume distribution and our definition of a potential (see also (3.1.23)).

Let $f: \mathbb{R}^n \mapsto \mathbb{R}$ be continuous at x^0. Then for each $\varepsilon > 0$ there exists an open neighbourhood (ball) $B(x^0, r)$ such that

$$|f(x) - f(x^0)| < \varepsilon \quad \text{for} \quad |x - x^0| < \delta(\varepsilon) = r. \tag{36}$$

This inequality is equivalent to $-\varepsilon < f(x) - f(x^0) < \varepsilon$ or

$$-\varepsilon + f(x^0) < f(x) \quad \text{for all} \quad x \in B(x^0, r) \tag{37}$$

and

$$f(x) < f(x^0) + \varepsilon \quad \text{for all} \quad x \in B(x^0, r). \tag{38}$$

Definition 3.4. A function $f: \mathbb{R}^n \mapsto \mathbb{R}$ satisfying the inequality (37) is called *lower semicontinuous (l.s.c.) at* x^0, a function f satisfying (38) *upper semicontinuous (u.s.c.) at* x^0. If $f(x^0) = -\infty$ then f is by definition l.s.c. If $f(x^0) = +\infty$ then f is l.s.c. if for each $\lambda < f(x^0)$ there exists a neighbourhood $B(x^0, r)$ such that $\lambda < f(x)$ for all $x \in B(x^0, r)$. A function f is called lower (upper) semicontinuous if it is lower (upper) semicontinuous at each x^0.

Let $(f_j)_{j \in J}$ be a family of functions. We define

$$f = \sup\{f_j, j \in J\} \quad \text{by} \quad f(x) := \sup\{f_j(x), j \in J\}. \tag{39}$$

Theorem 3.7. *The following properties hold for lower semicontinuous functions:*

(i) *If f is l.s.c. and $K \subset \mathbb{R}^n$ a compact set, then f attains its infimum on K, i.e., there exists at least one point $x^0 \in K$ such that $f(x^0) = \inf\{f(x), x \in K\}$.*
(ii) *Let $(f_j)_{j \in J}$ be a family of l.s.c. functions. Then $f = \sup\{f_j, j \in J\}$ is l.s.c.*
(iii) *If f is l.s.c., then there exists an increasing sequence of continuous functions f_k with $f = \lim_{k\to\infty} f_k$.*

If f is l.s.c. then $g := -f$ is u.s.c. Theorem 3.7 holds for u.s.c functions, if in (i) the infimum is replaced by supremum, in (ii) the supremum is replaced by infimum and in (iii) the increasing sequence is replaced by a decreasing sequence.

In classical potential theory (KELLOGG [1], TIKHONOV and SAMARSKI [1]) the potential $G_L\mu_\varrho$ of a volume distribution μ_ϱ is defined, if $x \in \operatorname{supp} \mu_\varrho$, by the limit

$$G_L\mu_\varrho(x) := \lim_{r\to 0} \int_{\Omega\setminus B(x^0,r)} G_L(x, y)\, \varrho(y)\, \mathrm{d}y. \tag{40}$$

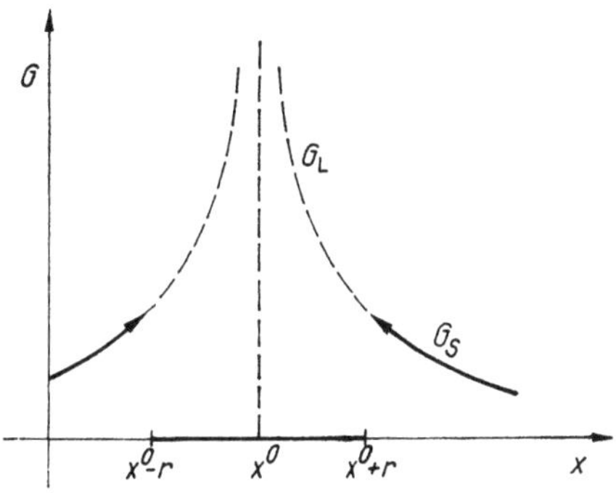

Fig. 3.8

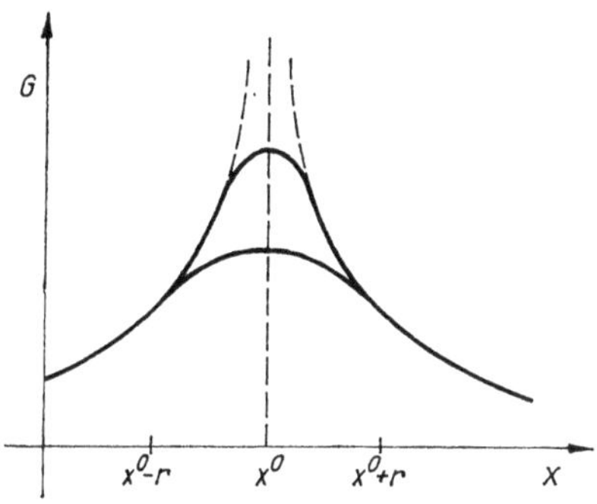

Fig. 3.9

Let $s = 1/r$ and

$$G_s(x, y) := \begin{cases} G_L(x, y) & \text{if} \quad |x - y| > \dfrac{1}{s}, \\ 0 & \text{if} \quad |x - y| \leqq \dfrac{1}{s}. \end{cases}$$

Then G_s is l.s.c. and $G_L(x, y) = \lim_{s\to\infty} G_s(x, y)$ is by Theorem 3.7 as the limit of an increasing sequence of l.s.c. functions a l.s.c. function (see Fig. 3.8). The approximation of G_L by continuous kernels G_k (see 3.1.20) is better because we use only continuous functions on the space $\mathbf{C}_0(\mathbb{R}^n)$ resp. $\mathbf{C}(K)$. With the aid of these spaces we get many results for potentials.

Let $\mathfrak{H}^u$ be the set of all l.s.c. functions, $\mathfrak{H}^0$ the set of all u.s.c. functions, $\mu\colon \mathbf{C}_0(\mathbb{R}^n) \mapsto \mathbb{R}$ a positive measure. We define for $h \in \mathfrak{H}^u$ (see ANGER and BAUER [1], BOURBAKI [1])

$$\mu^*(h) := \sup\,\{\mu(g),\ g \leqq h,\ g \in \mathbf{C}_0(\mathbb{R}^n)\} \tag{41}$$

and for $h \in \mathfrak{H}^0$

$$\mu_*(h) := \inf\,\{\mu(g),\ g \geqq h,\ g \in \mathbf{C}_0(\mathbb{R}^n)\} \tag{42}$$

(see Fig. 3.9 and Fig. 3.10). Then we define for a function $f\colon \mathbb{R}^n \mapsto \mathbb{R}$

$$\int^* f\, \mathrm{d}\mu := \inf\,\{\mu^*(h),\ h \in \mathfrak{H}^u,\ h \geqq f\} \tag{43}$$

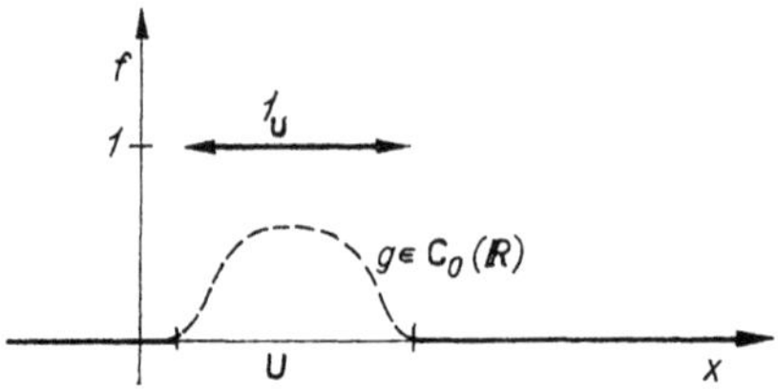

Fig. 3.10

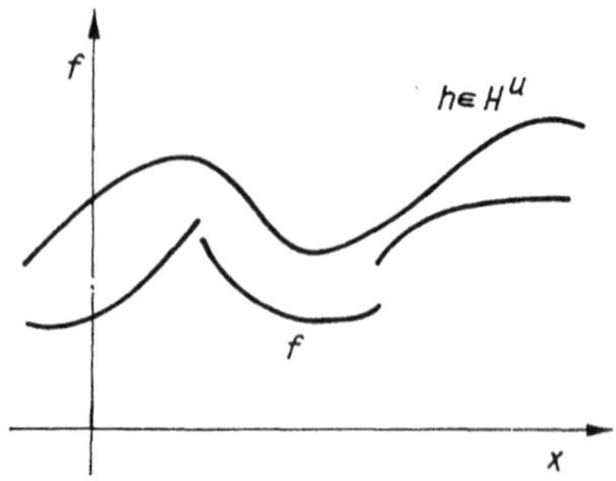

Fig. 3.11

and

$$\int_* f \, d\mu := \sup \{\mu^*(h), h \in \mathfrak{H}^0, h \leqq f\} \tag{44}$$

(see Fig. 3.11). If (43) and (44) are finite and equal, i.e.,

$$\int_* f \, d\mu = \int^* f \, d\mu =: \int f \, d\mu, \tag{45}$$

f is called μ-integrable. Each positive function $h \in \mathfrak{H}^u$ satisfying $\int {}^*h \, d\mu < \infty$ is μ-*integrable*. If $\mu = \mu^+ - \mu^-$ is the difference of two positive measures μ^+ and μ^- then

$$\int f \, d\mu := \int f \, d\mu^+ - \int f \, d\mu^-. \tag{46}$$

The set of all μ-integrable functions is denoted by $\mathbf{L}^1(\mu)$. In the case $\mu = \delta_{x^0}$ we get

$$\delta^*_{x^0}(h) = h(x^0) \quad \text{and} \quad \int^* f \, d\delta_{x^0} = f(x^0) = \int_* f \, d\delta_{x^0} \quad \text{if} \quad |f(x^0)| < \infty.$$

With respect to the measure δ_{x^0} each function bounded at x^0 is integrable. We remark that the measures μ_ϱ and μ_σ from Sect. 3.1. are special measures for which our theory also holds. Their (extended) integrals are called *Lebesgue integrals*. If f is Riemann integrable then relation (45) holds.

If $1_B \in \mathbf{L}^1(\mu)$, $B \subset \mathbb{R}^n$, then B is called μ-*integrable* and the value

$$\mu(B) := \int 1_B \, d\mu \tag{47}$$

is called the *measure* of B relative to μ.

Theorem 3.8. *The following results hold for* $\mu: \mathbf{C}_0(\mathbb{R}^n) \mapsto \mathbb{R}$:

(i) *Let f and g be μ-integrable. Then $f + g$ is μ-integrable and*

$$\int (f + g) \, d\mu = \int f \, d\mu + \int g \, d\mu. \tag{48}$$

(ii) *If f is μ-integrable and $c \in \mathbb{R}$ then (cf) is μ-integrable and*

$$\int (cf) \, d\mu = c \int f \, d\mu. \tag{49}$$

(iii) *If $\mu \geqq 0$ and f and g are μ-integrable functions satisfying $f \leqq g$ then*

$$\int f \, d\mu \leqq \int g \, d\mu. \tag{50}$$

Let f be a continuous function on the product space $\mathbb{R}^n \times \mathbb{R}^n$ with compact support, μ and ν measures on $\mathbb{R}^n$. Then there exists a uniquely determined measure $m\colon \mathbf{C}_0(\mathbb{R}^n \times \mathbb{R}^n) \mapsto \mathbb{R}$ such that

$$\int f(x, y) \,\mathrm{d}m(x, y) = \int \Big(\int f(x, y) \,\mathrm{d}\mu(y) \Big) \,\mathrm{d}\nu(x) = \int \Big(\int f(x, y) \,\mathrm{d}\nu(x) \Big) \,\mathrm{d}\mu(y). \tag{51}$$

The measure m is called the *product measure*. If $f \in \mathbf{L}^1(m)$ then (51) also holds. This relation is known as the *Fubini theorem*.

The further important theorems hold.

Theorem 3.9. *Let μ be a positive measure, $(f_k)_{k\in\mathbb{N}}$ an increasing sequence of μ-integrable functions. Then*

$$\lim_{k\to\infty} \int f_k \,\mathrm{d}\mu = \int^* f \,\mathrm{d}\mu, \quad \text{where} \quad f = \lim_{k\to\infty} f_k. \tag{52}$$

If f is μ-integrable then $\int^* f \,\mathrm{d}\mu = \int f \,\mathrm{d}\mu$.

Theorem 3.10. *If $(f_k)_{k\in\mathbb{N}}$ is a sequence of μ-integrable functions which converge to f and if g is a μ-integrable function such that $|f_k(x)| \leqq g(x)$ for all $k \in \mathbb{N}$, then f is μ-integrable and*

$$\lim_{k\to\infty} \int f_k \,\mathrm{d}\mu = \int f \,\mathrm{d}\mu. \tag{53}$$

One can find a proof of Theorem 3.9 and Theorem 3.10 in Anger and Bauer [1], Bauer [1], Bourbaki [1], Halmos [1].

3.4. Potentials, continuous potentials, C-capacity

We introduce the potential of a measure. Integrating the Newtonian potential on the whole $\mathbb{R}^n$ relative to the density $\varrho(y) = -\Delta\varphi(y)$, $\varphi \in \mathbf{C}_0^\infty(\mathbb{R}^n)$, one can prove the one-to-one correspondence between potential and measure. This theorem holds for each linear differential equation having a locally integrable fundamental solution and in a generalized sense for each linear differential equation. In order to prove the preceding fact the potential must be known at least on the support of the measure (sources). In applications the potential is only known outside the sources from which intrinsic ambiguities arise, i.e., different sources can give the same potential outside the sources. Further, $\partial\Omega$ is often not smooth like in geodesy, geophysics, space physics, solid-state physics, and so on. Therefore we have to develop a potential theory without any restrictions on $\partial\Omega$. The only condition we require is that the Green function be zero on $\partial\Omega$ (Theorem 4.5). Since we are working on the boundary and use the potential of a measure we have to introduce a concept of subsets which can be neglected integrating the potential with respect to a measure producing a continuous potential. In general the boundary of a bounded domain $\Omega \subset \mathbb{R}^n$ has n-dimensional Lebesgue measure zero, i.e., $\int \mathbf{1}_{\partial\Omega}(x)\,\mathrm{d}x = 0$. In the case of the Laplace equation this new notion of capacity comes from electrostatics and is the notion of capacity introduced in potential theory by N. Wiener in

1923. Our notion of $\mathbb{C}$-capacity can also be introduced relative to the heat equation (ANGER [4], DOOB [1], NETUKA and VESELY [1]), the equation (2.3.18) which has a bounded kernel, and to other equations (ANGER [4], MAZYA and SHAPOSHNIKOVA [1], SCHULZE and WILDENHAIN [1]). It seems that this notion is a central notion in mathematics. We define the $\mathbb{C}$-capacity by using continuous potentials. In this way problems between potentials are transformed into problems on the space of continuous functions and conversely.

Let $G: \mathbb{R}^n \times \mathbb{R}^n \mapsto \mathbb{R}$ be a mapping, such that

$$G^+ := \sup(G, 0), \qquad G^- := -\inf(G, 0) \tag{1}$$

are lower semicontinuous (l.s.c.). G is often called a *kernel*, G^+ and G^- the positive and negative part of G. The *adjoint kernel* is defined by

$$G^*(x, y) := G(y, x). \tag{2}$$

In integral equations kernels play an important role. It is well known that many concrete mappings in analysis can be represented by using kernels. Special kernels were defined in (1.2.4), (1.3.2), (1.3.18), (2.1.17)–(2.1.19), (2.1.38)–(2.1.40), (2.2.6), (2.3.19), (2.3.20), (2.3.25). A kernel can also be introduced relative to an open set $\Omega \subset \mathbb{R}^n$ or a compact set $K \subset \mathbb{R}^n$. More general kernels can be defined on locally compact spaces (DOOB [1], HELMS [1], LANDKOF [1]).

Definition 3.5. Let $G: \mathbb{R}^n \times \mathbb{R}^n \mapsto \mathbb{R}$ be a kernel such that G^+ and G^- are l.s.c., μ a positive measure. The potential $G\mu$ is defined at $x \in \mathbb{R}^n$ by

$$G\mu(x) := \int G^+(x, y)\,\mathrm{d}\mu(y) - \int G^-(x, y)\,\mathrm{d}\mu(y), \tag{3}$$

where $G\mu(x)$ assumes at most one of the values $+\infty$ und $-\infty$. If $\mu = \mu^+ - \mu^-$ the potential $G\mu$ is defined at x by

$$G\mu(x) := G^+\mu(x) - G^-\mu(x) \tag{4}$$

provided that $G^+\mu(x)$ and $G^-\mu(x)$ are defined. The adjoint potential is defined by

$$G^*\mu(x) := \int G^*(x, y)\,\mathrm{d}\mu(y) = \int G(y, x)\,\mathrm{d}\mu(y). \tag{5}$$

Let

$$G_\mathrm{D}(x, y) := H(x - y), \tag{6}$$

where H was defined in (1.3.3).

Remark 3.12. For the simplest differential equation $u' = 0$ we have already studied the potentials

$$G_\mathrm{D}\mu(x) := \int H(x - y)\,\mathrm{d}\mu(y) = \chi_1(x) + \chi_2(x) + \chi_3(x) \tag{7}$$

in Sect. 3.2. Here χ_1 was defined in (3.3.9), χ_2 in (3.3.11) and χ_3 is the singular part, i.e., $\chi_3' = 0$ almost everywhere.

Remark 3.13. The potentials of the classical measures μ_ϱ, μ_σ, μ_τ and the potentials of the special measures δ_{x^0}, $\mu_{\delta,\varrho}$, $\mu_{\delta,\sigma}$ are special cases of (3).

Remark 3.14. In order to define the potential $G\mu$ relative to a measure μ it is sufficient that G is measurable.

Theorem 3.11. *Let G be a kernel for which G^+ and G^- are l.s.c., μ a positive measure. Then the potential $G\mu$ is the difference of two l.s.c. functions.*

Proof. We prove the theorem in the case $G = G^+$. Since $G^+ : \mathbb{R}^n \times \mathbb{R}^n \mapsto \mathbb{R}$ is l.s.c. it can be represented as limit of an increasing sequence of continuous kernels $G_1, G_2, \ldots$ with compact support supp $G_k \subset \mathbb{R}^n \times \mathbb{R}^n$ (see Theorem 3.7). At first we prove that G_k is continuous.

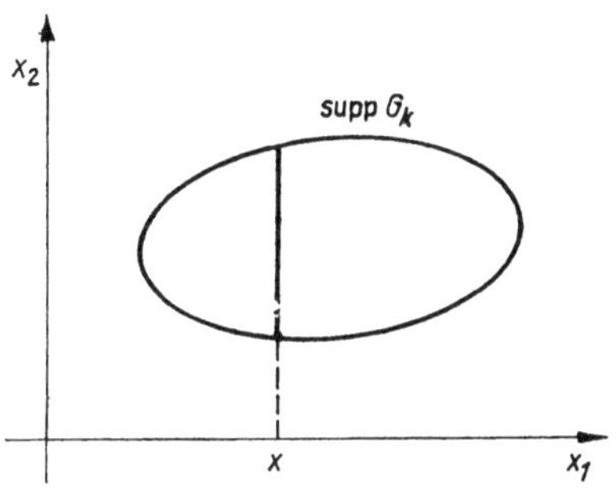

Fig. 3.12

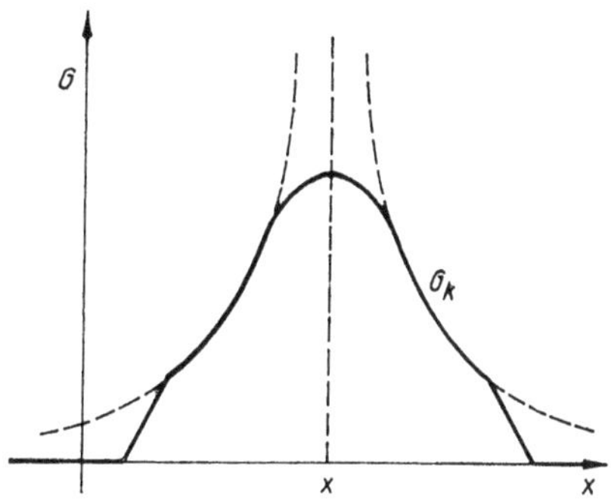

Fig. 3.13

Let $X := (x, y)$, $X' = (x', y')$. The continuity of $G_k : \mathbb{R}^n \times \mathbb{R}^n \mapsto \mathbb{R}$ means that to each $\varepsilon > 0$ there exists a $\delta(\varepsilon)$ such that

$$|G_k(x, y) - G_k(x', y')| < \varepsilon$$

for

$$|X - X'| = [(x_1 - x_1')^2 + \cdots + (x_n - x_n')^2 + (y_1 - y_1')^2 + \cdots + (y_n - y_n')^2]^{1/2} < \delta(\varepsilon).$$

The support of G_k is compact. Therefore the continuity is uniform, i.e., the quantity $\delta(\varepsilon)$ does not depend on $X = (x, y)$. Especially for $X' = (x', y)$ we get

$$|X - X'| = |x - x'| < \delta(\varepsilon).$$

The following estimates hold:

$$|G_k\mu(x) - G_k\mu(x')| = \left|\int \big(G_k(x, y) - G_k(x', y)\big)\, \mathrm{d}\mu(y)\right| \leqq \int |G_k(x, y) - G_k(x', y)|\, \mathrm{d}\mu(y) < \varepsilon \int \mathrm{d}\mu$$

for $|x - x'| < \delta(\varepsilon)$. From this inequality follows the continuity of $G_k\mu$. Further, from $G_k \leqq G_{k+1}$ and $\mu \geqq 0$ we get $G_k\mu \leqq G_{k+1}\mu$. Following Theorem 3.7 the limit

$$G\mu(x) = \lim_{k\to\infty} G_k\mu(x) \tag{8}$$

is lower semicontinuous. ∎

Remark 3.15. Let $G: \mathbb{R}^n \times \mathbb{R}^n \mapsto \mathbb{R}$ be a kernel continuous for $x \neq y$ and μ a measure with compact support. Then $G\mu$ is continuous for $x \notin \operatorname{supp} \mu$.

Proof. Similar to Theorem 3.11 one can prove

$$|G\mu(x) - G\mu(x')| \leqq \int |G(x, y) - G(x', y)| \,\mathrm{d}|\mu|\,(y),$$

where $|\mu| = \mu^+ + \mu^-$. Let $x, x' \in B(x^0, \delta_0)$, where $B(x^0, \delta_0) \cap \operatorname{supp} \mu = \emptyset$. Then $\overline{B(x^0, \delta_0)} \times \operatorname{supp} \mu \subset \mathbb{R}^n \times \mathbb{R}^n$ is a compact set on which the kernel G is uniformly continuous. From this it follows that

$$|G\mu(x^0) - G\mu(x)| < \varepsilon \quad \text{for} \quad |x^0 - x| < \delta(\varepsilon). \blacksquare$$

Now let G be continuously differentiable for $x \neq y$. This condition is fulfilled for the Newtonian kernel G_{L} and the kernel G_{H} of the heat equation, but not for the Kernel G_{D} defined in (2.3.19). We define

$$\partial_j G(x, y) := \begin{cases} \dfrac{\partial G(x, y)}{\partial x_j} & \text{if} \quad x \neq y, \\ 0 & \text{if} \quad x = y. \end{cases} \tag{9}$$

Then

$$\partial_j G: \mathbb{R}^n \times \mathbb{R}^n \setminus \{(x, x)\} \mapsto \mathbb{R}$$

is continuous.

Theorem 3.12. *Let G be the kernel defined in* (9) *and μ a measure with compact support. Then the relation*

$$\partial_j \int G(x, y) \,\mathrm{d}\mu(y) = \int \partial_j G(x, y) \,\mathrm{d}\mu(y) \tag{10}$$

holds for all points $x \notin \operatorname{supp} \mu$.

Proof. We prove this theorem for $j = 1$. Let

$$\xi = (\xi_1, x_2, \ldots, x_n), \qquad a = (a_1, x_2, \ldots, x_n); \qquad \xi, a \notin \operatorname{supp} \mu.$$

Then we get

$$G(x, y) = G(a, y) + \int_{a_1}^{x_1} \partial_{\xi_1} G(\xi, y) \,\mathrm{d}\xi_1. \tag{11}$$

The following relations are well known for the Riemann integral in the case $\mu = \mu_\varrho$. Using the Fubini theorem it follows that

$$\begin{aligned} G\mu(x) &= \int G(a, y) \,\mathrm{d}\mu(y) + \int \left(\int_{a_1}^{x_1} \partial_{\xi_1} G(\xi, y) \,\mathrm{d}\xi_1 \right) \mathrm{d}\mu(y) \\ &= \int G(a, y) \,\mathrm{d}\mu(y) + \int_{a_1}^{x_1} \left(\int \partial_{\xi_1} G(\xi, y) \,\mathrm{d}\mu(y) \right) \mathrm{d}\xi_1. \end{aligned} \tag{12}$$

The kernel $(\xi_1, y) \mapsto \partial_{\xi_1} G(\xi, y)$ is uniformly continuous for $\xi_1 \in [a_1 - \delta, a_1 + \delta]$ and $y \in \operatorname{supp} \mu$. Following Remark 3.15 one can prove that

$$\xi_1 \mapsto \int \partial_{\xi_1} G(\xi, y) \, \mathrm{d}\mu(y)$$

is continuous on $[a_1 - \delta, a_1 + \delta]$. By the first mean value theorem we get

$$\frac{1}{h_1} \left(G\mu(x) - G\mu(x') \right) = \int \partial_{\xi_1} G(\xi^*, y) \, \mathrm{d}\mu(y),$$

where $x_1 \leqq \xi_1^* \leqq x_1 + h_1$, $\xi^* = (\xi_1^*, x_2, \ldots, x_n)$. Therefore the limit of the left-hand side exists and is equal to

$$\partial_{x_1} G\mu(x) = \int \partial_{x_1} G(x, y) \, \mathrm{d}\mu(y). \blacksquare$$

Remark 3.16. Let G be the Newtonian kernel. Then relation (10) also holds on $\operatorname{supp} \mu$ provided that the right-hand side is continuous. In this case (11) holds only if ξ does not pass y. These conditions are fulfilled in the case $\mu = \mu_\varrho$ (see ANGER [4]).

Proposition 3.2. *Let μ be a measure with compact support, G one of the fundamental solutions defined in* (2.1.17)–(2.1.19), (2.1.38)–(2.1.40) *or* (2.2.6). *Then the potential $G\mu$ is a solution of the corresponding homogeneous differential equation $P(x, \partial)\, u = 0$ outside* $\operatorname{supp} \mu$.

Proof. If $x \notin \operatorname{supp} \mu$, $y \in \operatorname{supp} \mu$, the function $x \mapsto G(x, y)$ has derivatives of an arbitrary order. Therefore we get with the aid of Theorem 3.12

$$P(x, \partial)\, G\mu(x) = \int P(x, \partial)\, G(x, y) \, \mathrm{d}\mu(y) = \int 0 \, \mathrm{d}\mu(y) = 0. \blacksquare \tag{13}$$

Theorem 3.13. *The space $\mathrm{C}_0^\infty(\mathbb{R}^n)$ is dense in $\mathrm{C}_0(\mathbb{R}^n)$ in the following sense: to each $\varphi \in \mathrm{C}_0(\mathbb{R}^n)$ and each neighbourhood U of* $\operatorname{supp} \varphi$ *there exists a $\psi \in \mathrm{C}_0^\infty(\mathbb{R}^n)$ satisfying* $\operatorname{supp} \psi \subset U$ *and*

$$\|\varphi - \psi\|_{\mathrm{C}} < \varepsilon. \tag{14}$$

Proof. Let $\varphi \in C_0(\mathbb{R}^n)$ be given. We consider an open set U containing the closed set $\operatorname{supp} \varphi$ and the convolution integral

$$\psi(x) := (\varphi * \varphi_\delta)\,(x) = \int \varphi_\delta(x - y)\, \varphi(y) \, \mathrm{d}y$$

defined in (1.3.18). Following Theorem 3.12 one can prove that ψ has derivatives of an arbitrary order. If δ is sufficiently small the function $\varphi * \varphi_\delta$ is zero outside U. Then

$$|\varphi(x) - \psi(x)| = \int \left(\varphi(x) - \varphi(y) \right) \varphi_\delta(x - y) \, \mathrm{d}y$$

and

$$|\varphi(x) - \psi(x)| \leqq \max \left\{ |\varphi(x) - \varphi(y)|,\, y \in B(x, \delta) \right\}.$$

The given function φ has a compact support. Therefore it is uniformly continuous on supp φ. We get

$$|\varphi(x) - \psi(x)| < \varepsilon \quad \text{for} \quad \delta < \delta_0 \quad \text{and}$$

$$\|\varphi - \psi\|_{\mathrm{C}} = \max\{|\varphi(x) - \psi(x)|, x \in \mathbb{R}^n\} < \varepsilon. \blacksquare$$

Theorem 3.14. *Let G be a locally integrable fundamental solution, μ_1, μ_2 two measures with compact support. If $G\mu_1 = G\mu_2$ in $\mathbb{R}^n$ then $\mu_1 = \mu_2$.*

Proof. Multiplying $G\mu_1 = G\mu_2$ with $P^*(x, \partial)\, \psi(x)$, where $\psi \in \mathrm{C}_0^\infty(\mathbb{R}^n)$, and integrating this relation we obtain

$$\int \left(\int G(x, y)\, \mathrm{d}\mu_1(y)\right) P^*(x, \partial)\, \psi(x)\, \mathrm{d}x$$
$$= \int \left(\int G(x, y)\, \mathrm{d}\mu_2(y)\right) P^*(x, \partial)\, \psi(x)\, \mathrm{d}x. \tag{15}$$

Both integrals give (see, for instance, (2.1.22))

$$\int \left(\int G(x, y)\, P^*(x, \partial)\, \psi(x)\, \mathrm{d}x\right) \mathrm{d}\mu_j(y) = \int \psi(y)\, \mathrm{d}\mu_j(y), \qquad j = 1, 2.$$

From (15) it follows that

$$\int \psi\, \mathrm{d}\mu_1 = \int \psi\, \mathrm{d}\mu_2 \quad \text{for all} \quad \psi \in \mathrm{C}_0^\infty(\mathbb{R}^n).$$

Using Theorem 3.13 one can prove that

$$\int \varphi\, \mathrm{d}\mu_1 = \int \varphi\, \mathrm{d}\mu_2 \quad \text{for all} \quad \varphi \in \mathrm{C}_0(\mathbb{R}^n).$$

From the last relation it follows that $\mu_1 = \mu_2$. $\blacksquare$

We introduce the following sets (ANGER [1, 2, 3, 4], SCHULZE and WILDENHAIN [1]):

$$\begin{aligned} F^+(G) &:= \{\lambda \geqq 0\colon \text{supp}\, \lambda \text{ compact}, G^+\lambda \text{ and } G^-\lambda \text{ continuous}\}, \\ F(G) &:= \{\lambda = \lambda_1 - \lambda_2\colon \lambda_1, \lambda_2 \in F^+(G)\}, \\ D(G) &:= \{f = G\lambda\colon \lambda \in F(G)\}. \end{aligned} \tag{16}$$

Further, let $K \subset \mathbb{R}^n$ be a compact set. Often we have $K = \partial\Omega$, where $\Omega \subset \mathbb{R}^n$ is a bounded domain.

$$\begin{aligned} F^+(K; G) &:= \{\lambda \geqq 0\colon \text{supp}\, \lambda \subset K, G^+\lambda \text{ and } G^-\lambda \text{ continuous}\}, \\ F(K; G) &:= \{\lambda = \lambda_1 - \lambda_2\colon \lambda_1, \lambda_2 \in F^+(K; G)\}, \\ D(K; G) &:= \{f \in C(K)\colon f(x) = G\lambda(x), \lambda \in F(K; G)\}, \\ D_K(G) &:= \{f = G\lambda\colon \lambda \in F(K; G)\}. \end{aligned} \tag{17}$$

In the case of elliptic differential equations of second order and $G \geqq 0$ the following sets are introduced:

$$\begin{aligned} E^+(G) &:= \left\{\mu \geqq 0\colon \int G\mu\, \mathrm{d}\mu < \infty\right\}, \\ E(G) &:= \{\mu = \mu_1 - \mu_2\colon \mu_1, \mu_2 \in E^+(G)\}. \end{aligned} \tag{18}$$

It is possible that different G give the same $F(G)$. That depends on the singularity of G. For simplicity we sometimes write the symbols $F^+(K)$, $F(K)$, $D(K)$ and D_K for the sets introduced in (17).

Let μ be a measure. The restriction μ_K of μ on K is defined by

$$\mu_K(f) := \int 1_K(x)\, f(x)\, \mathrm{d}\mu(x), \tag{19}$$

where 1_K is the characteristic function of K.

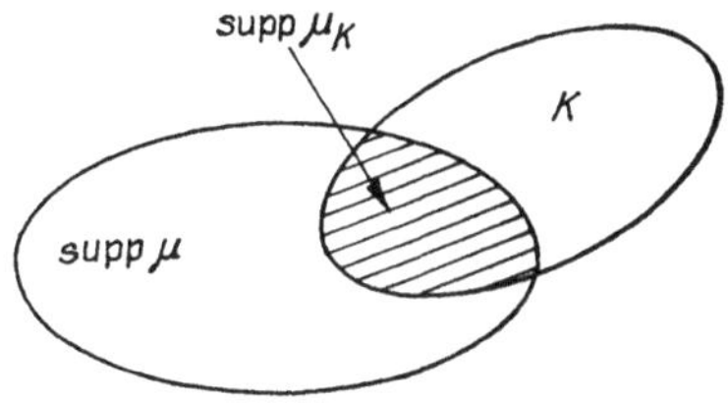

Fig. 3.14

In the case of the kernel G_D, $n = 1$, the set $F^+(G_D)$ can easily be characterized. The potential $\chi = G_D\mu$ is continuous if and only if $\mu(\{x^0\}) = 0$ for each $x^0 \in \mathbb{R}$, i.e., $\chi_2 = 0$. Of special interest are the functions χ_3 which are singular, i.e., $\chi_3' = 0$ almost everywhere. In Sect. 3.2 we constructed two special singular functions.

If $\mu = \mu_\varrho$, where the density ϱ is a bounded and integrable function, the Newtonian potential $G_L\mu_\varrho$ is continuous. Further, the potential $G_L\mu_\sigma$ of a single layer is continuous if the boundary $\partial\Omega$ is sufficiently smooth and σ continuous (Kellogg [1], Tikhonov and Samarski [1]).

Theorem 3.15. *Let $G \geqq 0$ be a lower semicontinuous kernel, $\lambda \in F^+(G)$. Then the restriction λ_K of λ on K is also an element of $F^+(G)$.*

Proof. Let $K' := \mathbb{R}^n \setminus K$, $\lambda \in F^+(G)$, $\lambda_1 := \lambda_K \geqq 0$, $\lambda_2 := \lambda_{K'} \geqq 0$. Then $\lambda = \lambda_1 + \lambda_2$. The potentials $G\lambda_1$ and $G\lambda_2$ are l.s.c. (Theorem 3.11) and the inequalities $G\lambda_1 \leqq G\lambda$ and $G\lambda_2 \leqq G\lambda$ hold. From $G\lambda = G\lambda_1 + G\lambda_2$ it follows that $G\lambda_1 = G\lambda + (-G\lambda_2)$. The righthand side as sum of two upper semicontinuous functions is a u.s.c. function. Since $G\lambda_1$ is l.s.c. and u.s.c. we obtain that $G\lambda_1$ is continuous. The same fact holds for $G\lambda_2$. ∎

In 1935 G. C. Evans and F. Vasilesco proved the following theorem.

Theorem 3.16. *Let G_L be the fundamental solution of the Laplace equation, $K \subset \mathbb{R}^n$ a compact set and μ a measure with* $\operatorname{supp} \mu = K$. *If the restriction* $\operatorname{rest}_K G_L\mu$ *of $G_L\mu$ on K is continuous, the $G_L\mu$ is continuous on the whole space $\mathbb{R}^n$.*

One can find a proof in Doob [1], Helms [1]. It depends on the inequality

$$\frac{1}{|x-y|} \leqq \frac{2}{|z_x - y|}, \tag{20}$$

where $y \in K$ and z_x a point of K nearest x. The potential is (by assumption) continuous at the inner points of K and continuous at the outer points of K. The Theorem 3.16 must be proved only at the boundary points of K. Some investigations on the continuity principle of Evans-Vasilesco relative to the heat equation can be found in DONT [1].

In electrostatics, the capacity cap (K) of an isolated conductor $K = \Omega \cup \partial\Omega \subset \mathbb{R}^2$ is defined as the ratio of the total charge Q in equilibrium on it to the constant value u_0 of the potential at its surface $\partial\Omega$. Let $G_L\lambda_0$ be the corresponding Newtonian potential, where $\lambda_0 \geqq 0$ and supp $\lambda_0 \subset \partial\Omega$. If $G_L\lambda_0 = 1$ on $\partial\Omega$ we get cap $(K) = Q = \int d\lambda_0$. The function $G_L\lambda_0$ is called conductor potential. It can be used to determine the total charge $\int d\lambda_0$ by computing the following surface integral extended over a ball $B(x^0, R)$ containing K in its interior (ANGER [4], KELLOGG [1]):

$$\text{cap}(K) = -\int\limits_{\partial B(x^0,R)} \frac{\partial G_L\lambda_0}{\partial n}\, dS. \tag{21}$$

In the case of a closed ball $K = \overline{B(x^0, R)} \subset \mathbb{R}^3$ the measure λ_0 has the form $\lambda_0 = 4\pi R\delta_{x^0,R}$. From this it follows that

$$\text{cap}(K) = 4\pi R. \tag{22}$$

In electrostatics the kernel has the form $1/|x - y|$. It differs from G_L by $1/4\pi$. Therefore our cap (K) has the factor 4π. There are various possibilities to introduce the capacity for a compact set $K \subset \mathbb{R}^n$ (ANGER [3, 4], FREHZE [1], MAZYA and SHAPOSHNIKOVA [1], SCHULZE and WILDENHAIN [1]). In 1923 N. Wiener introduced the capacity for a bounded set $B \subset \mathbb{R}^n$ by considering a sequence of smooth domains $\Omega_k \subset B$ tending to B. He constructs the conductor potential u_k relative to Ω_k using results of classical potential theory. By a limit N. Wiener obtains the conductor potential u_0 of $K = \overline{B}$ and introduces cap (K) by (21) (see KELLOGG [1]).

In 1932 Ch. de la Vallée Poussin defined cap (K) of the Newtonian kernel G_L by

$$\text{cap}(K) := \sup\left\{\int d\mu : \mu \geqq 0, \text{supp}\,\mu \subset K, G_L\mu(x) \leqq 1 \text{ on } K\right\}, \tag{23}$$

and in 1942 H. Cartan characterized the sets $B \subset \mathbb{R}^3$ of (inner) capacity zero by

$$\mu(B) = 0 \quad \text{for all} \quad \mu \in E^+(G_L) \tag{24}$$

(see ANGER [3], HELMS [1]).

In 1957 G. ANGER [1, 2] found an equivalent definition using continuous potentials in (23). Then a set B of (inner) capacity zero can be characterized by

$$\lambda(B) = 0 \quad \text{for all} \quad \lambda \in F^+(G_L). \tag{25}$$

Since $E^+(G_L) \supset F^+(G_L)$ from (24) it follows that (25) holds. In ANGER [4], HELMS [1] the following important theorem is proved relative to G_L, $n \geqq 2$. Let $\mu \in E^+(G_L)$ then there exists a sequence of measures $\lambda_j \in F^+(G_L)$ such that

$$\mu(f) = \sum_{j=1}^{\infty} \lambda_j(f). \tag{26}$$

If $f = 1_B$ one obtains

$$\mu(B) = \sum_{j=1}^{\infty} \lambda_j(B). \tag{27}$$

Using relation (27) one can prove that from (25) follows relation (24). H. Cartan introduced on $E(G_L)$ the scalar product (mutual energy)

$$(\mu, \nu) := \int G_L\mu \, d\nu \tag{28}$$

and proved basic theorems in potential theory by using results on Hilbert spaces (see DENY [1], HELMS [1], SCHULZE and WILDENHAIN [1]). But the measures δ_{x^0}, $\mu_{\delta,\varrho}$, $\mu_{\delta,\sigma}$ are no elements of $E(G_L)$. Further, the ideas of H. Cartan can be applied only for elliptic differential equations of second order.

The continuity of $G\lambda$ depends on the kernel G. Therefore, our capacity is adapted to the corresponding differential equation. In the following definition G can be a bounded discontinuous kernel like the kernel G_D defined in (2.3.19). For such a bounded kernel the definition of cap (K) following (23) gives to each point $x^0 \in \mathbb{R}^n$ a positive capacity while by using the continuous potentials the point x^0 has capacity zero. For differential equations of higher order, which often have a continuous fundamental solution one can require the existence of certain continuous derivatives, where the order depends on the special problem considered. In this direction further considerations can be found in ANGER [4, 6], SCHULZE and WILDENHAIN [1].

There are two further possibilities to introduce cap (K). Let $u \in C_0^\infty(\mathbb{R}^n)$ be a function satisfying $u = 1$ in a neighbourhood of K. Then (FREHZE [1])

$$\text{cap}\,(K) = 4\pi \inf_u \left(\int_\Omega (\text{grad}\, u)^2 \, dx\right). \tag{29}$$

Further, let

$$I(\mu) := \iint \frac{d\mu(x)\, d\mu(y)}{4\pi\, |x - y|} \tag{30}$$

be the energy of the positive measure μ satisfying supp $\mu \subset K$ and $\int d\mu = 1$. Then there exists a uniquely determined μ_0 satisfying (HELMS [1])

$$I(\mu_0) \leqq I(\mu). \tag{31}$$

This inequality can be proved by applying H. Cartan's ideas. The potential $G_L\mu_0$ is, if $\partial\Omega$ is sufficiently smooth, constant on $\bar{\Omega}$,

$$G_L\mu_0(x) = u_0 = \text{const}. \tag{32}$$

Then

$$\text{cap}\,(K) = \frac{4\pi}{u_0}. \tag{33}$$

The different definitions (23), (25) (29) and (33) give rise to different generalizations. For inverse problems definitions (23) and (25) seem to be the most suitable, since they are adapted to the special differential equation including different boundary and initial conditions.

Definition 3.6. Let $G = G^+ - G^-$ be a kernel such that G^+ and G^- are l.s.c. A compact set $K \subset \mathbb{R}^n$ has *positive* C-*capacity*, if there exists at least one $\lambda \in F^+(G)$, $\lambda \neq 0$, satisfying supp $\lambda \subset K$. In the other case K has C-*capacity zero*. The real number (depending on G)

$$\operatorname{cap}(K; G) := \sup \left\{\int \mathrm{d}\lambda : \lambda \in F^+(G), \operatorname{supp} \lambda \subset K, G^+\lambda(x) + G^-\lambda(x) \leqq 1 \text{ on } K\right\} \tag{34}$$

is called the C-*capacity of* K *relative* to G.

From the maximum principle of the Laplace equation it follows that the Newtonian potentials $G_L\lambda$ in Definition 3.6 satisfy $G_L\lambda \leqq 1$ on the whole $\mathbb{R}^n$, $n \geqq 3$ (Doob [1], Helms [1], Landkof [1]).

Following general principles in analysis, especially in measure theory, one can introduce an inner C-capacity.

Definition 3.7. The (*inner*) C_i-*capacity* of a set $B \subset \mathbb{R}^n$ is defined by

$$\operatorname{cap}_i(B; G) := \sup\{\operatorname{cap}(K; G), K \subset B\}, \tag{35}$$

where the supremum is extended over all compact subsets K of B. The (*outer*) C_e-*capacity* of B is defined by

$$\operatorname{cap}_e(B; G) := \inf\{\operatorname{cap}_i(U; G), U \supset B\}, \tag{36}$$

where the infimum is extended over all open sets U containing B. A set $B \subset \mathbb{R}^n$ is called C-*capacitable* if $\operatorname{cap}_i(B; G) = \operatorname{cap}_e(B; G)$. The common number is called the C-*capacity of* B and denoted by $\operatorname{cap}(B; G)$.

In 1953 G. Choquet [1] proved for the capacity $\operatorname{cap}(K; G_L)$, often denoted by $\operatorname{cap}(K)$, the following properties. Let $\mathfrak{K}$ be the system of all compact sets $K \subset \mathbb{R}^n$. Then

$$\operatorname{cap}(\emptyset) = 0 \quad \text{and} \quad \operatorname{cap}(K_1) \leqq \operatorname{cap}(K_2) \quad \text{for} \quad K_1 \subset K_2. \tag{37}$$

If $K_1, K_2, \ldots$ is a decreasing sequence of compact sets and $K = \bigcap_{j=1}^{\infty} K_j$ then

$$\operatorname{cap}(K) = \lim_{j\to\infty} \operatorname{cap}(K_j). \tag{38}$$

Further,

$$\operatorname{cap}(K_1 \cup K_2) + \operatorname{cap}(K_1 \cap K_2) \leqq \operatorname{cap}(K_1) + \operatorname{cap}(K_2). \tag{39}$$

For Borel subsets, or more generally for analytical sets, the Choquet Theorem 3.14 holds (Choquet [1], Doob [1], Helms [1]).

Theorem 3.17. *Let G_L be the Newtonian kernel. Then all Borel sets $B \subset \mathbb{R}^n$ are capacitable.*

Remark 3.17. The capacity can be introduced in metric spaces. In this case $\mathfrak{K}$ is the set of all compact subsets. A mapping $f\colon \mathfrak{K} \mapsto \mathbb{R}$ satisfying (37), (38) and (39) is called a *Choquet capacity*. Results concerning such abstract capacities can be found in Choquet [1], Doob [1], Frehze [1], Helms [1].

Now some properties of cap $(K; G)$ for different kernels G. Let $\Omega \subset \mathbb{R}^n$ be a domain, μ_ϱ the measure defined in (3.1.4). If ϱ is a bounded and integrable function then $G_L\mu_\varrho$ is continuous, i.e., $\bar{\Omega}$ is of positive C-capacity. Further, if $\partial\Omega$ is sufficiently smooth and σ a bounded and integrable function, then $G_L\mu_\sigma$ is continuous. From this it follows that $\partial\Omega$ has positive C-capacity. Further, it can be proved that the boundary $\partial\Omega$ of a bounded domain $\Omega \subset \mathbb{R}^n$ has positive C-capacity (Kellogg [1]).

Our C-capacity cap $(K; G)$ is more general than the capacities used in other publications. Every problem has it own capacity. Let us consider the kernel G_D defined in (6) or (2.3.19). We consider the Cantor set K_c constructed in Sect. 3.2 and its singular function χ_3. Following (3.3.13) one can assign a measure μ_3 to χ_3. The support of μ_3 is the Cantor set K_c. The potential $\chi_3 = G_D\mu_3$ is continuous and increasing. We have $\chi_3(x) = 0$ if $x \leqq 0$ and $\chi_3(x) = 1$ if $x \geqq 1$. From this it follows that cap $(K_c; G_D) = 1$. It is known that K_c has the Lebesgue measure zero. Now let $x^0 \in \mathbb{R}$ be an arbitrary point. A measure μ satisfying supp $\mu = \{x^0\}$ has the form $\mu = c\delta_{x^0}$. Its corresponding potential is discontinuous. Therefore we get cap $(\{x^0\}; G_D) = 0$. Let $[a, b] \subset \mathbb{R}$ be a closed interval containing x^0 in its interior. From $\mu \geqq 0$, supp $\mu \subset [a, b]$ and $\chi(x) = G_D\mu(x) = 1$ for $x \geqq b$ it follows that $G_D\mu(x) = G_D\mu(b) \leqq 1$. There exists a continuous potential $G_D\lambda_0(x)$ with this property. Therefore cap $([a, b]; G_D) = 1$ for each interval. The capacity of the kernel G_D does not have the property (38).

In Anger [4] the following theorem is proved. In Doob [1], Netuka and Vesely [1] one can find relations to other definitions of a capacity to the heat equation.

Theorem 3.18. *Let G_H be the fundamental solution of the heat equation, $I_n := \{(x, t) : x \in \mathbb{R}^n, t = T = \text{const}\}$ a hyperplane. Then each compact subset $K \subset I_n$ has C-capacity zero, i.e.,* cap $(K; G_H) = 0$.

For elliptic differential equations of second order such a property relative to I_n does not hold. Theorem 3.18 gives a difference between elliptic differential equations of second order and parabolic differential equations of second order. For our applications the following theorem plays an important role.

Theorem 3.19. *A set $B \subset \mathbb{R}^n$ has C_i-capacity zero if and only if $\lambda(B) = \int 1_B(x)\, d\lambda(x) = 0$ for all $\lambda \in F^+(G)$.*

Proof. a) The condition is necessary. Let $\text{cap}_i\,(B; G) = 0$. If there exists a $\lambda_0 \in F^+(G)$ satisfying $\lambda_0(B) \neq 0$ then there exists a compact set $K \subset B$ satisfying also $\lambda_0(K) \neq 0$. The restriction λ_1 of λ_0 on K is an element of $F^+(G)$ (see Theorem 3.16). Further,

$$\lambda_1(K) = \int 1_K(x)\, d\lambda_1(x) = \int 1_K(x)\, d\lambda_0(x) \neq 0.$$

Then $\text{cap}_i\,(B; G) \neq 0$, which is a contradiction.

b) The condition is sufficient. If $\lambda(B) = 0$ for all $\lambda \in F^+(G)$ then for each compact subset $K \subset B$ we have $\lambda(K) = 0$. There does not exist a $\lambda \in F^+(G)$, $\lambda \neq 0$, $\operatorname{supp} \lambda \subset K$. Therefore $\operatorname{cap}(K; G) = 0$ for each $K \subset B$, i.e., $\operatorname{cap}_i(B; G) = 0$. ∎

Proposition 3.3. *Let $B_k \subset \mathbb{R}^n$ be sets of C_i-capacity zero, $k = 1, 2, \ldots$ Then the union $B = \bigcup_{k=1}^{\infty} B_k$ has also C_i-capacity zero.*

Proof. By assumption

$$\lambda(B_k) = 0 \quad \text{for all} \quad \lambda \in F^+(G), \quad k = 1, 2, \ldots$$

From

$$\lambda(B) = \lambda\left(\bigcup_{k=1}^{\infty} B_k\right) \leqq \sum_{k=1}^{\infty} \lambda(B_k)$$

it follows that $\operatorname{cap}_i(B; G) = 0$. ∎

Proposition 3.4. *Let G_L be the Newtonian kernel. Each point $x \in \mathbb{R}^n$ and each denumerable set $B = \{x^1, x^2, \ldots\}$ have C_i-capacity zero. Further, since $\mu_\varrho \in F^+(G_L)$ each set of C_i-capacity zero has n-dimensional Lebesgue measure zero.*

Proof. Let $B = \{x^1, x^2, \ldots\} \subset \mathbb{R}^n$, $n \geqq 2$, be a denumerable set. Each measure $\mu \geqq 0$ with $\operatorname{supp} \mu \subset B$ has the form

$$\mu = \sum_{k=1}^{\infty} c_k \delta_x k, \qquad c_k \geqq 0. \tag{40}$$

Then

$$G_L\mu(x) = \sum_{k=1}^{\infty} c_k G_L(x, x^k) \tag{41}$$

is infinite at each x^k if $c_k \neq 0$. There does not exist a measure $\lambda \subset F^+(G_L)$ satisfying $\operatorname{supp} \lambda \subset B$ and $\lambda \neq 0$. The measures μ_ϱ are elements of $F^+(G_L)$ if ϱ is bounded. The Lebesgue measure can be calculated by setting $\varrho = 1$. Therefore from $\operatorname{cap}_i(B; G_L) = 0$ it follows that this set has Lebesgue measure zero. ∎

In 1935 G. C. Evans proved that for each set $B \subset \mathbb{R}^n$ satisfying $\operatorname{cap}_e(B; G_L) = 0$ there exists a positive measure μ such that $G_L\mu(x) = +\infty$ on B (Doob [1], Helms [1], Landkof [1]). This property is used to define sets of exterior capacity zero in axiomatic potential theory (Bauer [2], Bliedtner and Hansen [1], Netuka and Veseley (1]).

If G is a nonsymmetric kernel then $\int G\lambda \, d\mu = \int G^*\mu \, d\lambda$ holds. Therefore the following theorem is proved relative to $G^*\mu$.

Theorem 3.20. *If for all $\lambda \in F(K; G)$ the relation*

$$\int (G^*\mu_1 - G^*\mu_2) \, d\lambda = 0 \tag{42}$$

holds, then $G^*\mu_1 = G^*\mu_2$ on K except a set of C_i-capacity zero and conversely.

Proof. Let $B := \{x\colon G^*\mu_1(x) - G^*\mu_2(x) \neq 0\}$ be of positive $\mathrm{C_i}$-capacity. Then there exists a measure $\lambda_0 \in F^+(G)$, $\lambda_0 \neq 0$, such that $\operatorname{supp} \lambda_0 \subset B$ and

$$\int \left(G^*\mu_1(x) - G^*\mu_2(x)\right) \mathrm{d}\lambda_0(x) > 0,$$

which is a contradiction. If $G^*\mu_1(x) - G^*\mu_2(x) \neq 0$ except a set of $\mathrm{C_i}$-capacity zero, then (42) holds for all $\lambda \in F(K; G)$. ∎

In a certain sense it is possible to develop a potential theory for elliptic differential equations of higher order. One can introduce different notions of capacity and prove theorems like Theorem 3.17 (Anger [4], Schulze and Wildenhain [1], Wildenhain [1]).

We should like to derive some properties of the solution of the Dirichlet problem for smooth boundaries $\partial\Omega$. Let $\Omega = B(x^0, R) \subset \mathbb{R}^n$ be a ball and

$$u(x) = H_h(x) = \int_{\partial\Omega} h(y)\, P(x, y)\, \mathrm{d}S(y) \tag{43}$$

the solution of the Dirichlet problem. Here

$$P(x, y) = \frac{1}{RV_n} \frac{R^2 - |x - x^0|^2}{|x - y|^n}, \qquad n \geqq 2,$$

is the Poisson kernel. If $z \notin \Omega$ then $x \mapsto G_{\mathrm{L}}(x, z)$ is harmonic in Ω. From the maximum principle of the harmonic functions follows that the Dirichlet problem has a unique solution. Therefore we get

$$G_{\mathrm{L}}(x, z) = \int_{\partial\Omega} G_{\mathrm{L}}(y, z)\, P(x, y)\, \mathrm{d}S(y). \tag{44}$$

Let us write

$$\mathrm{d}\mu_x(y) := P(x, y)\, \mathrm{d}S(y) \tag{45}$$

and

$$G_{\mathrm{L}}(x, z) = \int G_{\mathrm{L}}(y, z)\, \mathrm{d}\delta_x(y).$$

Then (44) becomes

$$G_{\mathrm{L}}\delta_x(z) = G_{\mathrm{L}}\mu_x(z) \quad \text{for all} \quad z \notin \Omega. \tag{46}$$

The measure μ_x is called the *harmonic measure* of Ω. In modern potential theory the existence of μ_x is proved for each domain $\Omega \subset \mathbb{R}^n$, $n \geqq 2$. In general it does not have the special form (45), but the relation

$$H_h(x) = \int h(y)\, \mathrm{d}\mu_x(y) \tag{47}$$

holds (see also page 294, Bliedtner and Hansen [1], Doob [1], Helms [1]). From (3.1.18) or (3.1.22) it follows that the potential of a positive measure μ satisfies for each $B(x^0, R) \subset \mathbb{R}^n$

$$\int G_{\mathrm{L}}\mu(y)\, \mathrm{d}\delta_{x^0,R}(y) \leqq G_{\mathrm{L}}\mu(x^0). \tag{48}$$

Further, a harmonic function u satisfies for each $B(x^0, R) \subset \mathbb{R}^n$

$$\int u(y)\, \mathrm{d}\delta_{x^0,R}(y) = u(x^0), \tag{49}$$

and conversely, a function u satisfying (49) must be harmonic. Then $v := G_L\mu + u$ satisfies for each $B(x^0, R) \subset \mathbb{R}^n$

$$\int v(y) \, d\delta_{x^0,R}(y) \leqq v(x^0). \tag{50}$$

If $\mu = \mu_\varrho$, $\varrho \geqq 0$, sufficiently smooth, then $v = G_L\mu_\varrho + u$ satisfies

$$-\Delta v = \varrho \geqq 0. \tag{51}$$

These properties are characteristic for superharmonic functions (see also formula (8.3), Doob [1], Helms [1]).

Definition 3.8. Let $\Omega \subset \mathbb{R}^n$, $n \geqq 2$, be an open set, $u\colon \Omega \to \overline{\mathbb{R}}$ a l.s.c. function not identically $+\infty$ in each component of Ω. The function u is called *superharmonic*, if for each ball $B(x^0, R) \subset \mathbb{R}^n$ the inequality $\int u(y) \, d\mu_x(y) \leqq u(x)$ holds for all $x \in B(x^0, R)$. In $\mathbb{R}^n$ this definition is equivalent to $\int u(y) \, d\delta_{x^0,R}(y) \leqq u(x^0)$ or $\int u(y) \, dm_{x^0,R}(y) \leqq u(x^0)$. A function v is called *subharmonic* if $u := -v$ is superharmonic.

Remark 3.18. A function u is harmonic if and only if u is both superharmonic and subharmonic. With these notions the Dirichlet problem can be studied without any restriction on the domain $\Omega \subset \mathbb{R}^n$. Further, a theory of the Dirichlet problem is possible in Brelot spaces and Bauer spaces (see Chap. 8).

In Doob [1], Helms [1], Landkof [1] the famous *minimum principle* for superharmonic functions is proved (see also Theorem 8.1).

Theorem 3.21. *Let u be superharmonic in the domain $\Omega \subset \mathbb{R}^n$. Then u satisfies the minimum principle, i.e.,* $\inf \{u(x), x \in \Omega\}$ *is not attained unless u is constant. In this case from* $\liminf_{x \to z} u(x) \geqq 0$ *for all $z \in \partial\Omega$ it follows that $u \geqq 0$ on Ω.*

Further, the important theorem is proved.

Theorem 3.22. *Let $G_L^\Omega(x, y) := G_L\delta_x(y) - G_L\mu_x(y)$, $x, y \in \Omega$, be the Green function of Ω. For fixed x the Green function $y \mapsto G_L^\Omega(x, y)$ is superharmonic satisfying $G_L^\Omega(x, y) \geqq 0$ for all $x, y \in \Omega$.*

In 1930 F. Riezs [1] proved the famous *decomposition theorem* (Doob [1], Helms [1], Bliedtner and Hansen [1]).

Theorem 3.23. *Let $\Omega \subset \mathbb{R}^n$ be an open set and $u \geqq 0$ a superharmonic function defined on Ω. There is a unique measure $\nu \geqq 0$ on Ω such that $u = G_L\nu + h$, where h is harmonic on Ω.*

The following properties of superharmonic functions hold (Helms [1], Bliedtner and Hansen [1]).

Theorem 3.24. *If u and v are superharmonic on an open set Ω and $c > 0$, then*

(i) *cu is superharmonic on Ω,*
(ii) *$u + v$ is superharmonic on Ω,*
(iii) *$u \wedge v := \inf (u, v)$ is superharmonic on Ω.*

4. The space of continuous functions in inverse problems

4.1. A new transformation

At first for special domains $\Omega \subset \mathbb{R}^3$ we sketch some basic relations used in modern potential theory applying well known results of classical potential theory (formulas (4.1.6) through (4.1.15)). In modern potential theory these relations are proved for arbitrary domains by using the method of Perron-Wiener-Brelot based on superharmonic functions (HELMS [1]). That method holds for linear elliptic and parabolic differential equations of second order (DOOB [1]). As we have seen in Sect. 2.1 an axiomatic approach exists (BLIEDTNER and HANSEN [1]). In our book we need some of these results but do not prove them.

The new transformation is the following which we explain for Newtonian potentials. Instead of the potential $G_L\mu$ we shall consider the measure μ using the transformation

$$\mu(G_L\lambda) := \int G_L\lambda \, d\mu = \int G_L\mu \, d\lambda, \tag{1}$$

where $G_L\lambda$ is a continuous potential. The integrals in (1) are double integrals (energy integrals) in which the order of integration can be changed (Fubini theorem). Following Theorem 3.11 the correspondance $G_L\mu \leftrightarrow \mu$ is one-to-one if (1) is considered for all $\lambda \in F(G_L)$. The Newtonian potential $G_L\lambda$ tends to zero for $|x| \to \infty$. In this case one applies the space $\mathbf{C}_\omega(\mathbb{R}^n)$, the space of all continuous functions $f: \mathbb{R}^n \mapsto \mathbb{R}$ tending to zero for $x \to \omega$. Here ω is the point at infinity. The dual space $\mathbf{C}_\omega^*(\mathbb{R}^n)$ consists of all measures μ satisfying $\|\mu\| < \infty$. If the problems are studied on $K = \Omega \cup \partial\Omega \subset \mathbb{R}^n$, where Ω is a bounded domain, one applies the space $\mathbf{C}(K)$ and its dual space $\mathbf{C}^*(K)$ which consists of all measures μ concentrated on K. We do not need any restriction on Ω and μ, which is important in inverse theory, because the set of all solutions is to be studied.

If $\overline{D(\partial\Omega; G_L)} = \mathbf{C}(\partial\Omega)$ and $\operatorname{sup} \mu \subset \partial\Omega$ then μ can be uniquely determined by applying all $\lambda \in F(\partial\Omega; G_L)$ in (1). If $\overline{D(\partial\Omega; G_L)} \neq \mathbf{C}(\partial\Omega)$ then each continuous linear form L defined on $D(\partial\Omega; G_L)$ can be extended to a continuous linear form L' on $\mathbf{C}(\partial\Omega)$ (Hahn-Banach theorem). This extension is not unique. For simplicity we assume $\overline{D(\partial\Omega; G_L)} = \mathbf{C}(\partial\Omega)$ for all further considerations. This condition is always fulfilled in applications.

Let μ be a measure with compact support, which we assume for simplicity.

The potential $G_L\mu$ can be discontinuous, for instance, in the case $\mu = \delta_{x_0}$ or $\mu = \mu_\tau$ (Lebesgue spine, Example 3.2). In order to apply results of linear algebra and functional analysis it is more convenient to consider continuous linear forms on the space of continuous functions rather than potentials.

In applications the potential $G_L\mu$ or its gradient grad $G_L\mu$ is known only on the boundary of $\bar{\Omega} = \operatorname{supp}\mu$. Then

$$\int G_L\mu \,\mathrm{d}\lambda = \int G_L\lambda \,\mathrm{d}\mu = \mu(G_L\lambda) \tag{2}$$

can be calculated for each $\lambda \in F(\partial\Omega; G_L)$. We often have to consider

$$\int G_L\lambda \,\mathrm{d}\mu_1 = \int G_L\lambda \,\mathrm{d}\mu_2 \quad \text{for all} \quad \lambda \in F(\partial\Omega; G_L), \tag{3}$$

which defines an equivalence relation on $\mathbf{C}^*(\bar{\Omega})$. Following Theorem 3.20 we get

$$G_L\mu_1(x) = G_L\mu_2(x) \quad \text{on } \partial\Omega, \tag{4}$$

except a set of C-capacity zero and conversely. If $\overline{D(\partial\Omega; G_L)} = \mathbf{C}(\partial\Omega)$ then (4) holds for all $x \in \partial\Omega$ (see Theorem 4.5, (3.4.46), (19)).

From (3) the existence of further measures ν satisfying

$$\int G_L\lambda \,\mathrm{d}\nu = \int G_L\lambda \,\mathrm{d}\mu_1 \quad \text{for all} \quad \lambda \in F(\partial\Omega; G_L) \tag{5}$$

follows (see (12)). This fact is a consequence of the balayage principle which follows from the maximum principle. In gravimetry grad $G_L\mu$ can be measured on $\partial\Omega$. The existence of grad $G_L\mu$ is a restriction on μ. The special measures μ_ϱ defined in (3.1.4) fulfill this condition.

Now we should like to derive some concrete formulas for the balayage principle by using the Poisson integral. It has the form

$$u(x) = \int_{\partial\Omega} u(y)\, P(x, y) \,\mathrm{d}S(y), \tag{6}$$

where $P(x, y)$ was defined in (3.4.43). If $u(y) = G_L(y, z)$, $z \notin \Omega$, we obtain (see (3.4.44))

$$G_L(x, z) = \int_{\partial\Omega} G_L(y, z)\, P(x, y) \,\mathrm{d}S(y). \tag{7}$$

Integrating (7) with respect to $\lambda \in F(\partial\Omega; G_L)$ we get

$$G_L\lambda(x) = \int_{\partial\Omega} G_L\lambda(y)\, P(x, y) \,\mathrm{d}S(y). \tag{8}$$

Formula (8) is the solution of the Dirichlet problem with boundary-values $g(y) = G_L\lambda(y)$. By the first Harnack theorem we obtain – we have assumed $\overline{D(\partial\Omega; G_L)} = \mathbf{C}(\partial\Omega)$ –

$$u(x) = \int_{\partial\Omega} h(y)\, P(x, y) \,\mathrm{d}S(y), \qquad h \in \mathbf{C}(\partial\Omega). \tag{9}$$

Now let us integrate (7) with respect to μ, supp $\mu \subset \bar{\Omega}$. We have for $z \notin \Omega$

$$G_{\mathrm{L}}\mu(z) = \int_{\partial\Omega} G_{\mathrm{L}}(y, z) \left(\int P(x, y) \,\mathrm{d}\mu(x)\right) \mathrm{d}S(y)$$

$$= \int_{\partial\Omega} G_{\mathrm{L}}(y, z)\, \sigma(y) \,\mathrm{d}S(y) =: G_{\mathrm{L}}\mu_\sigma(z), \tag{10}$$

where

$$\sigma(y) := \int P(x, y) \,\mathrm{d}\mu(x). \tag{11}$$

The measures μ and μ_σ, where supp $\mu \subset \Omega$, give the same potential outside Ω. Let $g = \mathrm{rest}_{\partial\Omega}\, u$. Here u was defined in (9). Integrating (9) with respect to μ we get

$$\int u(x) \,\mathrm{d}\mu(x) = \int_{\partial\Omega} u(y)\, \sigma(y) \,\mathrm{d}S(y), \tag{12}$$

which defines an equivalence relation on the set of all functions $u \in \mathbf{C}(\bar{\Omega})$ harmonic in Ω. Relations (12) and (10) are equivalent. The measure μ_σ, where σ was defined in (11), is called the *swept-out measure* of μ. Especially if $\mu = \delta_x$, $x \in \Omega$, we obtain the solution of the Dirichlet problem. Further, we get

$$G_{\mathrm{L}}\mu(z) \geqq G_{\mathrm{L}}\mu_\sigma(z) \quad \text{for all} \quad z \in \mathbb{R}^n. \tag{13}$$

For $z \notin \Omega$ equality (13) was proved in (10). If $z \in \Omega$ then $z \mapsto G_{\mathrm{L}}\mu_\sigma(z)$ is harmonic. Since

$$v(z) := G_{\mathrm{L}}\mu(z) - G_{\mathrm{L}}\mu_\sigma(z) \tag{14}$$

is superharmonic in Ω and $v(z) = 0$ holds on $\partial\Omega$ from the minimum principle of the superharmonic functions it follows that $v(z) \geqq 0$ on Ω (Theorem 3.18).

In modern potential theory (13) is known as *domination principle*, an important inequality in the theory of the Laplace equation (Doob [1], Helms [1]). Integrating (13) with respect to $\lambda \in F^+(\partial\Omega; G_{\mathrm{L}})$ and changing the order of integration we obtain

$$\int G_{\mathrm{L}}\lambda \,\mathrm{d}\mu \geqq \int G_{\mathrm{L}}\lambda \,\mathrm{d}\mu_\sigma \quad \text{for all} \quad \lambda \in F^+(\bar{\Omega}; G_{\mathrm{L}}), \tag{15}$$

which is the *integral form of the domination principle.*

In the case of an arbitrary domain $\Omega \subset \mathbb{R}^n$ the solution $u = H_h$ of the Dirichlet problem is given by the harmonic measure μ_x, where μ_x is a positive measure satisfying supp $\mu_x \subset \partial\Omega$,

$$u(x) = \int h(y) \,\mathrm{d}\mu_x(y). \tag{16}$$

On the boundary we have

$$u(x) \to h(z) \text{ for } x \to z \in \partial\Omega, \text{except a set of } \mathbf{C}\text{-capacity zero.} \tag{17}$$

In this general case the swept-out measure ν of a given measure μ is given by Helms [1]:

$$\nu(f) := \int \mu_x(f) \,\mathrm{d}\mu(x), \qquad f \in \mathbf{C}_\omega(\mathbb{R}^n). \tag{18}$$

If $f(y) = G_L(y, z)$, $z \notin \bar{\Omega}$, we get

$$G_L\mu(z) = G_L\nu(z) \tag{19}$$

and on $\partial\Omega$

$$G_L\mu(z) = G_L\nu(z)\,, \text{ except a set of C-capacity zero.} \tag{20}$$

If $\mu \geqq 0$ we get the domination principle

$$G_L\mu(z) \geqq G_L\nu(z) \quad \text{for all} \quad z \in \mathbb{R}^n. \tag{21}$$

Integrating (20) with respect to $\lambda \in F^+(\partial\Omega; G_L)$ we get

$$\int G_L\mu \,\mathrm{d}\lambda = \int G_L\nu \,\mathrm{d}\lambda$$

or

$$\int G_L\lambda \,\mathrm{d}\mu = \int G_L\lambda \,\mathrm{d}\nu \quad \text{for all} \quad \lambda \in F(\partial\Omega; G_L). \tag{22}$$

By the first Harnack theorem we obtain

$$\int u \,\mathrm{d}\mu = \int u \,\mathrm{d}\nu \quad \text{for all} \quad u \in \overline{D_{\partial\Omega}(G_L)}, \tag{23}$$

where $\overline{D(\partial\Omega; G_L)} \neq C(\partial\Omega)$ is possible in the case of an arbitrary bounded $\partial\Omega$ (see, for instance, the Lebesgue spine). Inequality (21) is equivalent to

$$\int G_L\lambda \,\mathrm{d}\mu \geqq \int G_L\lambda \,\mathrm{d}\nu \quad \text{for all} \quad \lambda \in F^+(G_L). \tag{24}$$

The harmonic measure μ_x is uniquely determined for each $x \notin \partial\Omega$, also for an unbounded domain $\Omega \subset \mathbb{R}^n$, $n \geqq 2$. This follows from the domination principle.

In the case of the logarithmic potential ($n = 2$) one takes the Green function $G_L^{\Omega_0}$ of a sufficiently large ball $\Omega_0 = B(0, R) \subset \mathbb{R}^2$. Then $G_L^{\Omega_0} \geqq 0$ and all foregoing remarks hold relative to the kernel $G_L^{\Omega_0}$.

We have seen that the continuous potentials $G_L\lambda$ are an optimal set of potentials to develop a modern potential theory. In the case of an abstract potential theory the set of all $G_L\lambda$, $\lambda \in F^+(G_L)$, is replaced by a suitable cone (Bliedtner and Hansen [1], Doob (1]), i.e., if $f \in k$ and $r \in \mathbb{R}^+$ it follows that $fr \in k$. It is possible to do the same for the heat equation (Anger [4], Bauer [2]). But in this case some difficulties arise, because the solutions of the heat equation do not have all properties of the solutions of the Laplace equation. For instance, the hyperplanes I_n have C-capacity zero, i.e., cap $(K; G_H) = 0$ if $K \subset I_n$ is a compact subset (Anger [4]). Further, the convergence property K_D for solutions of parabolic equations is weaker than the convergence property K_B for solutions of elliptic differential equations (Bauer [2], Doob [1]).

4.2. The balayage principle as projection on the dual space of the space of continuous functions

In potential theory the extended balayage (sweeping-out principle) is proved relative to the Newtonian kernel (Doob [1], Helms [1]). In literature the properties (1) and (2) are called balayage. This operation has its origin in the work of H. Poincaré (Kellogg [1]). Here we call balayage the weaker property (1). In this form the balayage can be applied to other differential equations (Anger [4], Schulze and Wildenhain [1]) and formulated in dual systems using projections (Anger [4]) (see Sect. 4.4). Inequality (2) is then an additional property from which follows the unicity of the harmonic measure μ_x relative to an arbitrary domain $\Omega \subset \mathbb{R}^n$. In Corollary 4.1 we shall prove that the balayage principle holds relative to the adjoint kernel G^* if and only if the maximum principle is fulfilled in the form $|u(x)| \leqq M \sup \{|u(y)|, y \in \partial\Omega\}$ for all $u \in \overline{D_{\partial\Omega}(G)}$.

Theorem 4.1 (Extended balayage principle). *Let $\Omega \subset \mathbb{R}^n$ be a bounded domain, G_{L} the Newtonian kernel ($n \geqq 3$). To each positive measure μ (with compact support) there exists a positive measure ν on $\partial\Omega$ such that*

$$G_{\mathrm{L}}\mu(x) = G_{\mathrm{L}}\nu(x) \quad \text{on } \partial\Omega, \text{ except a set of } \mathbf{C}\text{-capacity zero,} \tag{1}$$

and

$$G_{\mathrm{L}}\mu(x) \geqq G_{\mathrm{L}}\nu(x) \quad \text{for all} \quad x \in \mathbb{R}^n. \tag{2}$$

If $\overline{D(\partial\Omega; G_{\mathrm{L}})} = \mathbf{C}(\partial\Omega)$ then relation (1) holds for all $x \in \partial\Omega$.

Remark 4.1. Let $\partial\Omega$ be sufficiently smooth. Using results of classical potential theory in (4.1.10) special swept-out measures were constructed.

Remark 4.2. If $\mu = \delta_x$, $x \notin \partial\Omega$, we obtain the uniquely determined harmonic measure μ_x solving the Dirichlet problem at x. It is an element of $E^+(G_{\mathrm{L}})$. Then

$$G_{\mathrm{L}}^{\Omega}(x, z) := G_{\mathrm{L}}(x, z) - G_{\mathrm{L}}\mu_x(z) \tag{3}$$

is the Green function. From (2) it follows that $G_{\mathrm{L}}^{\Omega}(x, z) \geqq 0$.

Remark 4.3. For the Helmholtz equation (2.1.34) and other linear elliptic differential equations of second-order relations (1) and (2) hold if Ω is sufficiently small (Miranda [1]). For the Helmholtz equation eigenfunctions exists, i.e., from $u(y) = 0$ on $\partial\Omega$ it follows that $u(x) \not\equiv 0$ in Ω. This can be proved by considering $u'' + k^2u = 0$. Further, the maximum principle has the following form: If there do not exist eigenfunctions the inequality

$$|u(x)| \leqq M \sup \{|u(y)| : y \in \partial\Omega\} \tag{4}$$

holds for all $x \in \Omega$ (Bliedtner and Hansen [1], Schulze and Wildenhain [1]).

In Sect 4.1 we proved for sufficiently smooth domains the extended balayage principle. Its general form is (see (4.1.23), (4.1.24)).

Theorem 4.2 (Integral form of the extended balayage principle). *Under the assumptions of Theorem 4.1 to each positive measure μ (with compact support) there exists a positive measure ν on $\partial\Omega$ such that*

$$\int u \, \mathrm{d}\mu = \int u \, \mathrm{d}\nu \quad \textit{holds for all} \quad u \in \overline{D_{\partial\Omega}(G_\mathrm{L})}, \tag{5}$$

where $\overline{D_{\partial\Omega}(G_\mathrm{L})}$ is the closure of $D_{\partial\Omega}(G_\mathrm{L})$ in the norm $\|\cdot\|_\mathbf{C}$. Additionally we have

$$\int G_\mathrm{L}\lambda \, \mathrm{d}\mu \geqq \int G_\mathrm{L}\lambda \, \mathrm{d}\nu \quad \textit{for all} \quad \lambda \in F^+(G_\mathrm{L}). \tag{6}$$

Let μ_x be the harmonic measure (relative to the Laplace equation). We define for $f \in \mathbf{C}(\bar{\Omega})$

$$\begin{aligned} Af(x) &:= \int f \, \mathrm{d}\mu_x, \qquad x \in \Omega, \\ Af(x) &:= f(x), \qquad x \in \partial\Omega. \end{aligned} \tag{7}$$

In the case of a smooth boundary (7) has the form (4.1.6). Therefore the following considerations can be understood in the special case of a ball $B(x^0, R) \subset \mathbb{R}^n$ using only Riemann integrals.

In the following we consider $D_{\partial\Omega}(G_\mathrm{L})$ as a subspace of $\mathbf{C}(\bar{\Omega})$, i.e., the potentials $G_\mathrm{L}\lambda$ are defined on $\bar{\Omega}$. Further, let

$$J(\partial\Omega) := \{f \in \mathbf{C}(\bar{\Omega}), f(y) = 0 \text{ on } \partial\Omega\}. \tag{8}$$

For simplicity, we put

$$X_1 := \overline{D_{\partial\Omega}(G_\mathrm{L})}, \qquad X_2 := J(\partial\Omega) \tag{9}$$

and define the direct sum of X_1 and X_2 by

$$X_1 + X_2 := \{f = f_1 + f_2, f_1 \in X_1, f_2 \in X_2\}. \tag{10}$$

Since the functions $f_1 \in X_1$ are harmonic in Ω we get

$$Af_1 = f_1 \quad \text{for all} \quad f_1 \in X_1. \tag{11}$$

The unicity of this representation follows from the maximum principle of the harmonic functions. Further, from $f_2(y) = 0$ on $\partial\Omega$ it follows that

$$Af_2 = 0 \quad \text{for all} \quad f_2 \in X_2. \tag{12}$$

The subspaces X_1 and X_2 have in common only the element 0. Therefore

$$X_1 \cap X_2 = \{0\}. \tag{13}$$

From this fact it follows that each $f \in X = \mathbf{C}(\bar{\Omega})$ (our assumption was $\overline{D(\partial\Omega; G_\mathrm{L})} = \mathbf{C}(\partial\Omega)$) can be uniquely represented in the form

$$f = f_1 + f_2, \qquad f_1 \in X_1, \qquad f_2 \in X_2, \tag{14}$$

symbolically written

$$X = X_1 \oplus X_2. \tag{15}$$

The space X is the direct sum of X_1 and X_2 (see Fig. 4.1).

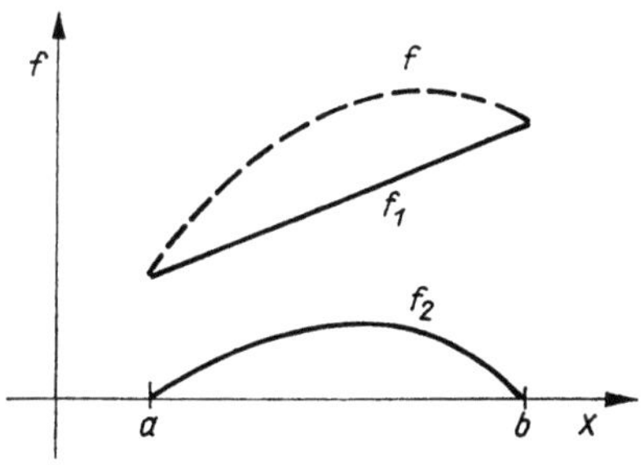

Fig. 4.1

The mapping $A: \mathbf{C}(\bar{\Omega}) \mapsto \mathbf{C}(\bar{\Omega})$ defined in (7) having the properties (11)–(15) is a *projection*, i.e.,

$$A^2 f = Af \quad \text{for all} \quad f \in X = \mathbf{C}(\bar{\Omega}). \tag{16}$$

One often writes P_{X_1} instead of A. We shall show that the *adjoint mapping* $A^*: \mathbf{C}^*(\bar{\Omega}) \mapsto \mathbf{C}^*(\bar{\Omega})$ describes the balayage principle.

At first the definition of A^*. Let

$$\langle f, \mu \rangle := \int f \, \mathrm{d}\mu, \qquad f \in \mathbf{C}(\bar{\Omega}), \qquad \mu \in \mathbf{C}^*(\bar{\Omega}). \tag{17}$$

The mapping $(f, \mu) \mapsto \langle f, \mu \rangle \in \mathbb{R}$ is a bilinear form, i.e., for arbitrary $f_1, f_2 \in \mathbf{C}(\bar{\Omega})$, $c_1, c_2 \in \mathbb{R}$ and $\mu \in \mathbf{C}^*(\bar{\Omega})$ we have

$$\langle c_1 f_1 + c_2 f_2, \mu \rangle = c_1 \langle f_1, \mu \rangle + c_2 \langle f_2, \mu \rangle \tag{18}$$

and for $\mu_1, \mu_2 \in \mathbf{C}^*(\bar{\Omega})$, $d_1, d_2 \in \mathbb{R}$ and $f \in \mathbf{C}(\bar{\Omega})$

$$\langle f, d_1 \mu_1 + d_2 \mu_2 \rangle = d_1 \langle f, \mu_1 \rangle + d_2 \langle f, \mu_2 \rangle. \tag{19}$$

In the case of the Euclidean space $\mathbb{R}^n$ the bilinear form is defined by

$$\langle x, y \rangle := \sum_{k=1}^{n} x_k y_k .$$

For better understanding let us calculate A^* for a domain with smooth boundary. The solution of the Dirichlet problem has then the form (2.1.45)

$$\begin{aligned} Af(x) &:= \int_{\partial\Omega} f(y) \, P(x, y) \, \mathrm{d}S(y), \qquad x \notin \Omega, \\ Af(x) &= f(x), \qquad x \in \partial\Omega, \end{aligned} \tag{20}$$

where in the case of a ball $B(x^0, R) \subset \mathbb{R}^n$, $n \geqq 2$, the kernel $P(x, y)$ was defined in (2.1.9)–(2.1.11). Let $\mu \in \mathbf{C}^*(\bar{\Omega})$. Then μ_1 denotes the restriction μ_Ω of μ on Ω, μ_2 the restriction $\mu_{\partial\Omega}$ of μ on $\partial\Omega$, $\mu \doteq \mu_1 + \mu_2$. By using (20) we obtain

$$\begin{aligned} \langle Af, \mu \rangle &:= \int Af(x) \, \mathrm{d}\mu(x) = \int \Bigg(\int_{\partial\Omega} f(y) \, P(x, y) \, \mathrm{d}S(y) \Bigg) \mathrm{d}\mu_1(x) + \int f(x) \, \mathrm{d}\mu_2(x) \\ &= \int_{\partial\Omega} \left(\int P(x, y) \, \mathrm{d}\mu_1(x) \right) f(y) \, \mathrm{d}S(y) + \int f(x) \, \mathrm{d}\mu_2(x) \\ &= \int_{\partial\Omega} f(y) \, \sigma(y) \, \mathrm{d}S(y) + \int f(x) \, \mathrm{d}\mu_2(x) = \langle f, A^* \mu \rangle, \end{aligned} \tag{21}$$

where

$$\begin{aligned}\sigma(y) &:= \int P(x, y)\, \mathrm{d}\mu_1(x),\\ \mathrm{d}A^*(y) &:= \sigma(y)\, \mathrm{d}S(y) + \mathrm{d}\mu_2(y).\end{aligned} \tag{22}$$

In this special case the adjoint mapping $A^*\colon \mathbf{C}^*(\bar{\Omega}) \mapsto \mathbf{C}^*(\bar{\Omega})$ is defined by

$$\int f(y)\, \mathrm{d}(A^*\mu(y)) = \int_{\partial\Omega} f(y)\, \sigma(y)\, \mathrm{d}S(y) + \int f(y)\, \mathrm{d}\mu_2(y). \tag{23}$$

For $\nu := A^*\mu$ one often writes $\Pi_{\partial\Omega}\mu$. One immediately sees that $A^*\mu$ is concentrated on $\partial\Omega$. Further, if $f \in \mathbf{C}(\bar{\Omega})$ is harmonic in Ω from (11) it follows that

$$\langle Af, \mu\rangle = \langle f, \mu\rangle = \int f\, \mathrm{d}\mu = \langle f, A^*\mu\rangle = \int f\, \mathrm{d}(A^*\mu) = \int f\, \mathrm{d}\nu, \tag{24}$$

the well known equivalence relation on $\mathbf{C}^*(\bar{\Omega})$ (see Anger [1, 2, 4]). Since we always assume that $\overline{D(\partial\Omega; G_\mathrm{L})} = \mathbf{C}(\partial\Omega)$, which is fulfilled in the case of a smooth boundary $\partial\Omega$, the swept-out measure $\nu = A^*\mu$ is uniquely determined.

Let $A\colon \mathbf{C}(\bar{\Omega}) \mapsto \mathbf{C}(\bar{\Omega})$ be a continuous linear mapping. Then the *adjoint mapping* $A^*\colon \mathbf{C}^*(\bar{\Omega}) \mapsto \mathbf{C}^*(\bar{\Omega})$ is defined by

$$\langle Af, \mu\rangle = \langle f, A^*\mu\rangle \quad \text{for all} \quad f \in \mathbf{C}(\bar{\Omega}) \quad \text{and all} \quad \mu \in \mathbf{C}^*(\Omega). \tag{25}$$

A^* is also a continuous linear mapping, i.e.,

$$\|A^*\mu\| \leqq \|A^*\|\, \|\mu\| \tag{26}$$

and

$$\|A^*\| = \|A\|. \tag{27}$$

For details see Bourbaki [2].

In general we have the following important theorem (Anger [4, 12]).

Theorem 4.3 (Projection theorem). *Let* $\overline{D(\partial\Omega; G_\mathrm{L})} = \mathbf{C}(\partial\Omega)$. *The mapping* $A\colon \mathbf{C}(\bar{\Omega}) \mapsto \mathbf{C}(\bar{\Omega})$ *defined in* (7) *solves the Dirichlet problem. It is a continuous projection. Its dual mapping* $A^*\colon \mathbf{C}^*(\bar{\Omega}) \mapsto \mathbf{C}^*(\bar{\Omega})$ *is also a continuous projection describing the balayage principle. It defines an equivalence relation on the set of all harmonic functions.*

Proof. a) We have assumed that $\overline{D(\partial\Omega; G_\mathrm{L})} = \mathbf{C}(\partial\Omega)$ holds. Therefore $\mathbf{C}(\bar{\Omega}) = X = X_1 \oplus X_2$, $X_1 = D_{\partial\Omega}(G_\mathrm{L})$, $X_2 = J(\partial\Omega)$. Each $f \in X$ can be uniquely represented in the form

$$f = f_1 + f_2, \qquad f_1 \in X_1, \qquad f_2 \in X_2.$$

From (11), (12), the property $|f_1(y)| = |f_1(y) + f_2(y)|$ for all $y \in \partial\Omega$, and the maximum principle it follows that

$$\begin{aligned}\|Af\|_\mathbf{C} &= \|Af_1\|_\mathbf{C} = \|f_1\|_\mathbf{C} = \sup\{|f_1(y)|,\, y \in \partial\Omega\}\\ &= \sup\{|f_1(y) + f_2(y)|,\, y \in \partial\Omega\} \leqq \sup\{|f_1(x) + f_2(x)|,\, x \in \bar{\Omega}\}\\ &= \|f_1 + f_2\|_\mathbf{C} = \|f\|_\mathbf{C}.\end{aligned} \tag{28}$$

The property $A^2f = Af$ for all $f \in \mathbf{C}(\bar{\Omega})$ was proved in (16).

b) The adjoint mapping A^* is defined by

$$\int (Af)\,\mathrm{d}\mu = \int f\,\mathrm{d}(A^*\mu) \quad \text{for all} \quad f \in X. \tag{29}$$

For $f_2 \in J(\partial\Omega)$ we get $Af_2 = 0$ and therefore

$$\int f_2\,\mathrm{d}(A^*\mu) = 0 \quad \text{for all} \quad f_2 \in J(\partial\Omega), \tag{30}$$

i.e., the support of the measure $A^*\mu$ is on $\partial\Omega$. Since the set of all restrictions of harmonic functions $f_1 \in X_1$ on $\partial\Omega$ is equal to $\mathbf{C}(\partial\Omega)$ the swept-out measure $\nu = A^*\mu$ is uniquely determined.

c) The equivalence relation follows from $Af_1 = f_1$ for all $f_1 \in X_1$,

$$\int Af_1\,\mathrm{d}\mu = \int f_1\,\mathrm{d}\mu = \int f_1\,\mathrm{d}(A^*\mu). \tag{31}$$

We now define the orthogonal spaces relative to X_1 and X_2. Let

$$X_1^\circ = \left(\overline{D_{\partial\Omega}(G_\mathrm{L})}\right)^\circ := \left\{\mu \in \mathbf{C}^*(\bar{\Omega}) : \mu(f) = 0 \text{ for all } f \in \overline{D_{\partial\Omega}(G_\mathrm{L})}\right\}, \tag{32}$$

$$X_2^\circ = \left(J(\partial\Omega)\right)^\circ := \{\mu \in \mathbf{C}^*(\bar{\Omega}) : \mu(f) = 0 \text{ for all } f \in J(\partial\Omega)\}. \tag{33}$$

The elements $\mu \in X_2^\circ$ consist of all $\mu \in \mathbf{C}^*(\Omega)$ whose support is on $\partial\Omega$, the elements of $\mu \in X_1^\circ$ of all $\mu \in \mathbf{C}^*(\bar{\Omega})$ satisfying the equivalence relation (31).

d) The continuity of A^* follows from (26). The next step is to prove

$$\mathbf{C}^*(\bar{\Omega}) = X_1^\circ \oplus X_2^\circ. \tag{34}$$

Let $\mu \in X_1^\circ \cap X_2^\circ \neq \{0\}$. Then $\mu \in \left(J(\partial\Omega)\right)^\circ$. From this it follows that $\operatorname{supp}\mu \subset \partial\Omega$. Since $\mu(f_1) = 0$ for all $f_1 \in X_1$ and $\operatorname{supp}\mu \subset \partial\Omega$ we get

$$0 = \int f_1\,\mathrm{d}\mu = \int \mathrm{rest}_{\partial\Omega} f_1\,\mathrm{d}\mu \quad \text{for all} \quad f_1 \in X_1.$$

The set of all $\mathrm{rest}_{\partial\Omega} f_1$ is dense in $\mathbf{C}(\partial\Omega)$. Therefore $\mu = 0$. We have the representation

$$\mu = (\mu - A^*\mu) + A^*\mu =: \mu_1 + \mu_2, \tag{35}$$

which is equivalent to

$$\int f\,\mathrm{d}\mu = \left(\int f\,\mathrm{d}\mu - \int Af\,\mathrm{d}\mu\right) + \int Af\,\mathrm{d}\mu.$$

From this it follows that $\mu_1 \in X_1^\circ$, $\mu_2 \in X_2^\circ$. Therefore (34) holds. In order to prove

$$A^*\mu_1 = 0 \quad \text{for all} \quad \mu_1 \in X_1^\circ \quad \text{and} \quad A^*\mu_2 = \mu_2 \quad \text{for all} \quad \mu_2 \in X_2^\circ \tag{36}$$

we use the definition of X_1° and X_2°. From $\mu_1 \in X_1^\circ$ it follows that

$$\int f_1\,\mathrm{d}\mu_1 = 0 \quad \text{for all} \quad f_1 \in X_1.$$

Therefore we obtain

$$A^*\mu_1(f) = \int f\,\mathrm{d}(A^*\mu_1) = \int Af\,\mathrm{d}\mu_1 = \int f_1\,\mathrm{d}\mu_1 = 0 \quad \text{for all} \quad f \in X,$$

i.e., $A^*\mu_1 = 0$. In a similar way we get for $\mu_2 \in X_2^\circ$, because $\int f_2 \, d\mu_2 = 0$ holds for all $f_2 \in X_2$,

$$A^*\mu_2(f) = \int f \, d(A^*\mu_2) = \int Af \, d\mu_2 = \int f_1 \, d\mu_2 = \int (f_1 + f_2) \, d\mu_2 = \int f \, d\mu_2 .$$

That means $A^*\mu_2 = \mu_2$ for all $\mu_2 \in X_2^\circ$. We have proved that

$$(A^*)^2 \mu = A^*(A^*\mu) = A^*\mu_2 = A^*\mu \tag{37}$$

holds for all $\mu \in \mathbf{C}^*(\bar{\Omega})$, i.e., A^* is a projection. ∎

Let $\Omega \subset \mathbb{R}^n$ be a bounded domain and μ a positive measure satisfying $\operatorname{supp} \mu \subset \bar{\Omega}$. In (24) or (31) we have seen that to μ there exist further measures ν satisfying

$$\int f \, d\nu = \int f \, d\mu \quad \text{for all} \quad f \in \mathbf{C}(\bar{\Omega}), \quad \Delta f = 0 \quad \text{in } \Omega .$$

The special measure $\nu = A^*\mu$, the swept-out measure concentrated on $\partial\Omega$, is uniquely determined. For a smooth boundary $\partial\Omega$ this equivalence relation has the form (21). In this case it is not necessary to know modern potential theory. In mathematics it is possible to work with sets of solutions, a strength of mathematics. To each $\mu \geqq 0$ with $\mu \in \mathbf{C}^*(\bar{\Omega})$ we introduce the set

$$\mathfrak{B}(\mu) := \left\{\nu \in \mathbf{C}^*(\bar{\Omega}) : \nu \geqq 0, \int f \, d\nu = \int f \, d\mu, f \in \overline{D_{\partial\Omega}(G_L)}\right\}. \tag{38}$$

From this definition it follows that $\mathfrak{B}(\mu) \subset \mathbf{C}^*(\bar{\Omega})$. In Sect. 4.1 we have proved, that (38) can be defined in the following equivalent manner:

$$\mathfrak{B}(\mu) := \{\nu \in \mathbf{C}^*(\bar{\Omega}) : \nu \geqq 0, G_L\nu(x) = G_L\mu(x) \text{ for all } x \notin \Omega\} . \tag{39}$$

The mapping $\mu \mapsto \mathfrak{B}(\mu)$ is a multi-valued mapping. Let $\mu \geqq 0$ be a measure satisfying $\operatorname{supp} \mu \subset \partial\Omega$. From (37) it follows that $A^*\mu = \mu$. We define

$$(A^*)^{-1} (\mu) := \{\nu \in \mathbf{C}^*(\bar{\Omega}) : A^*\nu = \mu\} . \tag{40}$$

If $\mu \in \mathbf{C}^*(\bar{\Omega})$ is an arbitrary positive measure and $A^*\mu = \mu_0$ we have to define (40) relative to μ_0. From (38) and (40) it follows that

$$\mathfrak{B}(\mu) \subset (A^*)^{-1} (\mu) .$$

In order to apply theorems of functional analysis to $\mathfrak{B}(\mu)$ it is necessary to introduce a notion of convergence. On $\mathbf{C}^*(\bar{\Omega})$ the norm

$$\|\mu\| := \sup \{|\mu(f)| : \|f\| \leqq 1\} \tag{41}$$

is considered. If $\mu \geqq 0$ then $\|\mu\| = \int d\mu$. With respect to this norm the space $\mathbf{C}^*(\bar{\Omega})$ becomes a Banach space. But this norm is too strong in applications. If $x^k \to x$ in $\mathbb{R}^n$ then the Dirac measures δ_{x^k} do not tend to δ_x in the norm (41). This follows from the fact that for

$$\nu_k := \delta_{x^k} - \delta_x = \nu_k^+ - \nu_k^- ,$$

where $\nu_k^+ = \delta_{x^k}$ and $\nu_k^- = \delta_x$, we have

$$\|\nu_k\| = \|\nu_k^+\| + \|\nu_k^-\| = \|\delta_{x^k}\| + \|\delta_x\| = 1 + 1 = 2 \quad \text{for all} \quad k \in \mathbb{N} .$$

For the solution of the Dirichlet problem we know that

$$\int f \, \mathrm{d}\mu_{x^k} \to \int f \, \mathrm{d}\delta_y = f(y) \quad \text{for} \quad x^k \to y \in \partial\Omega$$

holds for each $f \in \mathbf{C}(\bar{\Omega})$. That notion of convergence is the *weak convergence*. A sequence $(\mu_k)_{k\in\mathbb{N}}$ of measures $\mu_k \in \mathbf{C}^*(\bar{\Omega})$ *converges weakly* to $\mu \in \mathbf{C}^*(\bar{\Omega})$, if

$$\int f \, \mathrm{d}\mu_k \to \int f \, \mathrm{d}\mu \quad \text{for all} \quad f \in \mathbf{C}(\bar{\Omega}). \tag{42}$$

This convergence is generated by the following metric. Let

$$\{f_1, f_2, \ldots\} \subset \mathbf{C}(\bar{\Omega})$$

be a denumerable dense set, for instance, the set of all polynomials with rational coefficients. Then

$$d(\nu, \mu) := \sum_{j=1}^{\infty} \frac{1}{2^j} \frac{\left|\int f_j \, \mathrm{d}\nu - \int f_j \, \mathrm{d}\mu\right|}{1 + \left|\int f_j \, \mathrm{d}\nu - \int f_j \, \mathrm{d}\mu\right|} \tag{43}$$

is a metric on $\mathbf{C}^*(\bar{\Omega})$, which defines the *weak convergence.*

Definition 4.1. The *w^*-topology on* $\mathbf{C}^*(\bar{\Omega})$ is defined by the metric (43). A sequence $(\mu_k)_{k\in\mathbb{N}}$ of measures $\mu_k \in \mathbf{C}^*(\bar{\Omega})$ converges weakly to $\mu \in \mathbf{C}^*(\bar{\Omega})$, if and only if $\int f \, \mathrm{d}\mu_k \to \int f \, \mathrm{d}\mu$ for all $f \in \mathbf{C}(\bar{\Omega})$.

Remark 4.4. For each separable Banach space X the w^*-topology can be introduced on its dual space X^*. Further, the w-topology on X can be defined by changing the role of $f \in X$ and the continuous linear forms $L \in X^*$. For details see Bourbaki [2], Dunford and Schwartz [1].

4.3. Approximation theorems

Let $K \subset \Omega \subset \mathbb{R}^n$ be a compact subset. The question arises which $f \in \mathbf{C}(K)$ can be approximated by solutions $u = G_{\mathrm{L}}\lambda$, $\lambda \in F(\partial\Omega; G_{\mathrm{L}})$. Such results are important in order to prove that a measure $\nu \in \mathfrak{B}(\mu)$ satisfying $\operatorname{supp} \nu \subset K$ is uniquely determined by the values $G_{\mathrm{L}}\nu(x) = G_{\mathrm{L}}\mu(x)$ on $\partial\Omega$. Extremal elements of $\mathfrak{B}(\mu)$ have this property (see Sect. 5.3). As an example we consider the set $\mathfrak{B}(\delta_{x^0})$, where $x^0 \in \Omega$. The special measures $\delta_{x^0,r}$ and $m_{x^0,r}$, where $B(x^0, r) \subset \Omega$, are elements of $\mathfrak{B}(\delta_{x^0})$. The measures $\delta_{x^0,r}$ are extremal elements of $\mathfrak{B}(\delta_{x^0})$ because each $f \in \mathbf{C}(\partial B(x^0, r))$ can be uniformly approximated by potentials $G_{\mathrm{L}}\lambda$, $\lambda \in F(\partial\Omega; G_{\mathrm{L}})$ (Theorem 4.9 and Theorem 5.3). At the beginning of this section we prove that $\overline{D(\partial\Omega; G_{\mathrm{L}})} = \mathbf{C}(\partial\Omega)$ holds if and only if the Green function $G_{\mathrm{L}}^{\Omega}(x, z)$ is zero for all $z \in \partial\Omega$ and all $x \notin \partial\Omega$. These results are proved relative to the Laplace equation but they can be proved for other differential equations, too.

In order to prove some approximation theorems we shall use the following continuation property of harmonic functions u defined on a domain $\Omega \subset \mathbb{R}^n$: If $u(x) = 0$ on a ball $B(x^0, r) \subset \Omega$ then $u(x) \equiv 0$ on Ω. More generally this property holds for solutions of linear elliptic differential equations of second order and other differential equations (MIRANDA [1], DE LA PRADELLE [1]).

In the following we shall use the Green function

$$G_{\mathrm{L}}^{\Omega}(x, z) := G_{\mathrm{L}}\delta_x(z) - G_{\mathrm{L}}\mu_x(z),$$

where μ_x is the harmonic measure of Ω. In (3.4.43)–(3.4.46) we have seen that for smooth boundaries $\partial\Omega$ the Green function is zero on $\partial\Omega$. We shall give a further sufficient condition for $G_{\mathrm{L}}^{\Omega}(x, z) = 0$, $z \in \partial\Omega$, which is independent of the foregoing considerations.

Theorem 4.4. *Let $\Omega \subset \mathbb{R}^n$, $n \geqq 2$, be a domain and $z^0 \in \partial\Omega$ a boundary point, such that there exists a truncated closed solid cone in $\mathbb{R}^n \setminus \Omega$ with vertex at z^0. Then $G_{\mathrm{L}}^{\Omega}(x, z^0) = 0$ for all $x \in \Omega$.*

Proof. From (4.1.21) it follows that $G_{\mathrm{L}}^{\Omega}(x, z) \geqq 0$ for $x \in \Omega$ and $z \in \partial\Omega$. If $G_{\mathrm{L}}^{\Omega}(x^0, z) = 0$ for a special $x^0 \in \Omega$ then $G_{\mathrm{L}}^{\Omega}(x, z) = 0$ for all $x \in \Omega$, which is a consequence of the maximum principle of harmonic functions. We use a coordinate system such that the cone has the form (see Fig. 4.2)

$$\zeta_n^2 = cr^2 = c(\zeta_1^2 + \cdots + \zeta_{n-1}^2), \qquad \zeta_n \geqq 0.$$

Then $z^0 = 0$ and the inequality

$$|y - z| \geqq c^{1/2}\, |y - z^0|$$

holds for $z = (0, 0, \ldots, z_n)$ on the ζ_n-axis, where y is not an element of the cone.

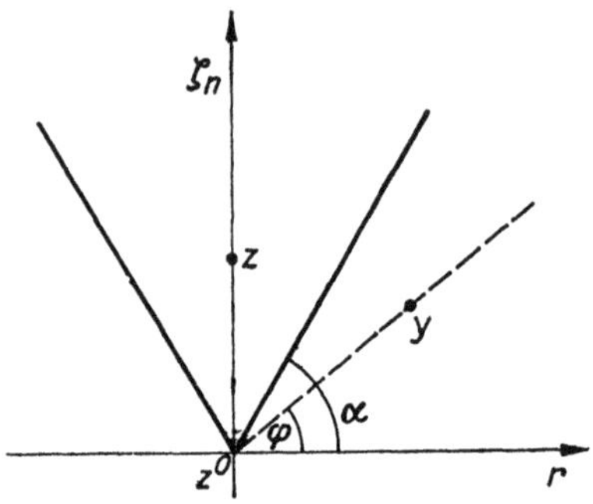

Fig. 4.2

In order to prove this we use the fact that the shortest distance from y to the ζ_n-axis is attained when $y_n = 0$. Since y is not an element of the cone we obtain

$$|y - z| \geqq (y_1^2 + \cdots + y_{n-1}^2)^{1/2} = |y| \cos\varphi \geqq |y| \cos\alpha = c^{1/2}\, |y|,$$

and because $z^0 = 0$

$$|y - z| \geqq c^{1/2}\, |y - z^0|.$$

Therefore we get

$$\frac{1}{|y - z|} \leqq c' \frac{1}{|y - z^0|}. \tag{1}$$

From (4.1.21) it follows that $G_L(x, z^0) \geqq G_L\mu_x(z^0)$. Using (1) and the fact that for $z \notin \Omega$ the function $x \mapsto G_L(x, z)$ is harmonic in Ω we obtain

$$G_L(x, z) = \int G_L(y, z) \, d\mu_x(y) \leqq c' \int G_L(y, z^0) \, d\mu_x(y) = c' G_L(x, z^0).$$

There exists an integrable majorant. Therefore

$$\begin{aligned} G_L(x, z^0) &= \lim_{z \to z^0} G_L(x, z) = \lim_{z \to z^0} \int G_L(y, z) \, d\mu_x(y) \\ &= \int \lim_{z \to z^0} G_L(y, z) \, d\mu_x(y) = \int G_L(y, z^0) \, d\mu_x(y), \end{aligned} \tag{2}$$

i.e.,

$$G_L^\Omega(x, z^0) = G_L(x, z^0) - \int G_L(y, z^0) \, d\mu_x(y) = 0. \blacksquare$$

Remark 4.5. The domain $\Omega := \{x \in \mathbb{R}^3 : G_L\mu_\tau(x) < 1 + c\}$ defined in Example 3.2 containing the point at infinity, does not satisfy the cone condition at $(0, 0, 0)$, i.e., there does not exist a truncated cone in $\Omega' := \{x \in \mathbb{R}^3 : G_L\mu_\tau(x) > 1 + c\}$ with vertex at $(0, 0, 0)$. In Example 3.2 we proved that for Ω the Dirichlet problem is not solvable at $(0, 0, 0)$.

Proposition 4.1. *Let $\Omega \subset \mathbb{R}^n$, $n \geqq 2$, be a bounded domain such that $G_L^\Omega(x, z) = 0$ for all $z \in \partial\Omega$. Then $\mu_x \in F^+(\partial\Omega; G_L)$.*

Proof. By assumption the equality

$$G_L(x, z) = G_L\mu_x(z) \text{ holds for all } z \in \partial\Omega, \text{ where } \operatorname{supp} \mu_x \subset \partial\Omega.$$

Then measure μ_x produces a continuous potential on its support. Using Theorem 3.16 we obtain the continuity of $G_L\mu_x$ on the whole space, i.e., $\mu_x \in F^+(\partial\Omega; G_L)$. ∎

Theorem 4.5. *Let $\Omega \subset \mathbb{R}^n$, $n \geqq 3$, be a domain such that for all $z \in \partial\Omega$ the Green function $G_L^\Omega(x, z)$ be zero for all $x \notin \partial\Omega$. Then $\overline{D(\partial\Omega; G_L)} = \mathbf{C}(\partial\Omega)$ and conversely.*

Proof. a) Let

$$G_L^\Omega(x, z) = G_L(x, z) - G_L\mu_x(z) = 0 \text{ for all } z \in \partial\Omega \text{ and all } x \notin \partial\Omega.$$

In order to prove $\overline{D(\partial\Omega; G_L)} = \mathbf{C}(\partial\Omega)$, we apply the Hahn-Banach theorem (Dunford and Schwartz [1]). This states that any bounded linear functional that is defined over some normed subspace X_1 of a Banach space X can be extended to the whole space without increasing its norm. From this theorem it follows that there exists a continuous linear functional $L: X \mapsto \mathbb{R}$ satisfying $L \neq 0$ and $L(f) = 0$ for all $f \in X$. Therefore, if $\overline{D(\partial\Omega; G_L)} \neq \mathbf{C}(\partial\Omega)$, then there exists a continuous linear functional $L: \mathbf{C}(\partial\Omega) \mapsto \mathbb{R}$ having the form $L = \mu^+ - \mu^-$, for which the relation

$$\int f \, d\mu = \int f \, d\mu^+ - \int f \, d\mu^- = 0 \quad \text{for all} \quad f \in \overline{D(\partial\Omega; G_L)} \tag{3}$$

holds. Here μ^+ and μ^- are positive measures (see (3.3.31)) concentrated on $\partial\Omega$. Let $f = G_L\mu_x$, $x \notin \partial\Omega$ fixed. Then

$$\int G_L\mu_x \,d\mu^+ = \int G_L\mu_x \,d\mu^- = \int G_L(x, y) \,d\mu^+(y) = \int G_L(x, y) \,d\mu^-(y)$$

for all $x \notin \partial\Omega$. Therefore we get

$$\int G_L(x, y) \,d\mu^+(y) = \int G_L(x, y) \,d\mu^-(y) \quad \text{for all} \quad x \notin \partial\Omega. \tag{4}$$

Further, relation (3) holds for all $f = G_L\lambda$. That means

$$\int G_L\lambda \,d\mu^+ = \int G_L\lambda \,d\mu^- \quad \text{or} \quad \int G_L\mu^+ \,d\lambda = \int G_L\mu^- \,d\lambda,$$

which by Theorem 3.20 is equivalent to

$$G_L\mu^+ = G_L\mu^- \text{ on } \partial\Omega, \text{ except a set of } \mathbf{C}\text{-capacity zero.} \tag{5}$$

From (4) and (5) it follows that

$$G_L\mu^+ = G_L\mu^- \text{ except a set of } \mathbf{C}\text{-capacity zero.} \tag{6}$$

Since sets of Lebesgue measure zero are of $\mathbf{C}$-capacity zero we can integrate (6) with respect to $(-\Delta\varphi)\,dx$ and obtain

$$\int \big(G_L\mu^+(x)\big)\big(-\Delta\varphi(x)\big)\,dx = \int \big(G_L\mu^-(x)\big)\big(-\Delta\varphi(x)\big)\,dx. \tag{7}$$

Using the Fubini theorem we get

$$\int \Big(\int G_L(x, y)\big(-\Delta\varphi(x)\big)\,dx\Big)\,d\mu^+(y) = \int \Big(\int G_L(x, y)\big(-\Delta\varphi(x)\big)\,dx\Big)\,d\mu^-(y),$$

and similar to the proof of Theorem 3.13

$$\int \varphi \,d\mu^+ = \int \varphi \,d\mu^-.$$

This means $\mu = \mu^+ - \mu^- = 0$ which is a contradiction. Therefore we have $\overline{D(\partial\Omega; G_L)} = \mathbf{C}(\partial\Omega)$.

b) Let $\overline{D(\partial\Omega; G_L)} = \mathbf{C}(\partial\Omega)$. We consider on $\partial\Omega$ the function f defined by $f(y) = G_L(y, z)$, $z \in \partial\Omega$ fixed. We can approximate f by an increasing sequence of function $f_k \in \mathbf{C}(\partial\Omega)$. For $x \notin \partial\Omega$ the solutions of the corresponding Dirichlet problem are

$$u_k(x) = \int f_k(y) \,d\mu_x(y), \tag{8}$$

where $u_k(x) \nearrow G_L(x, z)$. Therefore we obtain

$$0 = \lim_{k\to\infty} u_k(x) - \int \lim_{k\to\infty} f_k(x) \,d\mu_x(y) = G_L(x, z) - \int G_L(y, z) \,d\mu_x(y),$$

i.e., the Green function G_L^Ω is zero for $x \notin \partial\Omega$, $z \in \partial\Omega$. ∎

Remark 4.6. In the case $n = 2$ we have to study the problem relative to $\Omega_0 = B(x^0, R)$ and its Green function $G_L^{\Omega_0}$.

Corollary 4.1. *Let $\Omega \subset \mathbb{R}^n$ be a domain satisfying $G_L^\Omega(x, z) = 0$ for all $z \in \partial\Omega$. Then $G_L^\Omega(x, \zeta) \to 0$ for $\zeta \to z \in \partial\Omega$.*

Proof. In Proposition 4.1 we have proved $\mu_x \in F^+(\partial\Omega; G_L)$. Then $y \mapsto G_L\mu_x(y)$ is continuous. From this fact it follows that $G_L^\Omega(x, \zeta) \to 0$ for $\zeta \to z \in \partial\Omega$. ∎

Remark 4.7. In order to prove Corollary 4.1 it is sufficient that $G_L^\Omega(x, z) = 0$ on $\partial\Omega$ for a neighbourhood $U(z^0) \subset \partial\Omega$ of z^0. Then $G_L(x, z) = G_L\mu_x(z)$ for all $z \in U(z^0)$. Let μ_1 be the restriction of μ_x on $U(z^0)$, μ_2 the restriction of μ_x on $\partial\Omega \setminus U(z^0)$. Then $\mu_x = \mu_1 + \mu_2$, and $G_L\mu_1$ is continuous on $U(z^0)$. From Theorem 3.12 it follows that $\mu_1 \in F^+(\partial\Omega; G_L)$. Further, $G_L\mu_2$ is continuous on $U(z^0)$. Then $G_L^\Omega(x, \zeta) \to 0$ for $\zeta \to z \in U(z^0)$. This remark is important for Example 3.2. In this example the Green function G_L^Ω is not zero at (0, 0, 0).

Theorem 4.6. *Let $\Omega \subset \mathbb{R}^n$ be a bounded domain such that $\mathbb{R}^n \setminus \bar{\Omega}$ is simply connected and satisfies the cone property at each $y \in \partial\Omega$. Further, let $B(x^0, R)$ be a ball outside Ω. Then the set*

$$H_0(\partial\Omega) := \left\{ f \in \mathrm{C}(\partial\Omega) : f(x) = \sum_{k=1}^{m} c_k G_L(x, x^k),\ x^k \in B(x^0, R) \right\} \tag{9}$$

is dense in $\mathrm{C}(\partial\Omega)$ in the uniform topology.

Proof. If $\overline{H_0(\partial\Omega)} \neq \mathrm{C}(\partial\Omega)$ then there exists – like in the proof of Theorem 4.5 – a continuous linear form $L = \mu = \mu^+ - \mu^-$ on $\mathrm{C}(\partial\Omega)$ satisfying $L \neq 0$ and

$$\int f \,\mathrm{d}\mu = \int f \,\mathrm{d}\mu^+ - \int f \,\mathrm{d}\mu^- = 0 \quad \text{for all} \quad f \in H_0(\partial\Omega).$$

For $f(x) = G_L(x, x^k)$, $x^k \in B(x^0, R)$, we get

$$\int G_L(x, x^k) \,\mathrm{d}\mu^+(x) = \int G_L(x, x^k) \,\mathrm{d}\mu^-(x). \tag{10}$$

The potentials $G_L\mu^+$ and $G_L\mu^-$ are analytic outside $\bar{\Omega}$. Therefore from (10) it follows that this relation holds for all $z \notin \bar{\Omega}$, where $\mathbb{R}^n \setminus \Omega$ is simply connected. Let $z^0 \in \partial\Omega$ and z be an element of the cone with vertex at z^0. Applying inequality (1) to this special case we obtain

$$\int G_L(x, z) \,\mathrm{d}\mu^-(x) = \int G_L(x, z) \,\mathrm{d}\mu^+(x) \le c' \int G_L(x, z^0) \,\mathrm{d}\mu^+(x).$$

If $G_L\mu^+(z^0) < \infty$ we have a majorant. Following Theorem 3.10 we get

$$\lim_{z \to z^0} G_L\mu^+(z) = \lim_{z \to z^0} G_L\mu^-(z) = G_L\mu^+(z^0) = G_L\mu^-(z^0).$$

If $G_L\mu^+(z^0) = +\infty$ then from the l.s.c. it follows that

$$G_L\mu^+(z) \to G_L\mu^+(z^0) \quad \text{and similar} \quad G_L\mu^-(z) \to G_L\mu^-(z^0) \quad \text{for} \quad z \to z^0.$$

So we have

$$G_L\mu^+(z^0) = G_L\mu^-(z^0) \quad \text{for all} \quad z^0 \in \partial\Omega. \tag{11}$$

Integrating (11) with respect to $\lambda \in F^+(\partial\Omega; G_L)$ we obtain

$$\int G_L\lambda \, d\mu^+ = \int G_L\mu^+ \, d\lambda = \int G_L\mu^- \, d\lambda = \int G_L\lambda \, d\mu^-$$

for all $G_L\lambda \in D_{\partial\Omega}(G_L)$. From Theorem 4.4 and Theorem 4.5 it follows that $\overline{D(\partial\Omega; G_L)} = \mathbb{C}(\partial\Omega)$. Therefore $\mu^+ = \mu^-$, which is a contradiction to $\mu = \mu^+ - \mu^- \neq 0$. The set $H_0(\partial\Omega)$ is dense in $\mathbb{C}(\partial\Omega)$. ∎

Theorem 4.7. *Let*

$$H_1(\partial\Omega) := \left\{ f \in \mathbb{C}(\partial\Omega) : f(x) = \sum_{k=1}^{m} c_k G_L\lambda_k(x), \lambda_k \in F(G_L), \operatorname{supp} \lambda \subset \mathbb{R}^n \setminus \Omega \right\}. \quad (12)$$

Under the assumptions of Theorem 4.6 the set $H_1(\partial\Omega)$ is dense in $\mathbb{C}(\partial\Omega)$ in the uniform topology.

Proof. Let $B(x^0, r) \subset \mathbb{R}^n \setminus \Omega$, $\lambda = m_{x^0,r}$. Then for $x \in \partial\Omega$ we have

$$G_L\lambda(x) = G_L m_{x^0,r}(x) = G_L(x, x^0).$$

If x^0 is an element of a ball $B(z^0, R) \subset \mathbb{R}^n \setminus \Omega$ we can apply the proof of Theorem 4.6. ∎

In the following we should like to approximate continuous functions f defined on a compact subset $K \subset \Omega \subset \mathbb{R}^n$ by solutions u of the Dirichlet problem having h as boundary values. It is sufficient to vary h on a small subset $B \subset \partial\Omega$. In 1960 this problem was studied by H. Becker (see Anger [5], Göpfert et al. [1]. Hamann [1]) for the first time. We consider a domain Ω_0 containing Ω in its interior, such that $\mathbb{R}^n \setminus \Omega_0$ is simply connected and $\partial\Omega$ and $\partial\Omega_0$ coincide except the set $B = \bar{\Omega}_0 \cap \partial\Omega$ (see Fig. 4.3)

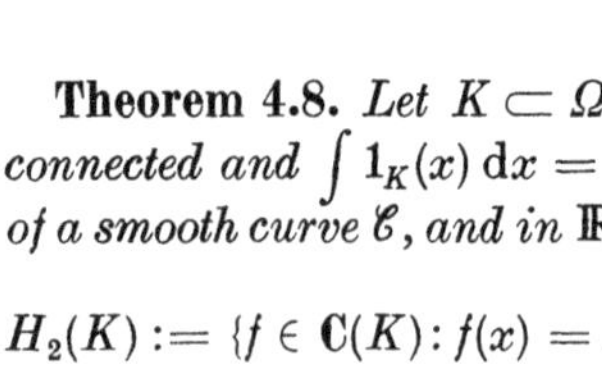

Fig. 4.3

Theorem 4.8. *Let $K \subset \Omega \subset \mathbb{R}^n$ be a compact set such that $\Omega \setminus K$ is simply connected and $\int 1_K(x) \, dx = 0$. This condition is fulfilled in $\mathbb{R}^2$ if K is the graph of a smooth curve $\mathscr{C}$, and in $\mathbb{R}^3$ if K is the graph of a smooth surface S. Then the set*

$$H_2(K) := \{f \in \mathbb{C}(K) : f(x) = H_h(x), h \in \mathbb{C}(\partial\Omega), h(y) = 0 \text{ for } y \in \partial\Omega \setminus B\} \quad (13)$$

is dense in $\mathbb{C}(K)$ in the uniform topology.

Proof. Let $\overline{H_2(K)} \neq \mathbb{C}(K)$. Like in the proof of Theorem 4.6 there exists a measure $\mu = \mu^+ - \mu^-$ with $\operatorname{supp} \mu \subset K$ satisfying

$$\int f \, d\mu^+ = \int f \, d\mu^- \quad \text{for all} \quad f \in H_2(K).$$

Let $G_L^{\Omega_0}$ be the Green function of Ω_0, where Ω_0 was defined before Theorem 4.8. It is zero on $\partial\Omega_0 \cap (\partial\Omega \setminus B)$. Let $B(x^0, r) \subset \Omega_0 \setminus \Omega$ and $x^k \in B(x^0, r)$. Then the set

$$H_3(K) := \left\{ f \in \mathbb{C}(K) : f(x) = \sum_{k=1}^{m} c_k G_L^{\Omega_0}(x, x^k),\ x^k \in B(x^0, r) \right\} \tag{14}$$

is of special interest. It produces special $f \in H_2(K)$. Then

$$\int G_L^{\Omega_0}(x, z)\, \mathrm{d}\mu^+(x) = \int G_L^{\Omega_0}(x, z)\, \mathrm{d}\mu^-(x) \tag{15}$$

for all $z \in B(x^0, r)$. The potentials $G_L^{\Omega_0}\mu^+$ and $G_L^{\Omega_0}\mu^-$ are analytic on $\Omega \setminus K$. Therefore

$$G_L^{\Omega_0}\mu^+(y) = G_L^{\Omega_0}\mu^-(y) \quad \text{for all} \quad y \in \Omega \setminus K. \tag{16}$$

Integrating (16) with respect to $(-\Delta\varphi)\,\mathrm{d}y$, $\varphi \in \mathbb{C}_0^\infty(\Omega)$, K has Lebesgue measure zero (in the case of a smooth graph the content zero), we obtain

$$\int G_L^{\Omega_0}\mu^+(y)\left(-\Delta\varphi(y)\right)\mathrm{d}y = \int G_L^{\Omega_0}\mu^-(y)\left(-\Delta\varphi(y)\right)\mathrm{d}y. \tag{17}$$

Since $G_L\mu_x$ is harmonic on Ω_0, where μ_x is the harmonic measure of Ω_0 at x, the relation

$$\int \left(G_L\mu_x(y)\right) \Delta\varphi(y)\, \mathrm{d}y = 0 \quad \text{holds for all} \quad \varphi \in \mathbb{C}_0^\infty(\Omega_0). \tag{18}$$

From (16) and (17) it follows that

$$\begin{aligned} -\int G_L^{\Omega_0}\mu^+(y)\, \Delta\varphi(y)\, \mathrm{d}y &= -\int \left(\int \left(G_L(x, y) - G_L\mu_x(y)\right) \mathrm{d}\mu^+(x)\right) \Delta\varphi(y)\, \mathrm{d}y \\ &= -\int \left(\int \left(G_L(x, y) - G_L\mu_x(y)\right) \Delta\varphi(y)\, \mathrm{d}y\right) \mathrm{d}\mu^+(x) \\ &= \int \varphi(x)\, \mathrm{d}\mu^+(x) = \int \varphi(x)\, \mathrm{d}\mu^-(x). \end{aligned}$$

By Theorem 3.13 we get $\mu^+ = \mu^-$, which is a contradiction to $\mu \neq 0$. Therefore $\overline{H_2(K)} = \mathbb{C}(K)$. ∎

Remark 4.8. Further approximation theorems and similar approximation theorems for differential equations of higher order were proved in Hamann [1], Schulze and Wildenhain [1].

As we often remarked the domains considered satisfy the condition

$$\overline{D(\partial\Omega; G_L)} = \mathbb{C}(\partial\Omega),$$

which is always true in applications.

Theorem 4.9. *Let Ω and Ω_1 be two bounded domains in $\mathbb{R}^n$ satisfying $\bar{\Omega}_1 \subset \Omega$. Further, we require that $\mathbb{R}^n \setminus \Omega$ and $\mathbb{R}^n \setminus \Omega_1$ were simply connected and that Ω_1 satisfies the cone property at each $y \in \partial\Omega_1$. Then the set of all restrictions $G_L\lambda|_{\partial\Omega_1}$, where $\lambda \in F(\partial\Omega; G_L)$, is dense in $\mathbb{C}(\partial\Omega_1)$.*

Proof. Let $B(x^0, r) \subset \mathbb{R}^n \setminus \bar{\Omega}$, $n \geq 3$, and $z \in B(x^0, r)$. Using Theorem 4.1 we can construct the harmonic measure μ_z concentrated on $\partial\Omega$ satisfying

$$G_L\mu_z(y) = G_L(z, y) \quad \text{for all} \quad y \in \partial\Omega .$$

From Proposition 4.1 it follows that $\mu_z \in F^+(\partial\Omega; G_L)$. In a similar way one can prove the existence of the harmonic measure $\tilde{\mu}_z$ concentrated on $\partial\Omega_1$ satisfying

$$G_L\tilde{\mu}_z(y) = G_L\mu_z(y) = G_L(z, y) \quad \text{for all} \quad y \in \partial\Omega_1, \tag{19}$$

where $\tilde{\mu}_z \in F(\partial\Omega_1; G_L)$. If the set of all $G_L\lambda|_{\partial\Omega_1}$ is not dense in $C(\partial\Omega_1)$ there exists, like in the proof of Theorem 4.6, a measure $\mu = \mu^+ - \mu^-$ concentrated on $\partial\Omega_1$ such that $\mu \neq 0$ and

$$\int f \, d\mu = \int f \, d\mu^+ - \int f \, d\mu^- = 0 \text{ for all } f = G_L\lambda|_{\partial\Omega_1}, \; \lambda \in F(\partial\Omega; G_L).$$

For $f(y) = G_L\tilde{\mu}_z(y)$ we get

$$\begin{aligned} \int G_L\tilde{\mu}_z(y) \, d\mu^+(y) &= \int G_L\mu_z(y) \, d\mu^+(y) = \int G_L(z, y) \, d\mu^+(y) = \int G_L\tilde{\mu}_z(y) \, d\mu^-(y) \\ &= \int G_L\mu_z(y) \, d\mu^-(y) = \int G_L(z, y) \, d\mu^-(y). \end{aligned} \tag{20}$$

Since the fundamental solution $z \mapsto G_L(z, y)$ is analytic for fixed $y \in \partial\Omega_1$ from (20) it follows that

$$\int G_L(z, y) \, d\mu^+(y) = \int G_L(z, y) \, d\mu^-(y) \quad \text{for all} \quad z \notin \bar{\Omega}_1 .$$

The further proof is similar to that in Theorem 4.6. We obtain $\mu^+ = \mu^-$, which is a contradiction. Therefore the set of all $G_L\lambda|_{\partial\Omega_1}$, $\lambda \in F(\partial\Omega; G_L)$, is dense in $C(\partial\Omega_1)$. If $n = 2$ we have to apply the Green function of a ball $\Omega_0 = B(0, R)$ containing Ω in its interior. ∎

In geodesy and other areas the harmonic polynomials are often used. Therefore we prove a special approximation theorem for these polynomials (LANDKOF [1]). We denote by h_k a harmonic polynomial of the order k. There exist $2k + 1$ linearly independent harmonic polynomials of the order k. In KELLOGG [1] it is proved that for fixed $x \notin B(0, r)$ and $y \in B(0, r)$

$$G_L(x, y) = \frac{1}{4\pi \, |x - y|} = \sum_{k=1}^{\infty} h_k(y) \tag{21}$$

uniformly holds on each compact set $K \subset B(0, r)$.

Proposition 4.2. *Let $\Omega \subset \mathbb{R}^n$, $n = 3$, be a bounded domain. Then on $\bar{\Omega}$ each harmonic polynomial h_k can be represented in the form*

$$h_k(x) = G_L\lambda_k(x), \qquad \lambda_k \in F^+(\partial\Omega; G_L), \qquad x \in \bar{\Omega}. \tag{22}$$

If $\partial\Omega$ is sufficiently smooth, the measure λ_k has the form

$$d\lambda_k(y) = \sigma(y) \, dS(y) = \left(\int_{\mathbb{R}^3 \setminus \Omega} P(x, y) \, \varrho_\varepsilon(x) \, dx \right) dS(y).$$

Proof. There exists an open ball $B(0, r)$ containing $\bar{\Omega}$ in its interior. Further, there exists an open set Ω_1 satisfying $\bar{\Omega} \subset \Omega_1 \subset B(0, r)$. Let

$$c_\varepsilon(x) := \int 1_{\Omega_1}(y)\, \varphi_\varepsilon(x - y)\, \mathrm{d}y, \tag{23}$$

where φ_ε was defined in (1.3.17). We have $\varphi_\varepsilon \in C_0^\infty(\mathbb{R}^n)$ and for sufficiently small ε

$$\operatorname{supp} c_\varepsilon \subset B(0, r), \qquad c_\varepsilon(x) = 1 \text{ on } \bar{\Omega}. \tag{24}$$

Let h_k be a harmonic polynomial. We define

$$u_k := h_k c_\varepsilon. \tag{25}$$

For

$$\varrho_\varepsilon := -\Delta u_k$$

we obtain

$$u_k(x) = \int G_{\mathrm{L}}(x, y)\, \varrho_\varepsilon(y)\, \mathrm{d}y = G_{\mathrm{L}}\mu_{\varrho_\varepsilon}(x). \tag{26}$$

The potential $G\mu_{\varrho_\varepsilon}$ is continuous, $\operatorname{supp} \mu_{\varrho_\varepsilon} \subset B(0, r) \setminus \Omega$. The swept-out measure λ_k of $\mu_{\varrho_\varepsilon}$ on $\partial\Omega$ is uniquely determined and represents h_k on $\bar{\Omega}$, i.e.,

$$u_k(x) = h_k(x) = G_{\mathrm{L}}\lambda_k(x), \qquad x \in \bar{\Omega}. \tag{27}$$

If $\partial\Omega$ is sufficiently smooth the measure λ_k has the form (see (4.1.10))

$$\mathrm{d}\lambda_k(y) = \sigma(y)\, \mathrm{d}S(y) = \left(\int_{\mathbb{R}^3 \setminus \Omega} P(x, y)\, \varrho_\varepsilon(x)\, \mathrm{d}x \right) \mathrm{d}S(y). \ ∎$$

Proposition 4.3. *Let $\Omega \subset \mathbb{R}^n$ be a bounded domain such that $\mathbb{R}^n \setminus \bar{\Omega}$ is simply connected satisfying the cone condition at each point $z \in \partial\Omega$. Then the set of all restrictions* $\operatorname{rest}_{\partial\Omega} h_k$ *of harmonic polynomials h_k is dense in* $C(\partial\Omega)$.

Proof. Let D be the set of all $\operatorname{rest}_{\partial\Omega} h_k$. If $\bar{D} \neq C(\partial\Omega)$ we can apply the ideas of the proof of Theorem 4.6. Then there exists a continuous linear form $\mu = \mu^+ - \mu^-$ such that $\mu \neq 0$ and

$$\int f\, \mathrm{d}\mu = \int f\, \mathrm{d}\mu^+ - \int f\, \mathrm{d}\mu^- = 0 \quad \text{for all} \quad f \in \bar{D}.$$

From (21) it follows that the Newtonian kernel is an element of $\bar{D}$. Therefore we get, if $B(0, r) \supset \bar{\Omega}$,

$$\int G_{\mathrm{L}}(x, y)\, \mathrm{d}\mu^+(y) = \int G_{\mathrm{L}}(x, y)\, \mathrm{d}\mu^-(y) \quad \text{for all} \quad x \notin B(0, r).$$

By continuation this equality holds for all $x \notin \bar{\Omega}$. Using the cone property we get the last equality on $\partial\Omega$. The further considerations correspond to those in the proof of Theorem 4.6. Therefore $\mu = \mu^+ - \mu^- = 0$, which is a contradiction. We have $\bar{D} = C(\partial\Omega)$. ∎

In applications the following approximation theorem is often used (Miranda [1], Moritz [1]). More generally it holds for linear elliptic differential equations of second order.

Theorem 4.10 (Runge approximation theorem). *Let $\Omega \subset \mathbb{R}^n$, $n \geqq 2$, be a bounded domain such that $\mathbb{R}^n \setminus \Omega$ is simply connected satisfying the cone condition at each point $z \in \partial\Omega$. Let further Ω_1 be a bounded domain having the same properties satisfying $\bar{\Omega} \subset \Omega_1$, then each function $u \in \mathbf{C}(\bar{\Omega})$ harmonic in Ω can be uniformly approximated by functions v harmonic in Ω_1,* i.e.,

$$|u(x) - v(x)| < \varepsilon \quad \textit{for all} \quad x \in \bar{\Omega}.$$

Especially, on Ω_1 the v can be restrictions of harmonic polynomials represented in the form $v = G_L\lambda$, $\lambda \in F(\partial\Omega_1; G_L)$.

Proof. Let $u \in \mathbf{C}(\bar{\Omega})$ be harmonic in Ω. Following Theorem 4.9 the set of all restrictions $v = \operatorname{rest}_{\partial\Omega} G_L\lambda$, $\lambda \in F(\partial\Omega_1; G_L)$, is dense in $\mathbf{C}(\partial\Omega)$. From

$$|u(x) - v(x)| < \varepsilon \quad \text{on } \partial\Omega \tag{28}$$

and from the maximum principle it follows that

$$|u(x) - v(x)| < \varepsilon \quad \text{on } \bar{\Omega}. \tag{29}$$

By Proposition 4.2 and Proposition 4.3 the functions v defined on $\bar{\Omega}_1$ can be harmonic polynomials having the form $v = G_L\lambda$, $\lambda \in F(\partial\Omega_1; G_L)$. ∎

Theorem 4.11. *Let $K \subset \Omega \subset \mathbb{R}^n$ be a compact set such that $\mathbb{R}^n \setminus K$ is simply connected and $\int 1_K(x)\,\mathrm{d}x = 0$. Further, let μ_1, μ_2 be two measures with* supp $\mu_j \subset K$, *$j = 1, 2$. From $G_L\mu_1(x) = G_L\mu_2(x)$ for all $x \notin \Omega$ it follows that $\mu_1 = \mu_2$.*

Proof. By analytic continuation the relation

$$G_L\mu_1(x) = G_L\mu_2(x) \quad \text{holds for all} \quad x \notin K. \tag{30}$$

Since $\int 1_K(x)\,\mathrm{d}x = 0$ we can multiply (30) by $\big(-\Delta\varphi(x)\big)\,\mathrm{d}x$ and integrate. Then

$$\int G_L\mu_1(x)\,\big(-\Delta\varphi(x)\big)\,\mathrm{d}x = \int G_L\mu_2(x)\,\big(-\Delta\varphi(x)\big)\,\mathrm{d}x.$$

Similarly to (17) and (18) we get

$$\int \varphi\,\mathrm{d}\mu_1 = \int \varphi\,\mathrm{d}\mu_2 \quad \text{for all} \quad \varphi \in \mathbf{C}_0(\mathbb{R}^n).$$

That means $\mu_1 = \mu_2$. ∎

Remark 4.9. The earth's gravitational potential $G_L\mu_0$ is approximatively computed by potentials of the form $G_L\mu_k$, where

$$\mu_k := \sum_{j=1}^{k} c_j\delta_{x^j}, \qquad x^j \notin \Omega_0. \tag{31}$$

From Theorem 4.11 it follows that the measures μ_k are uniquely determined by $G_L\mu_k(x)$, $x \notin \bar{\Omega}_0$ (see Barthelmes [1], Stromeyer and Ballani [1]).

Remark 4.10. If a measure $\nu \in \mathfrak{B}(\mu)$ (see formula (4.2.38)) has the property that for its support $K = \operatorname{supp} \nu$ the set $\mathbb{R}^n \setminus K$ is simply connected and

$\int 1_K(x)\,dx = 0$ holds, then ν is uniquely determined. Such measures $\nu \in \mathfrak{B}(\mu)$ are extremal elements of $\mathfrak{B}(\mu)$, i.e., if their support is known the measures are uniquely determined. The special measure ν_σ, where $\sigma = \psi_0$ was defined in (3.1.44), has this property.

4.4. A general form of the balayage principle

In Sect. 4.1 and Sect. 4.2 we have seen that the balayage principle relative to the Newtonian kernel depends on a Banach space X, for instance, $X = \mathbf{C}(\bar{\Omega})$, two subspaces X_1, X_2 satisfying

$$X = X_1 \oplus X_2 \tag{1}$$

and a continuous projection $A: X \mapsto X_1$,

$$\begin{aligned} Af_1 &= f_1 \quad \text{for all} \quad f_1 \in X_1, \\ Af_2 &= 0 \quad \text{for all} \quad f_2 \in X_2. \end{aligned} \tag{2}$$

Then the balayage principle is described by the adjoint mapping

$$A^*: X^* \mapsto X^*, \tag{3}$$

where A^* is defined by

$$\langle Af, f^* \rangle = \langle f, Af^* \rangle \quad \text{for all} \quad f \in X, f^* \in X^* \tag{4}$$

with

$$\langle f, f^* \rangle := f^*(f). \tag{5}$$

For the biharmonic equation $\Delta^2 u = 0$ the Dirichlet problem relative to an open set $\Omega \subset \mathbb{R}^n$ with boundary values

$$u(y) = g(y), \qquad \partial_{x_k} u(y) = g_k(y), \qquad y \in \partial\Omega, \qquad k = 1, 2, \ldots, n, \tag{6}$$

can be studied (see Sect. 1.6). In that case we consider the space

$$\mathbf{C}^1(\bar{\Omega}) = \{f \in C(\bar{\Omega}) : \partial_{x_k} f \in C(\bar{\Omega}),\, k = 1, 2, \ldots, n\}, \tag{7}$$

on which the norm

$$\|f\|_{\mathbf{C}^1} := \max\{|f(x)|, x \in \bar{\Omega}\} + \sum_{k=1}^{n} \max\{|\partial_{x_k} f(x)|, x \in \bar{\Omega}\} \tag{8}$$

is introduced. With this norm $\mathbf{C}^1(\bar{\Omega})$ is a Banach space. Further, let

$$X_1 := \{f \in \mathbf{C}^1(\bar{\Omega}) \cap \mathbf{C}^4(\Omega), \Delta^2 f = 0 \text{ in } \Omega\} \tag{9}$$

and

$$X_2 = J_1(\partial\Omega) := \{f \in \mathbf{C}^1(\bar{\Omega}) : f(y) = \partial_{x_k} f(y) = 0 \text{ on } \partial\Omega,\, k = 1, 2, \ldots, n\}. \tag{10}$$

With this pair a balayage theory can be developed (see Anger [4] and Schulze and Wildenhain [1]). In a certain sense these ideas can be applied to the Cauchy problem of hyperbolic equations. Using the fundamental solution, potentials of distributions of a finite order (Hörmander [2]), a potential theory for hyperbolic equations is to be developed (Anger [4]). In that way one can find finer structures for inverse source problems of such equations using results of functional analysis. In this section we develop a balayage theory for a Banach space X and prove the fundamental Theorem 4.12. It is applied to the concrete case $X = \mathrm{C}(\bar{\Omega})$. Further, the balayage principle is introduced for a general kernel $G^*: \Omega \times \Omega \mapsto \overline{\mathbb{R}}$.

Proposition 4.4. *A projection $A: X_1 \oplus X_2 \mapsto X_1$ is continuous if and only if the inequality $\|f_1\| \leqq M \inf \{\|f_1 + f_2\|, f_2 \in X_2\}$ holds for all $f_1 \in X_1$.*

Proof. From $X = X_1 \oplus X_2$ it follows that each $f \in X$ can be uniquely represented in the form

$$f = f_1 + f_2, \qquad f_1 \in X_1, \qquad f_2 \in X_2.$$

Then $A(f_1 + f_2) = f_1$.

a) The condition is necessary. Let A be continuous. Then for each $f \in X$ the inequality (Dunford and Schwartz [1])

$$\|f_1\| \leqq \|Af\| \leqq M \|f\| \leqq M \|f_1 + f_2\| \tag{11}$$

holds. From (11) it follows that

$$\|f_1\| \leqq M \inf \{\|f_1 + f_2\|, f_2 \in X_2\}.$$

b) The condition is also sufficient. By assumption the following inequalities hold:

$$\|Af\| = \|f_1\| \leqq M \inf \{\|f_1 + f_2\|, f_2 \in X_2\} \leqq M \|f_1 + f_2\| = M \|f\|.$$

From this follows the continuity of A. ∎

Proposition 4.5. *Let $X = \mathrm{C}(\bar{\Omega})$ and $X_2 = J(\partial\Omega)$. Then*

$$\inf \{\|f_1 + f_2\|_{\mathrm{C}}, f_2 \in J(\partial\Omega)\} = \max \{|f_1(x)|, x \in \partial\Omega\}, \tag{12}$$

where the right-hand side is the norm of $\mathrm{C}(\partial\Omega)$.

Proof. a) Let $f_1 \in X_1$, $g = \mathrm{rest}_{\partial\Omega} f_1$. To each $g \in \mathrm{C}(\partial\Omega)$ there exists at least one continuation $\tilde{f} \in \mathrm{C}(\bar{\Omega})$ such that (Tietz's theorem, see Dunford and Schwartz [1])

$$\mathrm{rest}_{\partial\Omega} \tilde{f} = g \text{ and } \max \{|\tilde{f}(x)|, x \in \bar{\Omega}\} = \max \{|g(x)|, x \in \partial\Omega\}. \tag{13}$$

We consider the function $f_2 := f_1 - \tilde{f}$, where on $\partial\Omega$

$$f_2(y) = f_1(y) - \tilde{f}(y) = g(y) - g(y) = 0$$

holds, i.e., $f_2 \in J(\partial\Omega)$. Then $f_1 = \tilde{f} + f_2$. For $\tilde{f}_2 := -f_2$ we get

$$\inf\{\|f_1 + f_2\|_C, f_2 \in J(\partial\Omega)\} \leqq \|f_1 + \tilde{f}_2\|_C = \|f_1 - f_2\|_C$$
$$= \|\tilde{f} + f_2 - f_2\|_C = \|\tilde{f}\|_C = \max\{|f_1(x)|, x \in \partial\Omega\}. \tag{14}$$

The last relation follows from (13).

b) Further, for all $f_2 \in J(\partial\Omega)$ the following inequalities hold:

$$\max\{|f_1(x)|, x \in \partial\Omega\} = \max\{|f_1(x) + f_2(x)|, x \in \partial\Omega\}$$
$$\leqq \max\{|f_1(x) + f_2(x)|, x \in \bar{\Omega}\} = \|f\|_C.$$

For all $f_2 \in J(\partial\Omega)$ we obtain

$$\inf\{\|f_1 + f_2\|_C, f_2 \in J(\partial\Omega)\} \geqq \max\{|f_1(x)|, x \in \partial\Omega\}. \tag{15}$$

Finally, from (14) and (15) we get

$$\max\{|f_1(x)|, x \in \partial\Omega\} = \inf\{\|f_1 + f_2\|_C, f_2 \in J(\partial\Omega)\}. \blacksquare$$

As in Sect. 4.2 we introduce the sets

$$X_1^\circ := \{f^* \in X^* : f^*(f_1) = 0 \text{ for all } f_1 \in X_1\}, \tag{16}$$

$$X_2^\circ := \{f^* \in X^* : f^*(f_2) = 0 \text{ for all } f_2 \in X_2\}. \tag{17}$$

Proposition 4.6. *Let $X = X_1 \oplus X_2$. Then X_1° and X_2° are closed subspaces of the dual space X^* satisfying $X_1^\circ \cap X_2^\circ = \{0\}$.*

Proof. a) Let $f_k^* \in X_1^\circ \subset X^*$ converging to $f_0^* \in X_0^*$, i.e., $\|f_k^* - f_0^*\| < \varepsilon$ for $k > k_0(\varepsilon)$. From $f_k^*(f_1) = 0$ it follows for all $f_1 \in X_1$ that

$$|f_0^*(f_1)| = |f_k^*(f_1) - f_0^*(f_1)| = |(f_k^* - f_0^*)(f_1)| \leqq \|f_k^* - f_0^*\| \|f_1\| < \varepsilon'$$

for $k > k_0'(\varepsilon')$, i.e., $f_0^* \in X_1^\circ$. A similar proof holds for X_2°.

b) We assume $g^* \in X_1^\circ \cap X_2^\circ$, $g^* \neq 0$. From $X = X_1 \oplus X_2$ it follows that $f \in X$ can be uniquely represented as

$$f = f_1 + f_2, \quad f_1 \in X_1, \quad f_2 \in X_2.$$

Then

$$g^*(f) = g^*(f_1 + f_2) = g^*(f_1) + g^*(f_2).$$

If $g^* \in X_1^\circ \cap X_2^\circ$ then $g^*(f_1) = 0$ and $g^*(f_2) = 0$. Therefore $g^*(f) = 0$ for all $f \in X$, i.e., $g^* = 0$, which is a contradiction. $\blacksquare$

The following definition is the abstract form of the conrete balayage principle studied in Sect. 4.2.

Definition 4.2 (Abstract balayage principle). Let X be a Banach space, X_1, X_2 two subspaces of X satisfying $X = X_1 \oplus X_2$, $A: X_1 \oplus X_2 \mapsto X_1$ the projection of X onto X_1. The *balayage principle* holds relative to X_1 if to each $f^* \in X^*$

there exists a $f_2^* \in X_2^\circ$ such that

$$f^*(Af) = f_2^*(f) \quad \text{for all} \quad f \in X. \tag{18}$$

Proposition 4.7. *The following relations are equivalent:*

$$\text{(i)}\ f^*(Af) = f_2^*(f) \quad \textit{for all} \quad f \in X,$$

$$\text{(ii)}\ f^*(f_1) = f_2^*(f_1) \quad \textit{for all} \quad f_1 \in X_1.$$

Proof. From $f^*(Af) = f_2^*(f)$, $Af = A(f_1 + f_2) = f_1$, and $f_2^*(f_2) = 0$ for all $f_2 \in X_2$ it follows that

$$f^*(Af) = f^*(f_1) = f_2^*(f_1 + f_2) = f_2^*(f_1) \quad \text{for all} \quad f_1 \in X_1$$

and conversely. ∎

Proposition 4.8. *The abstract balayage principle holds relative to* X_1 *if and only if* $X^* = X_1^\circ \oplus X_2^\circ$.

Proof. a) The condition is necessary. Let the balayage principle hold relative to X_1, i.e., to each $f^* \in X^*$ there exists an $f_2^* \in X_2^\circ \subset X^*$ satisfying

$$f^*(Af) = f_2^*(f) \quad \text{for all} \quad f \in X.$$

Let $f^* \in X^*$ be an arbitrary element of X^*. Then we define

$$f_1^*(f) := f^*(f) - f_2^*(f).$$

For $f_1 \in X_1$ we get

$$f_1^*(f_1) = f^*(f_1) - f_2^*(f_1) = 0,$$

i.e., $f_1^* \in X_1^\circ$. By assumption $f_2^* \in X_2^\circ$. From Proposition 4.6 it follows that $X_1^\circ \cap X_2^\circ = \{0\}$. Each $f^* \in X^*$ can be uniquely determined in the form

$$f^* = f_1^* + f_2^*, \qquad f_1^* \in X_1^\circ, \qquad f_2^* \in X_2^\circ.$$

That means $X^* = X_1^\circ \oplus X_2^\circ$.

b) The condition is sufficient. Let $X^* = X_1^\circ \oplus X_2^\circ$. Then each $f \in X^*$ has the unique representation

$$f^* = f_1^* + f_2^*, \qquad f_1^* \in X_1^\circ, \qquad f_2^* \in X_2^\circ.$$

From $f_1^* \in X_1^\circ$ it follows that $f_1^*(f_1) = 0$ and

$$f^*(f_1) = f_1^*(f_1) + f_2^*(f_1) = f_2^*(f_1) \quad \text{for all} \quad f_1 \in X_1.$$

From Proposition 4.7 we get

$$f^*(Af) = f_2^*(f) \quad \text{for all} \quad f \in X,$$

i.e., the abstract balayage principle holds relative to X_1. ∎

Theorem 4.12. *The abstract balayage principle holds relative to* X_1 *if and only if the projection defined in* (2) *is continuous.*

Proof. a) The condition is sufficient. For $f^* \in X^*$ the mapping $f \mapsto f^*(Af)$ is linear satisfying

$$|f^*(Af)| \leq \|f^*\| \, \|Af\| \leq \|f^*\| \, \|A\| \, \|f\|.$$

Therefore the linear form $f \mapsto f^*(Af)$ is continuous which can be represented by means of a continuous linear form $f_2^* \in X^*$. From

$$f^*(Af) = f_2^*(f) =: A^*f^*(f)$$

it follows that

$$f^*(Af_1) = f^*(f_1) = f_2^*(f_1) \quad \text{for all} \quad f_1 \in X_1$$

and

$$f^*(Af_2) = f_2^*(f_2) = 0 \quad \text{for all} \quad f_2 \in X_2,$$

i.e., the balayage principle holds to X_1.

b) The continuity of A is necessary. We assume that the balayage principle holds to X_1. From Proposition 4.8 it follows that

$$X^* = X_1^0 \oplus X_2^0.$$

To each $f^* \in X^*$ there exists an $f_2^* \in X_2^0$ such that

$$f^*(Af) = f_2^*(f).$$

That means that the adjoint mapping $A^*: X^* \mapsto X^*$ exists, where by definition (4)

$$\langle Af, f^* \rangle = f^*(Af) = \langle f, A^*f^* \rangle =: \langle f, f_2^* \rangle = f_2^*(f).$$

Similar to the proof of Theorem 4.3 one can show that

$$A^*: X_1^0 \oplus X_2^0 \mapsto X_2^0 \text{ is a projection onto } X_2^0. \tag{19}$$

Since X_1^0 and X_2^0 are closed subspaces (Proposition 4.6) satisfying $X^* = X_1^0 \oplus X_2^0$ it follows from the theorem of Banach-Schauder that the projection $A^*: X_1^0 \oplus X_2^0 \mapsto X_2^0$ is continuous (see Köthe [1]). We now consider the dual space X^{**} of X^*. On X^{**} the norm

$$\|f^{**}\| = \sup \{|f^{**}(f^*)|, \|f^*\| \leq 1\}$$

is introduced. The space X can be identified as a subspace of X^{**}, if

$$f^{**}(f^*) := f^*(f) \quad \text{for fixed} \quad f \in X.$$

Then (see Bourbaki [2])

$$\|f^{**}\| = \|f\|. \tag{20}$$

The mapping A^* is continuous on X^*, Its adjoint mapping

$$A^{**}: X^{**} = (X_1^0)^0 + (X_2^0)^0 \mapsto (X_1^0)^0$$

is also continuous and satisfies

$$\|f_1^{**}\| = \|A^{**}(f_1^{**} + f_2^{**})\| \leqq M \ \|f_1^{**} + f_2^{**}\|.$$

From (20) we get

$$\|f_1\| \leqq M \ \|f_1 + f_2\|, \qquad f_1 \in X_1, \qquad f_2 \in X_2,$$

from which follows the continuity of A. If $X = \mathbf{C}(K)$ then $\mathbf{C}(K)$ is a proper subspace of $\mathbf{C}^{**}(K)$, i.e., $\mathbf{C}(K) \subset C^{**}(K)$. For reflexive Banach spaces like $\mathbf{L}^p(K)$, $p \geqq 1$, the relation $X^{**} = X$ holds (DUNFORD and SCHWARTZ [1]). ▮

From Proposition 4.4 and Theorem 4.12 we get

Corollary 4.2. *The abstract balayage principle holds relative to X_1 if and only if*

$$\|f_1\| \leqq M \inf \{\|f_1 + f_2\|, f_2 \in X_2\} \quad \textit{for all} \quad f_1 \in X_1.$$

In the case $X = \mathbf{C}(\bar{\Omega})$ and $X_2 = J(\partial\Omega)$ we obtain

$$\|f_1\|_{\mathbf{C}} \leqq M \max \{|f_1(x)|, x \in \partial\Omega\} \quad \textit{for all} \quad f_1 \in X_1 \tag{21}$$

or

$$|f_1(z)| \leqq M \max \{|f_1(x)|, x \in \partial\Omega\} \quad \textit{for all} \quad z \in \bar{\Omega}.$$

Definition 4.3. Let $G: \bar{\Omega} \times \bar{\Omega} \mapsto \mathbb{R}$ be a kernel for which exists at least one measure $\lambda \geqq 0$, $\operatorname{supp} \lambda \subset \partial\Omega$, with continuous potentials $G^+\lambda$ and $G^-\lambda$, i.e., $F^+(\partial\Omega, G) \neq \emptyset$. The *balayage principle* holds relative to G^* and $\partial\Omega$ if to each measure μ with $\operatorname{supp} \mu \subset \bar{\Omega}$ there exists a measure ν satisfying $\operatorname{supp} \nu \subset \partial\Omega$ such that

$$G^*\mu(x) = G^*\nu(x) \quad \text{on} \quad \partial\Omega, \quad \textit{except a set of } \mathbf{C}\textit{-capacity zero.} \tag{22}$$

Theorem 4.13. *The balayage principle holds relative to G^* and $\partial\Omega$ if and only if*

$$|G\lambda(x)| \leqq M \max \{|G\lambda(y)|, y \in \partial\Omega\} \quad \textit{for all} \quad G\lambda \in D_{\partial\Omega}(G) \subset \mathbf{C}(\bar{\Omega}). \tag{23}$$

Proof. a) In Theorem 3.20 we proved that

$$G^*\mu(y) = G^*\nu(y) \quad \text{on} \quad \partial\Omega, \quad \text{except a set of } \mathbf{C}\text{-capacity zero,}$$

is equivalent to

$$\int G\lambda \, d\mu = \int G\lambda \, d\nu \quad \text{for all} \quad G\lambda \in D_{\partial\Omega}(G). \tag{24}$$

Let X_1 be the closure of $D_{\partial\Omega}(G)$ in the norm of the space $\mathbf{C}(\bar{\Omega})$, $X_2 = J(\partial\Omega)$. From (24) it follows that

$$\int u \, d\mu = \int u \, d\nu \quad \text{for all} \quad u \in X_1 = \overline{D_{\partial\Omega}(G)}. \tag{25}$$

Especially, let $\mu = \delta_x$. Then there exists a measure $\nu = \mu_x$, which is uniquely determined if $\overline{D(\partial\Omega; G)} = \mathbf{C}(\partial\Omega)$, satisfying

$$u(x) = \int u \, \mathrm{d}\mu_x .$$

Therefore $X_1 \cap X_2 = \{0\}$. In the following we consider $X = X_1 \oplus X_2$. From (25) it follows that to each $f^* = \mu$ there exists an $f_2^* = \nu$ such that for $f := u$ the relation

$$f^*(f) = f_2^*(f)$$

holds. We notice that we have to consider the continuous linear forms $f^* \in X^*$. If $X \neq \mathbf{C}(\bar{\Omega})$ then to one $f^* \in X^*$ there exists at least one measure μ satisfying $f^*(f) = \int f \, \mathrm{d}\mu$ (consequence of the theorem of Hahn-Banach). Proposition 4.7 gives the existence of the projection $A: X_1 \oplus X_2 \mapsto X_1$ such that $f^*(Af) = f_2^*(f)$ for all $f \in X$. From Theorem 4.12 it follows that A is continuous. Following Corollary 4.2 we get the inequality (23).

b) Conversely, let inequality (23) be fulfilled. Then this inequality holds for all $f_1 \in X_1$. We define on X_1 the linear form

$$f_1 \mapsto g^*(f_1) := \int f_1 \, \mathrm{d}\mu .$$

From (23) it follows that

$$|g^*(f_1)| \leqq \|\mu\| \max \{|f_1(x)|, x \in \bar{\Omega}\} \leqq M^* \max \{|f_1(x)|, x \in \partial\Omega\} .$$

We define

$$g^*(f_2) = 0 \quad \text{for all} \quad f_2 \in X_2 .$$

Then $X = X_1 \oplus X_2$ and

$$\begin{aligned} |g^*(f_1 + f_2)| = |g^*(f_1)| &\leqq M^* \max \{|f_1(x)|, x \in \partial\Omega\} \\ &= M^* \max \{|f_1(x) + f_2(x)|, x \in \partial\Omega\} \\ &\leqq M^* \max \{|f_1(x) + f_2(x)|, x \in \bar{\Omega}\} . \end{aligned}$$

On $X \subset \mathbf{C}(\bar{\Omega})$ the linear form $f \to g^*(f)$ is continuous. By means of the theorem of Hahn-Banach this linear form can be continued onto $\mathbf{C}(\bar{\Omega})$, such that

$$\tilde{g}^* \in \mathbf{C}^*(\bar{\Omega}) \quad \text{and} \quad \tilde{g}^*(f) = g^*(f) \text{ on } X .$$

From a theorem of F. Riesz and J. Radon (see Theorem 3.6 and Remark 3.9) it follows that g^* can be represented by a measure $\nu \in \mathbf{C}^*(\bar{\Omega})$ concentrated on $\partial\Omega$. The last relation follows from $\int f_2 \, \mathrm{d}\nu = g^*(f_2) = 0$ for all $f_2 \in X_2$. Therefore to each $f^* \in X^*$ there exists a $g^* = f_2^* \in X^*$ satisfying

$$f^*(f_1) = f_2^*(f_1) \quad \text{for all} \quad f_1 \in X_1 ,$$

$$f^*(f_2) = 0 \quad \text{for all} \quad f_2 \in X_2 .$$

Following Proposition 4.7 these relations are equivalent to

$$f^*(Af) = f_2^*(f) = A^*(f) ,$$

i.e., the balayage principle holds relative to X_1 or all G satisfying (23). ∎

5. Inverse source problems

5.1. Some special inverse source problems in $\mathbb{R}^1$

We should like to discuss some special inverse source problems in $\mathbb{R}^1$ in order to get a first feeling for such problems. The mathematical tools used are simple. The extremal elements of the sets considered are of special interest. If the support of an extremal element is known it is uniquely determined.

At first some geometrical considerations. Let M be the set of all points of a closed triangle in $\mathbb{R}^2$. Then each $x \in M$ can be uniquely represented in the form

$$x = \sum_{k=1}^{3} c_k x^k, \qquad c_k \geqq 0, \qquad \sum_{k=1}^{3} c_k = 1, \tag{1}$$

where x^k are the corners of the triangle (see Fig. 3.3). The x^k are *extremal elements* of M, i.e., from $x^k = tz^1 + (1 - t)\, z^2$, $z^1, z^2 \in M$, $0 \leqq t \leqq 1$, it follows that $t = 0$ or $t = 1$. The points x^1, x^2, x^3 are uniquely determined. A closed ball $K(x^0, r) \subset \mathbb{R}^n$ is a convex set M whose extremal points consist of the whole boundary $\partial K(x^0, r)$. In this case the representation of $x \in M$ by means of extremal points is not unique (see Fig. 3.2). If M is a convex set of an infinite-dimensional functional space the representation (1) holds using a special integration procedure (Choquet [1]). In the concrete cases (3.1.19) and (3.1.45) the integration over extremal elements is simple. In the general case the theorem of Krein-Milman-Choquet proves this representation (Dunford and Schwartz [1], Köthe [1]).

Example 5.1. Let G be the fundamental solution of $P(\partial)\, u := u'$ defined in (1.3.3), i.e., $G(x, y) = H(x - y)$, where H is the Heaviside function. Following (3.3.13) the potentials $G\mu$, $\mu \geqq 0$, supp μ compact, have the form

$$G\mu(x) = \chi_1(x) + \chi_2(x) + \chi_3(x), \tag{2}$$

where χ_1 has a density ϱ, χ_2 is a step function and χ_3 a singular function. Let

$$\mathfrak{B}(\delta_{x^0}) := \{\mu \geqq 0 \colon \operatorname{supp} \mu \leqq x^0,\ G\mu(x) = G(x, x^0) \text{ for } x > x^0\}. \tag{3}$$

If $\mu \in \mathfrak{B}(\delta_{x^0})$ we get

$$G\mu(x) = \chi_1(x) + \chi_2(x) + \chi_3(x) = 1 \quad \text{for} \quad x > x^0 \tag{4}$$

and therefore $\int d\mu = 1$ (see Fig. 5.1). In the case (2) the potentials $G\mu$ are completely known.

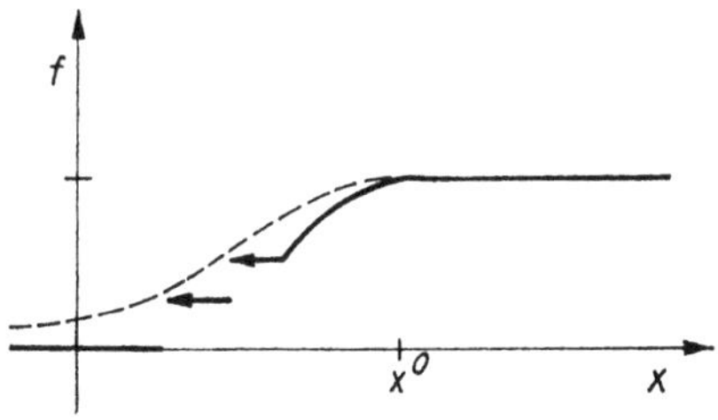

Fig. 5.1

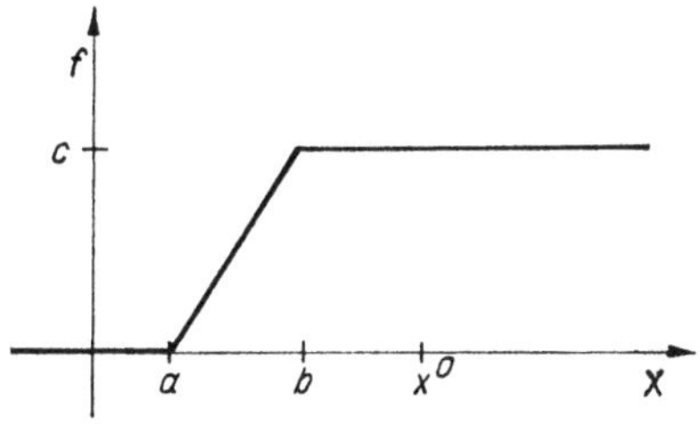

Fig. 5.2

The Dirac measures δ_x, $x < x^0$, are extremal elements of $\mathfrak{B}(\delta_{x^0})$. In order to prove this we assume that δ_x is not extremal. Then we have the representation

$$\delta_x = t\nu_1 + (1 - t)\,\nu_2\,,\; 0 < t < 1\,,\; \nu_1 \neq \nu_2,\; \nu_1,\; \nu_2 \in \mathfrak{B}(\delta_{x^0})\,. \tag{5}$$

From

$$\delta_x(f) = f(x) = 0 \quad \text{for} \quad x \notin \operatorname{supp} f,\; \nu_1 \geqq 0\,,\; \nu_2 \geqq 0$$

$$\text{and} \quad \delta_x \geqq t\nu_1\,,\; \delta_x \geqq (1 - t)\,\nu_2$$

it follows that $\nu_1 = a_1\delta_x$, $\nu_2 = a_2\delta_x$. From (4) we get

$$\int \mathrm{d}\nu_1 = \int \mathrm{d}\nu_2 = 1$$

and therefore $a_1 = a_2 = 1$, i.e., $\nu_1 = \delta_x$, $\nu_2 = \delta_x$. The measure δ_x, $x < x^0$, is uniquely determined if its support $\operatorname{supp} \delta_x = \{x\}$ is known.

Remark 5.1. In integration theory it is proved that the Dirac measures δ_x are the extremal elements of the set K^+ consisting of all positive measures μ satisfying $\int \mathrm{d}\mu = 1$. Each measure $\mu \in K^+$ is then the weak limit (Bourbaki [1])

$$\mu(f) = \lim_{k\to\infty} \sum_{j=1}^{k} c_j\delta_{x^j}(f)\,, \qquad c_j \geqq 0\,, \qquad \sum_{j=1}^{k} c_j = 1\,, \qquad f \in \mathbf{C}_0(\mathbb{R}^n)\,. \tag{6}$$

This result is well known for Riemann integrals. In our case we have to consider measures μ satisfying $\operatorname{supp} \mu \leqq x^0$.

Remark 5.2. In (6), (3.1.19), and (3.1.45) we have seen that certain elements of a convex set can be represented by an integration procedure extended over extremal elements.

Example 5.2. Let $a < b \leqq x^0$,

$$\begin{aligned} &\chi_1(x) = 0 \quad \text{for} \quad x \leqq a < x^0\,, \qquad \chi_1(x) = c \quad \text{for} \quad b \leqq x\,, \\ &\chi_1(x) = \frac{x - a}{b - a}\, d \quad \text{for} \quad a \leqq x \leqq b \end{aligned} \tag{7}$$

and ν the measure satisfying $G\nu(x) = \chi_1(x)$, where G is the kernel of Example 5.1. From the value $G\nu(x^0) = c$ the function χ_1 can be uniquely determined (see

Fig. 5.2). Further, let χ_1^* be the function defined by

$$\chi_1^*(x) = 0 \quad \text{for} \quad x \leqq a < x^0, \qquad \chi_1^*(x) = c \quad \text{for} \quad x \geqq b_c,$$
$$\chi_1^*(x) = (x - a)\, d \quad \text{for} \quad a \leqq x \leqq b_c \leqq x_0. \tag{8}$$

If $(b_c - a)\, d = c$ the point $b_c = x^0$ can be uniquely determined from the value $G\nu(x^0) = c$.

Remark 5.3. There are other possibilities to uniquely determine a function $\chi = G\nu$, $\nu \in \mathfrak{B}(\delta_{x^0})$. In the case of the Laplace equation function (7) corresponds to the problem where the earth's shape is given and the density is to be determined. Function (8) corresponds to the problem where the density is given and the shape is to be determined.

Example 5.3. Let χ_c be a strictly increasing function with $\chi_c(x) = c$ for all $x > x^0$ (see Fig. 5.3). Let $\mathfrak{B}_c(\delta_{x^0})$ be the set of all measures $\nu \in \mathfrak{B}(\delta_{x^0})$ satisfying

$$G\nu(x) \leqq \chi_c(x) \quad \text{for all} \quad x \leqq x^0. \tag{9}$$

Here $\mathfrak{B}(\delta_{x^0})$ was introduced in Example 5.1.

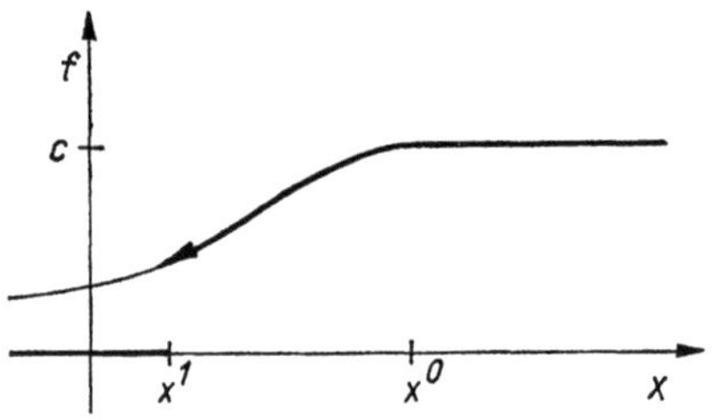

Fig. 5.3

The potentials of extremal elements $\nu_e \in \mathfrak{B}_c(\delta_{x^0})$, where supp $\nu_e \leqq x^0$, have the form

$$G\nu_e(x) = 0 \quad \text{for} \quad x \leqq x^1, \qquad G\nu_e(x) = \chi_c(x) \leqq c \quad \text{for} \quad x^1 < x. \tag{10}$$

In order to prove this we assume that there exist two different $\nu_1, \nu_2 \in \mathfrak{B}_c(\delta_{x^0})$ satisfying

$$\nu_e = t\nu_1 + (1 - t)\, \nu_2, \qquad 0 < t < 1.$$

Then

$$G\nu_e(t) = tG\nu_1(x) + (1 - t)\, G\nu_2(x).$$

Since $\nu_1 \geqq 0$, $\nu_2 \geqq 0$, $G\nu_e \geqq tG\nu_1$, $G\nu_e \geqq (1 - t)\, G\nu_2$ it follows from $G\nu_e(x) = 0$ for $x \leqq x^1$ always $G\nu_1(x) = G\nu_2(x) = 0$. If $G\nu_j(x) < \chi_c(x)$, $j = 1$ or 2, we obtain

$$G\nu_e(x) = tG\nu_1(x) + (1 - t)\, G\nu_2(x) < \chi_c(x),$$

which is a contradiction. Therefore $\nu_j = \nu_e$ for $j = 1, 2$, i.e., ν_e is uniquely determined, also an extremal element of $\mathfrak{B}_c(\delta_{x^0})$. The following question is of special interest: which representation of χ_c holds with respect to the extremal elements of $\mathfrak{B}(\delta_{x^0})$?

Example 5.4. Let

$$\begin{aligned} G_L(x, y) &= x(1 - y) \quad \text{for} \quad 0 \leqq x \leqq y, \\ G_L(x, y) &= (1 - x)\, y \quad \text{for} \quad y \leqq x \leqq 1 \end{aligned} \tag{11}$$

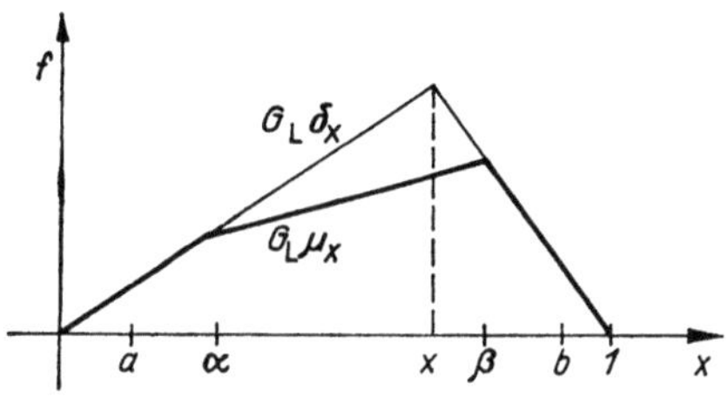

Fig. 5.4

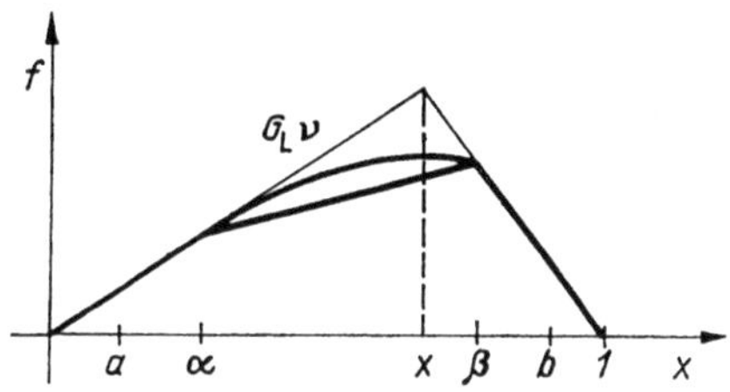

Fig. 5.5

be a special fundamental solution of the one-dimensional Laplace operator $P(\partial)\, u := -u''$. The function G_L defined in (11) is the Green function of the open interval $]0, 1[\subset \mathbb{R}$ satisfying $G_L(x, y) = G_L(y, x)$. The following considerations hold for a fixed closed interval $[a, b] \subset\,]0, 1[$. Let $[\alpha, \beta] \subset [a, b]$ be a further interval. The solution of the Dirichlet problem relative to $]\alpha, \beta[$ has the form (see Fig. 5.4)

$$u(x) = g(\alpha)\, \frac{\beta - x}{\beta - \alpha} + g(\beta)\, \frac{x - \alpha}{\beta - \alpha} = \int g(y)\, d\mu_x(y), \tag{12}$$

where

$$\mu_x := \frac{\beta - x}{\beta - \alpha}\, \delta_\alpha + \frac{x - \alpha}{\beta - \alpha}\, \delta_\beta. \tag{13}$$

Let $\nu \geqq 0$ be a measure satisfying $\operatorname{supp} \nu \subset [a, b]$. Then

$$\int G_L\nu(y)\, d\mu_x(y) = \int G_L\mu_x(z)\, d\nu(z) \leqq \int G_L(x, z)\, d\nu(z) = G_L\nu(x) \tag{14}$$

for all $x \in [\alpha, \beta]$ and each $[\alpha, \beta] \subset [a, b]$ (see Fig. 5.5). As a solution of the Dirichlet problem the left-hand side of (14) is a straight line segment. Therefore $G_L\nu$ is a concave function on $[0, 1]$, i.e., a superharmonic function. In the following we consider for $x^0 \in [a, b]$ the set

$$\mathfrak{B}(\delta_{x^0}) := \{\nu \geqq 0 \colon \operatorname{supp} \nu \subset [a, b],\, G_L\nu(x) = G_L\delta_{x^0}(x),\, x \notin [a, b]\}. \tag{15}$$

On the interval $[\alpha, \beta]$ there exists an *equilibrium distribution* $\lambda_0 \geqq 0$, i.e., $G_L\lambda_0(x) = 1$ on $[\alpha, \beta]$. It has the form

$$\lambda_0 = \frac{1}{\alpha}\, \delta_\alpha + \frac{1}{1 - \beta}\, \delta_\beta.$$

Let $\nu_1, \nu_2 \in \mathfrak{B}(\delta_{x^0})$. From $G_L\nu_1(x) = G_L\nu_2(x)$, $x \notin [a, b]$, it follows that ($\alpha = a$, $\beta = b$)

$$\int G_L\nu_1 \, d\lambda_0 = \int G_L\lambda_0 \, d\nu_1 = \int G_L\nu_2 \, d\lambda_0 = \int G_L\lambda_0 \, d\nu_2,$$

or $\int d\nu_1 = \int d\nu_2$. Since ν_2 can be the measure δ_{x^0} we get

$$\int d\nu = 1 \quad \text{for all} \quad \nu \in \mathfrak{B}(\delta_{x^0}). \tag{16}$$

Now we prove that the measures $\nu_0 := \mu_{x^0}$ with $\operatorname{supp} \mu_{x^0} = \{\alpha, \beta\} \subset [a, b]$ are extremal elements of $\mathfrak{B}(\delta_{x^0})$. If ν_0 is not extremal, we have

$$\nu_0 = t\nu_1 + (1 - t)\,\nu_2, \qquad 0 < t < 1, \qquad \nu_1 \neq \nu_2, \qquad \nu_1, \nu_2 \in \mathfrak{B}(\delta_{x^0}).$$

From $\nu_0 \geqq t\nu_1$, $\nu_0 \geqq (1 - t)\,\nu_2$ and $\nu_0(f) = 0$ for all f satisfying $(\operatorname{supp} f) \cap \{\alpha, \beta\} = \emptyset$, it follows that

$$\nu_1 = c_1\delta_\alpha + c_2\delta_\beta, \qquad \nu_2 = d_1\delta_\alpha + d_2\delta_\beta, \qquad c_j \geqq 0, \qquad d_j \geqq 0.$$

The potentials $G_L\nu_j$ are continuous, equal to $G_L\nu_0$ outside $[\alpha, \beta]$ and harmonic in $]\alpha, \beta[$. From these facts it follows that

$$G_L\nu_0 = G_L\nu_1 = G_L\nu_2 \quad \text{on} \quad \mathbb{R},$$

i.e., $\nu_0 = \nu_1 = \nu_2$. Therefore $\nu_0 = \mu_{x^0}$ is an extremal element of $\mathfrak{B}(\delta_{x^0})$.

5.2. Some special extremal elements of $\mathfrak{B}(\delta_{x^0})$

In formula (4.2.39) we have introduced the set

$$\mathfrak{B}(\mu) := \{\nu \geqq 0 : \operatorname{supp} \nu \subset \bar{\Omega}, G_L\nu(x) = G_L\mu(x), x \notin \Omega\}. \tag{1}$$

Here μ is a given positive measure satisfying $\operatorname{supp} \mu \subset \bar{\Omega}$. Following (4.2.38) the set $\mathfrak{B}(\mu)$ can be defined in the following manner:

$$\mathfrak{B}(\mu) := \left\{\nu \geqq 0 : \operatorname{supp} \nu \subset \bar{\Omega}, \int u \, d\nu = \int u \, d\mu, u \in C(\bar{\Omega}) \cap C^2(\Omega), \Delta u = 0 \text{ in } \Omega\right\}. \tag{2}$$

Definition (2) is used if we apply theorems of functional analysis. Formula (1) resp. (2) is an equivalence relation on the dual space $C^*(\bar{\Omega})$, i.e., on all measures concentrated on $\bar{\Omega}$.

In the following let $\Omega = B(x^0, R) \in \mathbb{R}^2$, $x^0 = 0$, be an open ball. We shall study some special extremal measures of $\mathfrak{B}(\delta_{x^0})$. The swept-out measure $\Pi_{\partial\Omega}\delta_z$ of δ_z, $z \in B(x^0, R)$, on $\partial\Omega$ is uniquely determined and has the form (see (4.1.10) and (4.2.23)), where $y = (R, \varphi)$,

$$\Pi_{\partial\Omega}\delta_z(f) = \int_0^{2\pi} f(y)\, P(z, y) \, ds(y). \tag{3}$$

If $z = x^0$ we obtain

$$\Pi_{\partial\Omega}\delta_{x^0}(f) = \int_0^{2\pi} f(R, \beta) \frac{1}{2\pi} \, d\beta. \tag{4}$$

We now consider the measure (see Fig. 5.6)

$$\nu_a := a\delta_z + \Pi_{\partial\Omega}(\delta_{x^0} - a\delta_z), \qquad a \geqq 0. \tag{5}$$

From (3) and (4) it follows that ν_a has the form

$$\begin{aligned}\nu_a(f) &= af(z) + \int_0^{2\pi} \left(\frac{1}{2\pi R} - aP(z, y)\right) f(R, \beta)\, R\, \mathrm{d}\beta \\ &= a\delta_z(f) + \gamma_a(f).\end{aligned} \tag{6}$$

Here

$$\gamma_a(f) = \int_0^{2\pi} g_a(\beta)\, f(R, \beta)\, \mathrm{d}\beta, \qquad g_a(\beta) := \frac{1}{2\pi}\left(1 - a\,\frac{R^2 - |z|^2}{|z - y|^2}\right), \tag{7}$$

where g_a is analytic on $\partial B(x^0, R)$. If a is sufficiently small we get $g_a(\beta) \geqq 0$ (see Fig. 5.7). The same formulas hold for $B(x^0, R)$, $x^0 \in R^2$. Then z resp. y is to be replaced by $z - x^0$ resp. $y - x^0$.

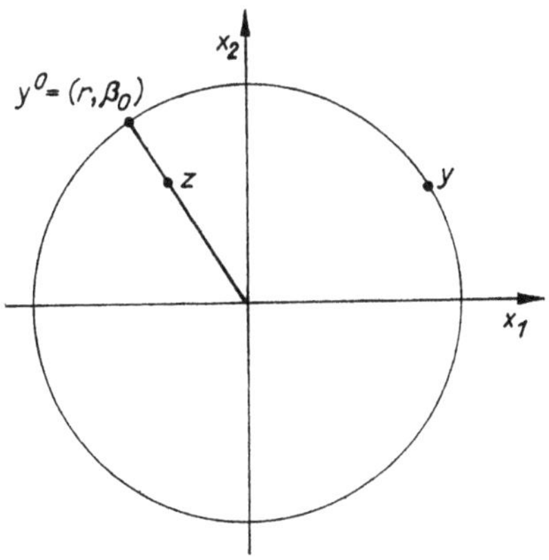

Fig. 5.6

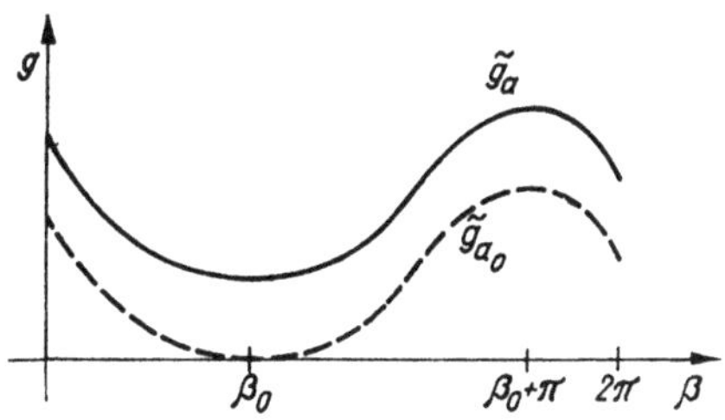

Fig. 5.7

Proposition 5.1. *Let g_a be defined in (7). In $\mathbb{R}^2$ there exists a uniquely determined number $a_0 = \dfrac{R - |z|}{R + |z|}$ such that $g_{a_0}(\beta_0) = 0$ and $g_{a_0}(\beta) > 0$ for $\beta \neq \beta_0$. The corresponding ν_{a_0} is an extremal element of $\mathfrak{B}(\delta_{x^0})$.*

Proof. The function g_a defined in (7) attains its minimum for $y^0 = (R, \beta_0)$, where $z = (|z|, \beta_0)$. Then $|z - y^0|^2 = (R - |z|)^2$ and

$$\frac{R^2 - |z|^2}{|z - y^0|^2} = \frac{(R + |z|)(R - |z|)}{(R - |z|)^2} = \frac{R + |z|}{R - |z|}. \tag{8}$$

If

$$a_0 = \frac{R - |z|}{R + |z|}$$

one gets

$$g_{a_0}(\beta_0) = \frac{1}{2\pi}\left(1 - a_0\,\frac{R^2 - |z|^2}{|z - y^0|^2}\right) = 0 \tag{9}$$

and $g_{a_0}(\beta) > 0$ otherwise. The measure ν_{a_0} is an extremal element of $\mathfrak{B}(\delta_{x^0})$. In order to prove this we assume that ν_{a_0} is not extremal. Then it has the form

$$\nu_{a_0} = t\nu_1 + (1 - t)\,\nu_2\,, \qquad 0 < t < 1\,, \qquad \nu_1 \neq \nu_2\,, \qquad \nu_1, \nu_2 \in \mathfrak{B}(\delta_{x^0})\,. \tag{10}$$

From $\nu_1 \geqq 0$ and $\nu_2 \geqq 0$ it follows that $\nu_{a_0} \geqq t\nu_1$, $\nu_{a_0} \geqq (1 - t)\,\nu_2$. Let $(\operatorname{supp} f) \cap \partial\Omega = \emptyset$. If additionally $(\operatorname{supp} f) \cap \partial\Omega \cup \{z\} = \emptyset$ we get the representation $\nu_j = c_j\delta_z$, $j = 1, 2$. For all f we obtain

$$\nu_j = c_j\delta_z + \gamma_j\,, \qquad \gamma_j \geqq 0, \qquad \operatorname{supp}\gamma_j \subset \partial\Omega\,, \qquad j = 1, 2\,. \tag{11}$$

As $\nu_j \in \mathfrak{B}(\delta_{x^0})$ we have for $x \notin B(x^0, R)$

$$G_{\mathrm{L}}\gamma_j(x) = G_{\mathrm{L}}\delta_{x^0}(x) - c_jG_{\mathrm{L}}\delta_z(x)\,. \tag{12}$$

The swept-out measure of $\delta_{x^0} - c_j\delta_z$, which is uniquely determined, is equal to γ_j and has the form (see (7))

$$\gamma_j(f) = \int_0^{2\pi} g_j(\beta)\, f(R, \beta)\, \mathrm{d}\beta\,, \quad g_j \text{ continuous.} \tag{13}$$

From $\gamma_j \geqq 0$ it follows that $g_j \geqq 0$, and for all f vanishing at z from (10)

$$\gamma_{a_0}(f) = t\gamma_1(f) + (1 - t)\,\gamma_2(f)\,, \qquad 0 < t < 1\,,$$

which is equivalent to

$$\int_0^{2\pi} \big(g_{a_0}(\beta) - tg_1(\beta) - (1 - t)\,g_2(\beta)\big)\, f(R, \beta)\, \mathrm{d}\beta = 0\,.$$

By a limit, $f(R, \beta)$ tends to the Dirac measure, we obtain

$$g_{a_0}(\beta) = tg_1(\beta) + (1 - t)\,g_2(\beta)\,, \qquad 0 < t < 1\,. \tag{14}$$

From $g_{a_0}(\beta_0) = 0$ and $g_j \geqq 0$ it follows that $g_j(\beta_0) = 0$, $j = 1, 2$. If $c_j \neq a_0$ we obtain either $g_j(\beta_0) > 0$ or $g_j(\beta_0) < 0$. Therefore $c_j = a_0$, $j = 1, 2$. The representation (10) is not possible for ν_{a_0}, i.e., ν_{a_0} is an extremal element. ∎

Similar considerations are possible in $\mathbb{R}^3$ and more generally in $\mathbb{R}^n$, $n > 3$. Here we consider a ball $\Omega = B(x^0, R) \subset \mathbb{R}^3$, $x^0 = 0$. Let $z \in B(x^0, R)$ be fixed and

$$\nu_a := a\delta_z + \Pi_{\partial\Omega}(\delta_{x^0} - a\delta_z)\,, \qquad a \geqq 0\,. \tag{15}$$

In this case ν_a has the form

$$\nu_a(f) = a\delta_z(f) + \gamma_a(f)\,,$$

where

$$\gamma_a(f) = \int\limits_{\vartheta_2=0}^{2\pi} \int\limits_{\vartheta_1=0}^{\pi} \frac{1}{4\pi R} \left(\frac{1}{R} - a \frac{R^2 - |z|^2}{|z - y|^3}\right) f(y) \,\mathrm{d}S(y) = \int\limits_{\partial B(x^0,R)} g_{a_0}(y) f(y) \,\mathrm{d}S(y), \tag{16}$$

where

$$g_a(y) = \frac{1}{4\pi R} \left(\frac{1}{R} - a \frac{R^2 - |z|^2}{|z - y|^3}\right). \tag{17}$$

If $B(x^0, R)$ is an arbitrary ball, z resp. y are to be replaced by $z - x^0$ resp. $y - x^0$.

Proposition 5.2. *Let g_a be defined in* (17). *In* $\mathbb{R}^3$ *there exists a uniquely determined $a_0 = (R - |z|)^2/R(R + |z|)$ such that $g_{a_0}(y^0) = 0$ and $g_{a_0}(y) > 0$ otherwise. Here $y^0 \in \partial B(x^0, R)$. The corresponding ν_{a_0} is an extremal element of $\mathfrak{B}(\delta_{x^0})$.*

Proof. Let $z = (|z|, \vartheta_1^0, \vartheta_2^0) \in B(x^0, R)$. The minimum of $|z - y|^3$ is attained for $y^0 = (R, \vartheta_1^0, \vartheta_2^0)$. Then $|z - y^0| = R - |z|$ and

$$g_{a_0}(y^0) = \frac{1}{4\pi R} \left(\frac{1}{R} - a_0 \frac{R^2 - |z|^2}{(R - |z|)^3}\right) = 0$$

and $g_{a_0}(y) > 0$ otherwise. The extremality of ν_{a_0} is proved in a similar way as in Proposition 5.1. ∎

We now shall study another set of extremal elements of $\mathfrak{B}(\delta_{x^0})$ containing two Dirac measures. Let $\Omega = B(x^0, R)$, $x^0 = 0$, and

$$\nu_{a_1,a_2} := a_1\delta_{z^1} + a_2\delta_{z^2} + \gamma(f), \tag{18}$$

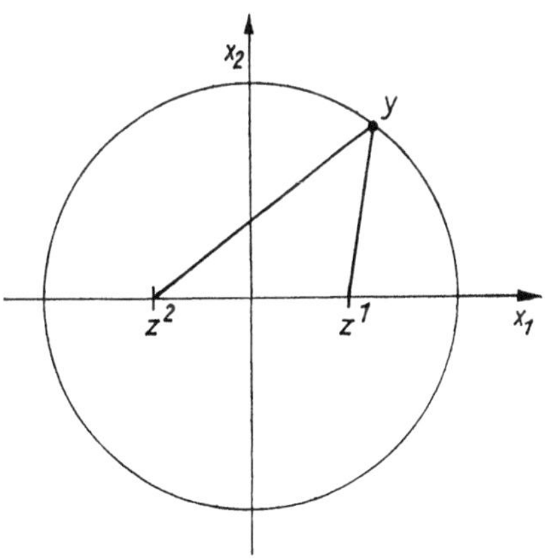

Fig. 5.8

where $a_1 \geqq 0$, $a_2 \geqq 0$, $z^1 = (|z|, 0)$, $z^2 = (-|z|, 0)$. If $\nu_{a_1,a_2} \in \mathfrak{B}(\delta_{x^0})$ we get

$$\nu_{a_1,a_2} = a_1\delta_{z^1} + a_2\delta_{z^2} + \Pi_{\partial\Omega}(\delta_{x^0} - a_1\delta_{z^1} - a_2\delta_{z^2}). \tag{19}$$

From (18) and (19) it follows that, $y = (R, \beta)$,

$$\gamma(f) = \int_0^{2\pi} \left(\frac{1}{2\pi R} - a_1 P(z^1, y) - a_2 P(z^2, y) \right) f(R, \beta) \, R \, \mathrm{d}\beta$$

$$= \int_0^{2\pi} g_{a_1,a_2}(\beta) \, f(R, \beta) \, \mathrm{d}\beta \,, \tag{20}$$

where

$$g_{a_1,a_2}(\beta) = \left(\frac{1}{2\pi R} - a_1 P(z^1, y) - a_2 P(z^2, y) \right) R \,. \tag{21}$$

The inequality $g_{a_1,a_2} \geqq 0$ follows from $\gamma \geqq 0$.

Proposition 5.3. *In* $\mathbb{R}^2$ *there exists a uniquely determined* $a_0 = 1 \Big/ \left(\frac{R + |z|}{R - |z|} + \frac{R - |z|}{R + |z|} \right)$ *such that the corresponding measure* ν_{a_0,a_0} *defined in* (18) *is an extremal element of* $\mathfrak{B}(\delta_{x^0})$.

Proof. a) At first we consider the function $g_{a,a}$ defined in (21):

$$g_{a,a}(\beta) = \frac{1}{2\pi} \left(1 - a \frac{R^2 - |z|^2}{|z^1 - y|^2} - a \frac{R^2 - |z|^2}{|z^2 - y|^2} \right). \tag{22}$$

We shall prove that the minimum of $g_{a,a}$ is attained at $\beta = 0$ and $\beta = \pi$; the maximum at $\beta = \pi/2$ and $\beta = 3\pi/2$. Further, we have

$$g_{a,a}(\beta) = g_{a,a}(-\beta) \quad \text{and} \quad g_{a,a}\left(\frac{\pi}{2} + \beta \right) = g_{a,a}\left(\frac{\pi}{2} - \beta \right).$$

In order to prove this let $\beta = 0$. Then $|z^1 - y| = R - |z|$ and $|z^2 - y| = R + |z|$. In this case we get

$$\begin{aligned} \frac{R^2 - |z|^2}{|z^1 - y|^2} &= \frac{R^2 - |z|^2}{(R - |z|)^2} = \frac{R + |z|}{R - |z|}, \\ \frac{R^2 - |z|^2}{|z^2 - y|^2} &= \frac{R^2 - |z|^2}{(R + |z|)^2} = \frac{R - |z|}{R + |z|} \end{aligned} \tag{23}$$

and

$$\frac{R^2 - |z|^2}{|z^1 - y|^2} + \frac{R^2 - |z|^2}{|z^1 - y|^2} = \frac{R + |z|}{R - |z|} + \frac{R - |z|}{R + |z|}. \tag{24}$$

If $\beta = \pi/2$ we have $|z^1 - y|^2 = |z^2 - y|^2 = R^2 + |z|^2$ and

$$\frac{R^2 - |z|^2}{|z^1 - y|^2} + \frac{R^2 - |z|^2}{|z^2 - y|^2} = 2 \frac{(R^2 - |z|^2)}{R^2 + |z|^2}. \tag{25}$$

The following inequality holds for $0 \leqq |z| < R$:

$$\frac{1}{(R^2 + |z|^2)^2} \leqq \frac{1}{(R^2 - |z|^2)^2}.$$

From this inequality it follows that

$$\frac{2}{(R^2 + |z|^2)} \leqq 2\,\frac{(R^2 + |z|^2)}{(R^2 - |z|^2)^2} = \frac{1}{(R + |z|)^2} + \frac{1}{(R - |z|)^2}.$$

Finally, we get

$$\frac{2(R^2 - |z|^2)}{R^2 + |z|^2} \leqq \frac{R^2 - |z|^2}{(R + |z|)^2} + \frac{R^2 - |z|^2}{(R - |z|)^2} = \frac{R + |z|}{R - |z|} + \frac{R - |z|}{R + |z|}, \tag{26}$$

that means

$$g_{a,a}(0) < g_{a,a}\left(\frac{\pi}{2}\right).$$

Further, $g_{a,a}$ is increasing for $0 \leqq \beta \leqq \dfrac{\pi}{2}$ since

$$|z^1 - y|^2 = R^2 + |z|^2 - 2R\,|z| \cos \beta, \qquad |z^2 - y|^2 = R^2 + |z|^2 + 2R\,|z| \cos \beta$$

and

$$g'_{a,a}(\beta) = c\,(\sin \beta)\left(\frac{1}{|z^1 - y|^4} - \frac{1}{|z^2 - y|^4}\right) > 0 \quad \text{for} \quad 0 < \beta < \frac{\pi}{2}.$$

From (24), (25), (26), and the last inequality we obtain

$$g_{a,a}(0) < g_{a,a}(\beta) < g_{a,a}\left(\frac{\pi}{2}\right) \quad \text{for} \quad 0 < \varphi < \frac{\pi}{2}.$$

If

$$a = a_0 = 1\Big/\left(\frac{R + |z|}{R - |z|} + \frac{R - |z|}{R + |z|}\right),$$

we have

$$g_{a,a}(0) = g_{a,a}(\pi) = 0, \qquad g_{a,a}(\beta) > 0 \text{ otherwise}.$$

b) In order to prove that ν_{a_0,a_0} is an extremal element we assume like in (18)

$$\nu_{a,a} = t\nu_1 + (1 - t)\,\nu_2, \qquad \nu_1 \neq \nu_2, \qquad \nu_1, \nu_2 \in \mathfrak{B}(\delta_{x^0}), \qquad 0 < t < 1.$$

Then $\nu_{a,a} \geqq t\nu_1$, $\nu_{a,a} \geqq (1 - t)\,\nu_2$. From this it follows that

$$\nu_j = c_j\delta_{z^1} + d_j\delta_{z^2} + \gamma_j,$$

where

$$\gamma_j(f) = \int_0^{2\pi} g_j(\beta)\, f(R, \beta)\, d\beta, \qquad j = 1, 2.$$

Similarly to (14) we can prove that

$$g_{a_0,a_0}(\beta) = tg_j(\beta) + (1-t)\, g_2(\beta), \qquad g_j \geqq 0.$$

From $g_{a_0,a_0}(0) = g_{a_0,a_0}(\pi) = 0$ it follows that

$$g_1(0) = g_2(0) = 0, \qquad g_1(\pi) = g_2(\pi) = 0.$$

The problem is symmetric with respect to $\beta = \pi/2$. If $c_1 \neq d_j$ then $g_j(0) \neq g_j(\pi)$, which is a contradiction to $g_j(0 = g_j(\pi) = 0$. Therefore ν_{a_0,a_0} is extremal. ▮

In $\mathbb{R}^3$ we shall construct a uniquely determined $\nu_{a,a} \in \mathfrak{B}(\delta_{x^0})$ by considering

$$\nu_{a_1,a_2} := a_1\delta_{z^1} + a_2\delta_{z^2} + \gamma(f), \tag{27}$$

where

$$z^1 = (|z|, 0, 0), \qquad z^2 = (-|z|, 0, 0).$$

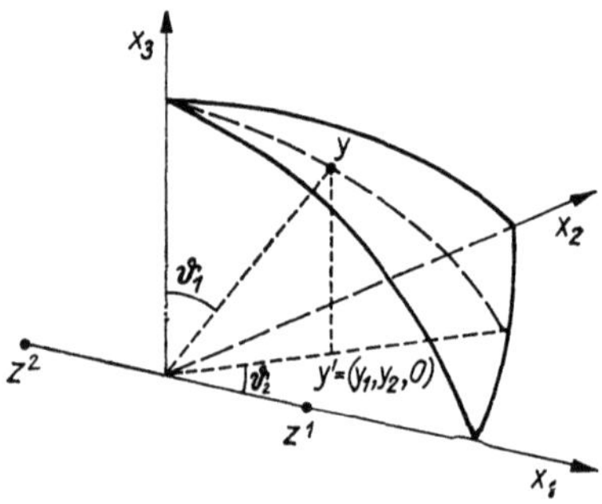

Fig. 5.9

Similarly to (20) we get

$$\gamma(f) = \int\limits_{\partial B(0,R)} \left(\frac{1}{4\pi R^2} - a_1 P(z^1, y) - a_2 P(z^2, y)\right) f(y)\, \mathrm{d}S(y)$$
$$= \int\limits_{\partial B(0,R)} g_{a_1,a_2}(y)\, f(y)\, \mathrm{d}S(y). \tag{28}$$

The function $y \mapsto g_{a_1,a_2}(y)$ only depends on $(y_1 - |z|)^2 + \varrho^2$ and $(y_1 + |z|)^2 - \varrho^2$, where $\varrho^2 = y_2^2 + y_3^2$. It is sufficient to study g_{a_1,a_2} in the (x_1, x_2)-plane. Then we get for $y = (R, \pi/2, \vartheta_2)$

$$|z^1 - y|^2 = R^2 + |z|^2 - 2R\,|z| \cos\vartheta_2, \qquad |z^2 - y|^2 = R^2 + |z|^2 + 2R\,|z| \cos\vartheta_2$$

and

$$g_{a_1,a_2}(\vartheta_2) = \frac{1}{4\pi R}\left(\frac{1}{R} - a_1 \frac{R^2 - |z|^2}{|z^1 - y|^3} - a_2 \frac{R^2 - |z|^2}{|z^2 - y|^3}\right). \tag{29}$$

Proposition 5.4. *In* $\mathbb{R}^3$ *there exists a uniquely determined*

$$a_0 = 1 \bigg/ \left(\frac{(R + |z|)\, R}{(R - |z|)^2} + \frac{(R - |z|)\, R}{(R + |z|)^2}\right)$$

such that the corresponding measure $\nu_{a,a}$ *defined in* (27) *is an extremal element of* $\mathfrak{B}(\delta_{x^0})$.

Proof. a) From (29) it follows that

$$g_{a,a}(0) = \frac{1}{4\pi R}\left(\frac{1}{R} - a\,\frac{(R^2 - |z|^2)}{(R - |z|)^3} - a\,\frac{(R^2 - |z|^2)}{(R + |z|)^3}\right)$$
$$= \frac{1}{4\pi R}\left(\frac{1}{R} - a\,\frac{(R + |z|)}{(R - |z|)^2} - a\,\frac{(R - |z|)}{(R + |z|)^2}\right)$$

and

$$g_{a,a}\left(\frac{\pi}{2}\right) = \frac{1}{4\pi R}\left(\frac{1}{R} - 2a\,\frac{(R^2 - |z|^2)}{(R^2 + |z|^2)^{3/2}}\right).$$

In order to prove the inequality

$$\frac{2(R^2 - |z|^2)}{(R^2 + |z|^2)^{3/2}} \leqq \frac{(R^2 - |z|^2)}{(R + |z|)^3} + \frac{(R^2 - |z|^2)}{(R - |z|)^3}, \tag{30}$$

we at first consider the inequalities

$$(R^2 - |z|^2)^3 \leqq (R^2 + 3\,|z|^2)^2\,R^2 \tag{31}$$

and

$$\frac{1}{(R^2 + |z|^2)^{3/2}} \leqq \frac{1}{(R^2 - |z|^2)^{3/2}}. \tag{32}$$

From (31) and (32) we obtain

$$\frac{1}{(R^2 + |z|^2)^{3/2}\,(R^2 + 3\,|z|^2)\,R} \leqq \frac{1}{(R^2 - |z|^2)^3}$$

and

$$\frac{2}{(R^2 + |z|^2)^{3/2}} \leqq \frac{(R^2 + 3\,|z|^2)\,2R}{(R^2 - |z|^2)^3} \leqq \frac{1}{(R + |z|)^3} + \frac{1}{(R - |z|)^3},$$

from which follows (30). Therefore

$$g_{a,a}(0) < g_{a,a}\left(\frac{\pi}{2}\right).$$

The function $g_{a,a}(\vartheta_2)$ attains its minimum at $\vartheta_2 = 0$ since

$$g'_{a,a}(\vartheta_2) = c \sin\vartheta_2\left(\frac{1}{|z^1 - y|^5} - \frac{1}{|z^2 - y|^5}\right) > 0 \quad \text{for} \quad 0 < \vartheta_2 < \frac{\pi}{2}.$$

If

$$a_0 = 1\bigg/\left(\frac{(R + |z|)\,R}{(R - |z|)^2} + \frac{(R - |z|)\,R}{(R + |z|)^2}\right)$$

we get

$$g_{a,a}(0) = g_{a,a}(\pi) = 0, \qquad g_{a,a}(\vartheta_2) > 0 \text{ otherwise.}$$

b) To prove that ν_{a_0,a_0} is an extremal element we can apply the same ideas as in the proof of Proposition 5.3. ▮

5.3. General theorems on inverse source problems

In this section we shall study the set $\mathfrak{B}(\mu)$ of all positive measures concentrated on the closure $\bar{\Omega}$ of a bounded domain $\Omega \subset \mathbb{R}^n$ producing the same Newtonian potential outside Ω from a systematic point of view. It corresponds to the information content of the problem considered. In general not more information is known in geophysics by using the gravitational field. The set $\mathfrak{B}(\mu)$ is convex and weakly compact. It can be characterized by an equivalence relation on the dual space $C^*(\bar{\Omega})$. Theorem 5.3 does not depend on the special differential equation. This theorem characterizes a general property of subspaces in functional analysis. We also apply it to inverse source problems of the heat equation. In this way we can use results on convex sets studied in infinite-dimensional spaces, especially for subsets of $C^*(\bar{\Omega})$. In Section 5.4 one can find concrete results for the extremal elements of $\mathfrak{B}(\mu)$ which are uniquely determined if the support of such a measure is known. In a certain sense these results are geometric conditions. The inverse gravitational problem in geophysics, where the gradient of the potential is measured, can be reduced to the inverse source problem considered. In all considerations of this book we assume $\overline{D(\partial\Omega; G_L)} = C(\partial\Omega)$. This condition is equivalent that the Green function is zero at the boundary. Non-smooth boundaries like in geophysics also fulfill this condition. In order to uniquely determine an element $\nu_0 \in \mathfrak{B}(\mu)$ additional information is necessary. Each additional mathematical piece of information must be clarified by the experiment.

Following (4.2.38) the set $\mathfrak{B}(\mu)$ can be defined as

$$\mathfrak{B}(\mu) := \left\{\nu \geqq 0 \colon \operatorname{supp} \nu \subset \bar{\Omega}, \int f \, d\nu = \int f \, d\mu, f \in C(\bar{\Omega}) \cap C^2(\Omega), \Delta f = 0 \text{ in } \Omega\right\}. \tag{1}$$

Using the mapping A defined in (4.2.7), which gives the solution of the Dirichlet problem relative to $\bar{\Omega}$, the set $\mathfrak{B}(\mu)$ can be introduced in the equivalent form

$$\mathfrak{B}(\mu) := \left\{\nu \geqq 0 \colon \operatorname{supp} \nu \subset \bar{\Omega}, \int (Af) \, d\nu = \int f \, d(A^*\mu) \text{ for all } f \in C(\bar{\Omega})\right\}. \tag{2}$$

Theorem 5.1. *Let $\nu \in \mathfrak{B}(\mu)$. Then $\int d\nu = \int d\mu$.*

Proof. The measure

$$\delta_{x^0,R}(f) = \frac{1}{4\pi R^2} \int\limits_{\partial B(x^0,R)} f(y) \, dS(y) \tag{3}$$

yields on $B(x^0, R)$ the potential $G_L\delta_{x^0,R}(x) = 1/4\pi R$ (see Example 3.1). Therefore

$$\lambda_0 := 4\pi R\delta_{x^0,R} \tag{4}$$

gives the potential

$$G_L\lambda_0(x) = 1 \quad \text{on} \quad B(x^0, R). \tag{5}$$

Let $\Omega \subset B(x^0, R)$, $\mu \geqq 0$ satisfying $\operatorname{supp} \mu \subset \bar{\Omega}$ and $\nu \in \mathfrak{B}(\mu)$. Following the balayage principle (Theorem 4.1) there exists on $\partial\Omega$ a measure $\lambda_0' \geqq 0$ satisfying

$$G_L\lambda_0'(x) = 1 \quad \text{on } \partial\Omega. \tag{6}$$

From the maximum principle it follows that (6) holds on the closed domain $\bar{\Omega}$. The measure λ_0' is called the *equilibrium distribution* and $G_L\lambda_0'$ the *equilibrium potential*. From

$$G_L\nu(x) = G_L\mu(x) \quad \text{on } \partial\Omega$$

it follows that

$$\int G_L\nu \,\mathrm{d}\lambda_0' = \int G_L\mu \,\mathrm{d}\lambda_0' \quad \text{or} \quad \int G_L\lambda_0' \,\mathrm{d}\nu = \int G_L\lambda_0' \,\mathrm{d}\mu,$$

from which wet get $\int \mathrm{d}\nu = \int \mathrm{d}\mu$. ∎

The dual space $\mathbf{C}^*(\bar{\Omega})$ consists of all measures μ satisfying $\operatorname{supp} \mu \subset \bar{\Omega}$. Therefore

$$\mathfrak{B}(\mu) \subset \mathbf{C}^*(\bar{\Omega}) \tag{7}$$

holds. On the space $\mathbf{C}^*(\bar{\Omega})$ the weak topology (often called w^*-topology) is introduced. A sequence $(\nu_k)_{k\in\mathbf{N}}$ of measures $\nu_k \in \mathbf{C}^*(\bar{\Omega})$ converges weakly to $\mu \in \mathbf{C}^*(\bar{\Omega})$ if and only if

$$\int f \,\mathrm{d}\nu_k \to \int f \,\mathrm{d}\mu \quad \text{for each} \quad f \in \mathbf{C}(\bar{\Omega}). \tag{8}$$

The w^*-topology can be defined by the metric (4.2.43). A set $M \subset \mathbf{C}^*(\bar{\Omega})$ is weakly compact if and only if each infinite subset of M contains a weakly convergent subsequence of measures ν_k, converging to a measure $\nu \in M$. In $\mathbf{C}^*(\bar{\Omega})$ each ball, i.e., the set of all $\nu \in \mathbf{C}^*(\bar{\Omega})$ satisfying $\|\nu\| \leqq c$, is weakly compact. This theorem was proved by F. Riesz [1] in 1930 and has been the base for many existence theorems, also in potential theory.

Theorem 5.2. *The set $\mathfrak{B}(\mu) \subset \mathbf{C}^*(\bar{\Omega})$ is convex and weakly compact.*

Proof. Let

$$\nu = t\nu_1 + (1 - t)\,\nu_2, \qquad \nu_1, \nu_2 \in \mathfrak{B}(\mu), \qquad 0 \leqq t \leqq 1.$$

From (1) it follows that

$$\int f \,\mathrm{d}\nu = t \int f \,\mathrm{d}\nu_1 + (1 - t) \int f \,\mathrm{d}\nu_2 = t \int f \,\mathrm{d}\mu + (1 - t) \int f \,\mathrm{d}\mu = \int f \,\mathrm{d}\mu, \tag{9}$$

where $f \in \mathbf{C}(\bar{\Omega}) \cap \mathbf{C}^2(\Omega)$, $\Delta f = 0$ in Ω. Therefore $\mathfrak{B}(\mu)$ is convex. We now prove that $\mathfrak{B}(\mu)$ is weakly closed. From

$$\int f \, d\nu_k \to \int f \, d\nu, \qquad f \in \mathbf{C}(\bar{\Omega}), \qquad \nu \in \mathbf{C}^*(\bar{\Omega}), \qquad \nu_k \in \mathfrak{B}(\mu),$$

and the property

$$\int f \, d\nu_k = \int f \, d\mu, \qquad \Delta f = 0 \quad \text{in } \Omega,$$

it follows that

$$\int f \, d\nu_k \to \int f \, d\nu = \int f \, d\mu, \qquad \Delta f = 0 \quad \text{in } \Omega,$$

i.e., $\nu \in \mathfrak{B}(\mu)$. Since $\int d\nu = \int d\mu$ (see Theorem 5.1) we have $\|\nu\| \leqq c$ for all $\nu \in \mathfrak{B}(\mu)$. The closed set $\mathfrak{B}(\mu)$ as closed subset of a (weakly) compact set is a (weakly) compact set (Bourbaki [1], Dunford and Schwartz [1], Köthe [1]). ▮

The set of all extremal elements of $\mathfrak{B}(\mu)$ is denoted by ex $\mathfrak{B}(\mu)$. Special extremal elements have been constructed in Example 3.1, Example 3.3 and in Sect. 5.2.

We define

$$H_G(\bar{\Omega}) := \left\{ f \in C(\bar{\Omega}) : f(x) = \sum_{k=1}^{m} c_k G_L(x, z^k),\ z^k \notin \bar{\Omega} \right\}. \tag{10}$$

On $\mathbf{L}^1(\nu)$, $\nu \geqq 0$, the norm

$$\|f\|_{\mathbf{L}^1(\nu)} := \int |f(x)| \, d\nu(x) \tag{11}$$

is introduced. The dual space $\mathbf{L}^\infty(\nu)$ of $\mathbf{L}^1(\nu)$ consists of all functions $\varrho : \operatorname{supp} \nu \mapsto \bar{\mathbb{R}}$ which are bounded, except a ν-negligible set (Dunford and Schwartz [1]). The norm on $\mathbf{L}^\infty(\nu)$ is given by

$$\|\varrho\|_{\mathbf{L}^\infty} := \operatorname{ess\,sup} \{|\varrho(x)|,\ x \in \operatorname{supp} \nu\}. \tag{12}$$

The supremum in (12) is to be extended over $\varrho(x)$, except a ν-negligible set. Using the space $\mathbf{L}^1(\nu)$ and its dual space $\mathbf{L}^\infty(\nu)$ one can prove the following theorem, which in general holds for a subspace $H \subset \mathbf{C}(\bar{\Omega})$ and an equivalence relation on $\mathbf{C}^*(\bar{\Omega})$ defined by H. The following fundamental theorem was proved by R. G. Douglas [1] (see also Anger [12], Schulze and Wildenhain [1]).

Theorem 5.3. *A measure $\nu \in \mathfrak{B}(\mu)$ is extremal if and only if $H_G(\bar{\Omega})$ is dense in $\mathbf{L}^1(\nu)$ with respect to the norm* (11).

Proof. a) Let $\nu \in \mathfrak{B}(\mu)$ be an extremal element. We have to prove that $\overline{H_G(\bar{\Omega})} = \mathbf{L}^1(\nu)$ holds. If we make the contrary assumption that $\overline{H_G(\bar{\Omega})} \neq \mathbf{L}^1(\nu)$ holds, then there exists a continuous linear form $l \neq 0$ on $\mathbf{L}^1(\nu)$ satisfying $l(f) = 0$ for all $f \in \overline{H_G(\bar{\Omega})}$ (consequence of the theorem of Hahn-Banach, see Dunford and Schwartz [1]). Because the dual space of $\mathbf{L}^1(\nu)$ is the space $\mathbf{L}^\infty(\nu)$, l has the form

$$l(f) = \int f(y) \, \varrho(y) \, d\nu(y), \tag{13}$$

where ϱ is ν-essentially bounded. The norm of ϱ, $\varrho \in \mathbf{L}^\infty(\nu)$, can be chosen less than 1. By the construction of l the relation holds:

$$G_{\mathrm{L}}l(x) := \int G_{\mathrm{L}}(x, y)\, \varrho(y)\, \mathrm{d}\nu(y) = 0 \quad \text{for} \quad x \notin \Omega. \tag{14}$$

The measures ($\varrho\nu := l$)

$$\gamma_1 := \nu + \varrho\nu, \qquad \gamma_2 := \nu - \varrho\nu \tag{15}$$

yield the same potential as ν, because (14) holds, i.e.,

$$\begin{aligned} G_{\mathrm{L}}\gamma_1(x) &:= \int G_{\mathrm{L}}(x, y)\, \mathrm{d}\nu(y) + \int G_{\mathrm{L}}(x, y)\, \varrho(y)\, \mathrm{d}\nu(y) \\ &= \int G_{\mathrm{L}}(x, y)\, \mathrm{d}\nu(y) = \int G_{\mathrm{L}}(x, y)\, \mathrm{d}\mu(y), \qquad x \notin \bar{\Omega}, \\ G_{\mathrm{L}}\gamma_2(x) &:= \int G_{\mathrm{L}}(x, y)\, \mathrm{d}\nu(y) - \int G_{\mathrm{L}}(x, y)\, \varrho(y)\, \mathrm{d}\nu(y) \\ &= \int G_{\mathrm{L}}(x, y)\, \mathrm{d}\nu(y) = \int G_{\mathrm{L}}(x, y)\, \mathrm{d}\mu(y), \qquad x \notin \bar{\Omega}. \end{aligned}$$

Then $\nu = \frac{1}{2}(\gamma_1 + \gamma_2)$, which contradicts the hypothesis that ν is extremal.

b) Let $\overline{H_G(\bar{\Omega})} = L^1(\nu)$. We have to prove that ν is extremal. If

$$\nu = t\nu_1 + (1 - t)\, \nu_2, \qquad 0 < t < 1, \qquad \nu_1, \nu_2 \in \mathfrak{B}(\mu),$$

then $\nu \geqq t\nu_1$ and $\nu \geqq (1 - t)\, \nu_2$. Let $\mathbf{1}_B$ be the characteristic function of a measurable subset $B \subset \bar{\Omega}$. As $H_G(\bar{\Omega})$ is dense in $\mathbf{L}^1(\nu)$ there exists a sequence $(f_k)_{k\in\mathbf{N}}$ of functions $f_k \in H_G(\bar{\Omega})$ such that

$$\|\mathbf{1}_B - f_k\|_{\mathbf{L}^1(\nu)} \to 0.$$

From the inequality

$$\|\mathbf{1}_B - f_k\|_{\mathbf{L}^1(\nu_1)} \leqq \frac{1}{t}\, \|\mathbf{1}_B - f_k\|_{\mathbf{L}^1(\nu)}$$

it follows that

$$\|\mathbf{1}_B - f_k\|_{\mathbf{L}^1(\nu_1)} \to 0.$$

As ν, $\nu_1 \in \mathfrak{B}(\mu)$, the relation $\int f_k\, \mathrm{d}\nu = \int f_k\, \mathrm{d}\nu_1$ holds for all $k = 1, 2, \ldots$, since $\Delta f_k = 0$ in Ω. Using the inequalities

$$\begin{aligned} \left|\int \mathbf{1}_B\, \mathrm{d}\nu - \int \mathbf{1}_B\, \mathrm{d}\nu_1\right| &\leqq \left|\int (\mathbf{1}_B - f_k)\, \mathrm{d}\nu\right| + \left|\int (\mathbf{1}_B - f_k)\, \mathrm{d}\nu_1\right| \\ &\leqq \int |\mathbf{1}_B - f_k|\, \mathrm{d}\nu + \int |\mathbf{1}_B - f_k|\, \mathrm{d}\nu_1 \\ &\leqq \frac{\varepsilon}{2} + \frac{\varepsilon}{2} = \varepsilon \quad \text{for} \quad k > k_0(\varepsilon), \end{aligned}$$

we get $\nu(B) = \nu_1(B)$ for all measurable sets $B \subset \bar{\Omega}$. This is equivalent to $\nu = \nu_1$. Similarly we can prove $\nu = \nu_2$. Therefore ν is an extremal element of $\mathfrak{B}(\mu)$. ∎

Theorem 5.4. *Let $\Omega \subset \mathbb{R}^n$ be a bounded domain and μ a positive measure satisfying* $\operatorname{supp} \mu = \bar{\Omega}$. *A measure $\nu_0 \in \mathfrak{B}(\mu)$ having the form $\nu_0 = \nu_1 + \nu_2$, where $d\nu_1(y) = \varrho(y)\, dy$, $0 < c \leqq \varrho$, and* $\operatorname{supp} \nu_1 = \bar{\Omega}$, *cannot be an extremal element of $\mathfrak{B}(\mu)$.*

Proof. By assumption there exists a ball $B(x^0, r) \subset \operatorname{supp} \nu_0$. Let $B := \{x\colon (3r/4) < |x - x^0| < r\}$ and $B(x, r') \subset B$ a further ball. From Gauss's theorem of the arithmetic mean it follows that the harmonic function $u \in H_G(\bar{\Omega})$ satisfies the relation

$$u(x) = \frac{1}{V_n(r')} \int\limits_{B(x,r')} u(y)\, dy, \tag{16}$$

where $V_n(r')$ is the volume of the ball. Using (16) we get the estimation

$$|u(x)| \leqq \frac{1}{c V_n(r')} \int\limits_{B(x,r')} |c u(y)|\, dy \leqq C \int\limits_{B(x,r')} |\varrho(y)\, u(y)|\, dy,$$

and from this inequality

$$|u(x)| \leqq C \int\limits_{B(x,r')} \varrho(y)\, |u(y)|\, dy \leqq C \int\limits_{B} \varrho(y)\, |u(y)|\, dy. \tag{17}$$

Following Theorem 5.3 a measure $\nu_0 \in \mathfrak{B}(\mu)$ is extremal if and only if for $f \in \mathbf{L}^1(\nu)$ and each $\varepsilon > 0$ exists a $u_k \in H_G(\bar{\Omega})$ such that

$$\|f - u_k\|_{\mathbf{L}^1(\nu_0)} = \int |f - u_k|\, d\nu_0 < \varepsilon. \tag{18}$$

Let $f(y) = 0$ on B and $f(y) = 1$ on $B(x^0, r/4)$. From (18) and $\nu_0 = \nu_1 + \nu_2 \geqq \nu_1$ we get

$$\int\limits_{B} |u_k(y)|\, \varrho(y)\, dy < \varepsilon \quad \text{and} \quad \int\limits_{B(x^0,r/4)} |1 - u_k(y)|\, \varrho(y)\, dy < \varepsilon \tag{19}$$

and from (17)

$$|u_k(x)| \leqq C \int\limits_{B} |u_k(y)|\, \varrho(y)\, dy < \varepsilon,$$

i.e.,

$$\lim_{k\to\infty} u_k(x) = 0 \quad \text{on } B',$$

where

$$B' := \{x\colon (3r/4) + \varepsilon' < |x - x^0| < r - \varepsilon'\}, \qquad \varepsilon' > 0,$$

is sufficiently small. Using the maximum principle of harmonic functions we get

$$\lim_{k\to\infty} u_k(x) = 0 \quad \text{uniformly on} \quad B(x^0, r - \varepsilon'). \tag{20}$$

On the other hand it follows from (19) that

$$\int_{B(x^0,r/4)} \varrho(y)\,\mathrm{d}y = \lim_{k\to\infty} \int_{B(x^0,r/4)} u_k(y)\,\varrho(y)\,\mathrm{d}y .$$

By (20) the right-hand side is zero, which is a contradiction. Therefore the measure ν_0 is not extremal.

Remark 5.4. If a planet, for instance, the earth has the property that its mass distribution ν_0 satisfies the conditions of Theorem 5.4, then ν_0 cannot be an extremal element of $\mathfrak{B}(\nu_0)$. The mass distribution ν_0 is built up by an integration procedure extended over the set ex $\mathfrak{B}(\nu_0)$. One can find such special integration procedures in (3.1.19) and (3.1.45). For the general case no investigations have been done up to now (Anger [16]).

5.4. The geometrical interpretation of extremal measures

In the following the inverse source problem for a bounded domain $\Omega \subset \mathbb{R}^2$ and the Dirac measure δ_{x^0}, $x^0 \in \Omega$, is studied. The set $\mathfrak{B}(\delta_{x^0})$ consists of all positive measures on $\bar{\Omega}$ which yield the fundamental solution $G_L(x, x^0)$ outside Ω. Special measures were constructed in Example 3.1 and Section 5.2. There exist two classes of extremal measures $\nu \in \mathfrak{B}(\delta_{x^0})$. The first class consists of all $\nu = \mu_{x^0}$, where μ_{x^0} is the harmonic measure which gives the solution of the Dirichlet problem at x^0 relative to a subdomain $\Omega' \subset \Omega$ containing x^0, $\mathbb{R}^n \setminus \Omega'$ simply connected. The measures of the second class have the form $\nu = \nu_K + \gamma$, where supp $\gamma \subset \partial\Omega'$ and supp $\nu_K \subset K \subset \Omega'$, K compact. As examples see (5.2.6) and (5.2.18). For simplicity we consider the case $\Omega' = B(x^0, r) \subset \mathbb{R}^2$. The measure γ has the form $\mathrm{d}\gamma = g_0(\beta)\,\mathrm{d}\beta$, $g_0 \geqq 0$. If $g_0 > 0$ then $\nu_K = 0$. If $\nu_K \neq 0$ then g_0 has at most $2m$ roots and supp ν_K consists at most of $2m$ points (Theorem 5.7). This property must have a physical interpretation, especially in astrophysics, electronics, continuum mechanics, solid-state physics, diffusion, and so on (Anger [12], [13], [16]).

At first some facts which follow from $g_0(\beta_0) = 0$. Let $u_k \in \mathrm{C}^2(\Omega) \cap \mathrm{C}(\bar{\Omega})$ be harmonic functions. In the case of a ball $\Omega = B(x^0, R) \subset \mathbb{R}^2$ the harmonic functions u_k can be represented by means of the Poisson integral

$$u_k(x) = \int_{\partial B(x^0,R)} u_k(y)\,P(x,y)\,\mathrm{d}s(y) = \int_0^{2\pi} u_k(y)\,P(x,y)\,R\,\mathrm{d}\beta, \qquad y = (R,\beta). \tag{1}$$

Of special interest is the point $y^0 = (R, \beta_0)$ where $g_0(\beta_0) = 0$, $g_0(\beta) \geqq 0$ for $\beta \neq \beta_0$. For g_0 holds the following representation:

$$g_0(\beta) = (\beta - \beta_0)^{2k}\,\tilde{R}_{2k}(\beta, \beta_0), \qquad \tilde{R}_{2k}(\beta_0, \beta_0) > 0. \tag{2}$$

In the following we assume $k = 1$, which is fulfilled in Proposition 5.1 through Proposition 5.4. It is an open problem that in the general case only $k = 1$ is possible. If we find a sequence $(u_{k'})$ defined on $\partial\Omega$ satisfying $\cap$ supp $u_{k'} = \{\beta_0\}$

and

$$\int_0^{2\pi} u_{k'}(y)\, R\, \mathrm{d}\beta \to d_1, \qquad \int_0^{2\pi} (\beta - \beta_0)\, u_{k'}(y)\, R\, \mathrm{d}\beta \to 0,$$
$$\int_0^{2\pi} (\beta - \beta_0)^2\, u_{k'}(y)\, R\, \mathrm{d}\beta \to 0 \tag{3}$$

and a second sequence $(u_{k''})$ defined on $\partial\Omega$ satisfying $\cap$ supp $u_{k''} = \{\beta_0\}$ and

$$\int_0^{2\pi} u_{k''}(y)\, R\, \mathrm{d}\beta \to 0, \qquad \int_0^{2\pi} (\beta - \beta_0)\, u_{k''}(y)\, R\, \mathrm{d}\beta \to d_2,$$
$$\int_0^{2\pi} (\beta - \beta_0)^2\, u_{k''}(y)\, R\, \mathrm{d}\beta \to 0, \tag{4}$$

from $u_k = u_{k'} + u_{k''}$ and

$$u_k(x) = P(x, y^0) \int_0^{2\pi} u_k(y)\, R\, \mathrm{d}\beta + \frac{\partial P(x, y^0)}{\partial \beta} \int_0^{2\pi} (\beta - \beta_0)\, u_k(y)\, R\, \mathrm{d}\beta + R(x, u_k) \tag{5}$$

it follows that

$$u_k(x) \to d_1 P(x, y^0) + d_2 \frac{\partial P(x, y^0)}{\partial \beta} \quad \text{in } \Omega. \tag{6}$$

Here $u_k(x)$ is the solution of the Dirichlet problem relative to the boundary values $u_k(y) = u_{k'}(y) + u_{k''}(y)$. The sequence $(u_{k'})$ tends to the point measure $d_1 \delta_{\beta_0}$, $(u_{k''})$ to the measure $d_2 \delta_{\beta_0}$. Relation (6) will be proved in the following theorems. Similar investigations are possible for a ball $\Omega = B(x^0, R) \subset \mathbb{R}^n$. They should also be possible for smooth domains $\Omega \subset \mathbb{R}^n$. It is an open question which results hold for an arbitrary domain $\Omega \subset \mathbb{R}^n$, for instance, the earth $\Omega_0 \subset \mathbb{R}^3$. In this case $\partial\Omega_0$ is not smooth, the measure γ cannot be represented in the form $\mathrm{d}\gamma(y) = g_0(\beta)\, \mathrm{d}\beta$ with g_0 analytic (see ANGER [16]).

Let

$$\nu = \nu_K + \gamma \in \mathfrak{B}(\delta_{x^0}), \qquad \operatorname{supp} \gamma \subset \partial\Omega, \qquad \operatorname{supp} \nu_K \subset K \subset \Omega. \tag{7}$$

Special ν were constructed in Sect. 5.2. From (4.2.22) it follows that

$$\Pi_{\partial\Omega} \nu_K = A^* \nu_K, \qquad \mathrm{d}A^* \nu_K = \sigma(y)\, \mathrm{d}S(y), \qquad \sigma(y) = \int P(x, y)\, \mathrm{d}\nu_K(x). \tag{8}$$

Relation (4.2.38) yields for $f \in \mathrm{C}^2(\Omega) \cap \mathrm{C}(\bar{\Omega})$, $\Delta f = 0$ in Ω,

$$\int f\, \mathrm{d}\delta_{x^0} = \int f\, \mathrm{d}\mu_{x^0} = \int f\, \mathrm{d}\nu_K + \int f\, \mathrm{d}\gamma, \tag{9}$$

from which follow the relations

$$\int f(y)\, \mathrm{d}\gamma(y) = \int f(y) \left(P(x^0, y) - \sigma(y)\right) \mathrm{d}S(y) = \int f(y)\, g_0(y)\, \mathrm{d}S(y), \tag{10}$$

where

$$g_0(y) = P(x^0, y) - \int P(x, y)\, \mathrm{d}\nu_K(x). \tag{11}$$

Since the set of all $\mathrm{rest}_{\partial\Omega} f$ is equal to $\mathbf{C}(\partial\Omega)$, the density g_0 is uniquely determined. We introduce for $a > 0$ the function (see Fig. 5.7)

$$\tilde{g}_a(y) := P(x^0, y) - a \int P(x, y)\, \mathrm{d}\nu_K(x). \tag{12}$$

If $a = 1$ we get $\tilde{g}_1(y) = g_0(y)$. Then

$$\tilde{g}_a(y) = P(x^0, y)\,\big(1 - ah(y)\big), \qquad h(y) = \int \big(P(x, y)/P(x^0, y)\big)\, \mathrm{d}\nu_K(x). \tag{13}$$

For a ball $B(x^0, R) \subset \mathbb{R}^2$ the function h has the form

$$h(y) = \int \frac{R^2 - |x - x^0|^2}{|x - y|^2}\, \mathrm{d}\nu_K(x). \tag{14}$$

If $0 < a < 1$ the inequality $\tilde{g}_a > g_0 \geqq 0$ holds on $\partial B(x^0, R)$. From $g_0 > 0$ we obtain the inequality $1 > h(y) > 0$. There exists at least one $y^0 = (R, \beta_0)$ such that

$$\sup\,\{h(y), y \in \partial\Omega\} = h(y^0).$$

Since $y \mapsto h(y)$ is an analytic function with respect to β the relation $h(y) = h(y^0)$ holds at most at the points $y^1, \ldots, y^m$.

The following inequality is important in later considerations. In the case $\nu = \nu_K + \gamma \in \mathrm{ex}\, \mathfrak{B}(\delta_{x^0})$ and $g_0 > 0$ from this inequality it follows that the measure ν_K must be zero.

Proposition 5.5. *Let $g_0\colon \partial\Omega \mapsto \mathbb{R}$ be strictly positive. Then there exists a constant $b > 0$ such that on $\partial\Omega$ holds the inequality*

$$g_0(y) \geqq b \int P(x, y)\, \mathrm{d}\nu_K(x). \tag{15}$$

Proof. We consider the function

$$\tilde{g}_a(y) = P(x^0, y)\,\big(1 - ah(y)\big), \qquad a > 0.$$

If $a_0 = 1/h(y^0)$ then $\tilde{g}_{a_0}(y^0) = 0$. From $g_0 = \tilde{g}_1 > 0$ it follows that $a_0 = 1 + b$, $b > 0$, and $\tilde{g}_{a_0} \geqq 0$. Finally, we obtain

$$\begin{aligned} g_0(y) &= \tilde{g}_1(y) = P(x^0, y)\,\big(1 - a_0 h(y) + bh(y)\big) \\ &= \tilde{g}_{a_0}(y) + b \int P(x, y)\, \mathrm{d}\nu_K(x) \geqq b \int P(x, y)\, \mathrm{d}\nu_K(y). \ \blacksquare \end{aligned}$$

Proposition 5.6. *Let $g_0\colon \partial\Omega \mapsto \mathbb{R}$ be strictly positive and $u_k \in H_G(\bar{\Omega})$ satisfying*

$$\int_0^{2\pi} |u_k(y)|\, g_0(y)\, R\, \mathrm{d}\beta \to 0 \quad \textit{for} \quad k \to \infty.$$

Then

$$\int |u_k(x)|\, \mathrm{d}\nu_K(x) \to 0 \quad \textit{and} \quad u_k(x^0) \to 0 \quad \textit{for} \quad k \to \infty. \tag{16}$$

Proof. Using (15) the following inequality holds:

$$\int |u_k(x)| \,\mathrm{d}\nu_K(x) = \int \left| \int_0^{2\pi} P(x, y)\, u_k(y)\, R\, \mathrm{d}\beta \right| \mathrm{d}\nu_K(x)$$

$$\leqq \int \left(\int_0^{2\pi} P(x, y)\, |u_k(y)|\, R\, \mathrm{d}\beta \right) \mathrm{d}\nu_K(x)$$

$$= \int_0^{2\pi} \left(\int P(x, y)\, \mathrm{d}\nu_K(x) \right) |u_k(y)|\, R\, \mathrm{d}\beta \leqq \frac{1}{b} \int_0^{2\pi} g_0(y)\, |u_k(y)|\, R\, \mathrm{d}\beta. \tag{17}$$

From

$$\int_0^{2\pi} g_0(y)\, |u_k(y)|\, R\, \mathrm{d}\beta \to 0 \quad \text{it follows that} \quad \int |u_k(x)|\, \mathrm{d}\nu_K(x) \to 0$$

for $k \to \infty$. Now we consider the relation

$$g_0(y) + \int P(x, y)\, \mathrm{d}\nu_K(x) = P(x^0, y) = \frac{1}{2\pi R}.$$

Then

$$|u_k(x^0)| = \left| \int_0^{2\pi} P(x^0, y)\, u_k(y)\, R\, \mathrm{d}\beta \right| = \left| \int_0^{2\pi} \left(g_0(y) + \int P(x, y)\, \mathrm{d}\nu_K(x) \right) u_k(y)\, R\, \mathrm{d}\beta \right|$$

$$\leqq \left| \int_0^{2\pi} g_0(y)\, u_k(y)\, R\, \mathrm{d}\beta \right| + \left| \int u_k(x)\, \mathrm{d}\nu_K(x) \right|$$

$$\leqq \int_0^{2\pi} g_0(y)\, |u_k(y)|\, R\, \mathrm{d}\beta + \int |u_k(x)|\, \mathrm{d}\nu_K(x).$$

From the last inequality follows $u_k(x^0) \to 0$ for $k \to \infty$. ∎

Theorem 5.5. *Let* $g_0 : \partial\Omega \mapsto \mathbb{R}$ *be nonnegative and* $\nu = \nu_K + \gamma$ *an extremal element of* $\mathfrak{B}(\delta_{x^0})$. *If* $g_0 > 0$ *then* $\nu_K = 0$; *if* $\nu_K \neq 0$ *then* g_0 *must have at least one root.*

Proof. Let $g_0 > 0$ and $\nu_K \neq 0$. We consider the following function $f_1 \in \mathbf{L}^1(\nu)$ satisfying

$$f_1(y) = 0 \quad \text{on } \partial\Omega, \qquad f_1(x) = 1 \quad \text{on supp } \nu_K.$$

Since ν is an extremal element from Theorem 5.3 it follows that to f_1 there exists a sequence of functions $u_k \in H_G(\bar{\Omega})$ such that

$$\|f_1 - u_k\|_{\mathbf{L}^1(\nu)} = \int |1 - u_k(x)|\, \mathrm{d}\nu_K(x) + \int_0^{2\pi} |u_k(x)|\, g_0(y)\, R\, \mathrm{d}\beta < \varepsilon \tag{18}$$

for $k > k_0(\varepsilon)$. From

$$\left|\int \mathrm{d}\nu_K - \int u_k(x)\,\mathrm{d}\nu_K(x)\right| \leqq \int |1 - u_k(x)|\,\mathrm{d}\nu_K(x) < \varepsilon$$

it follows that

$$0 \neq \int \mathrm{d}\nu_K = \lim_{k\to\infty} \int u_k(x)\,\mathrm{d}\nu_K(x). \tag{19}$$

Further, from (18) and (15) we obtain

$$\begin{aligned}
\varepsilon &> \int_0^{2\pi} |u_k(y)|\, g_0(y)\, R\,\mathrm{d}\beta \geqq \left|\int_0^{2\pi} u_k(y)\, g_0(y)\, R\,\mathrm{d}\beta\right| \\
&\geqq \left| b \int_0^{2\pi} u_k(y) \int P(x, y)\,\mathrm{d}\nu_K(x)\, R\,\mathrm{d}\beta\right| \\
&= \left| b \int \left(\int_0^{2\pi} u_k(y)\, P(x, y)\, R\,\mathrm{d}\beta\right) \mathrm{d}\nu_K(x)\right| = \left| b \int u_k(x)\,\mathrm{d}\nu_K(x)\right|,
\end{aligned} \tag{20}$$

i.e.,

$$\lim_{k\to\infty} \int u_k(x)\,\mathrm{d}\nu_K(x) = 0,$$

which contradicts (19). From $g_0 > 0$ and $\nu = \nu_K + \gamma \in \operatorname{ex} \mathfrak{B}(\delta_{x^0})$ it follows that $\nu_K = 0$. If $\nu_K \neq 0$ the function g_0 must have at least one root. ∎

In order to prove (3) we construct the following functions. Let

$$\begin{aligned}
v_k(\beta) &:= \frac{c}{\varepsilon} \exp\left(-\frac{\varepsilon^2}{\varepsilon^2 - |\beta - \beta_0|^2}\right) \quad \text{if} \quad |\beta - \beta_0| < \varepsilon = \frac{1}{k}, \\
v_k(\beta) &= 0 \text{ otherwise.}
\end{aligned} \tag{21}$$

The constant is to be determined such that

$$c \int_0^{2\pi} \exp\left(-\frac{1}{1 - |\beta - \beta_0|^2}\right) \mathrm{d}\beta = 1.$$

The following relations hold:

$$v_k(\beta_0) = \frac{c}{\varepsilon\,\mathrm{e}}, \qquad v_k(\beta_0 + \beta^*) = v_k(\beta_0 - \beta^*), \tag{22}$$

and

$$\int_{\beta_0-\pi}^{\beta_0+\pi} v_k(\beta)\,\mathrm{d}\beta = \int_{\beta_0-\varepsilon}^{\beta_0+\varepsilon} v_k(\beta)\,\mathrm{d}\beta = 1. \tag{23}$$

Then

$$(\beta - \beta_0)\, v_k(\beta) = \beta^* v_k(\beta_0 + \beta^*) \tag{24}$$

and

$$\int_0^{2\pi} (\beta - \beta_0)\, v_k(\beta)\, \mathrm{d}\beta = 0. \tag{25}$$

Further,

$$\left|\int_0^{2\pi} (\beta - \beta_0)^2\, v_k(\beta)\, R\, \mathrm{d}\beta\right| = \left|\int_{\beta_0-\varepsilon}^{\beta_0+\varepsilon} (\beta - \beta_0)^2\, v_k(\beta)\, R\, \mathrm{d}\beta\right|$$
$$\leqq R\varepsilon^2 \int_{-\varepsilon}^{+\varepsilon} v_k(\beta^* + \beta_0)\, \mathrm{d}\beta^* \leqq R\varepsilon^2 \to 0. \tag{26}$$

In Fig. 5.10 one sees the function v_k, in Fig. 5.11 a further function used in Theorem 5.6.

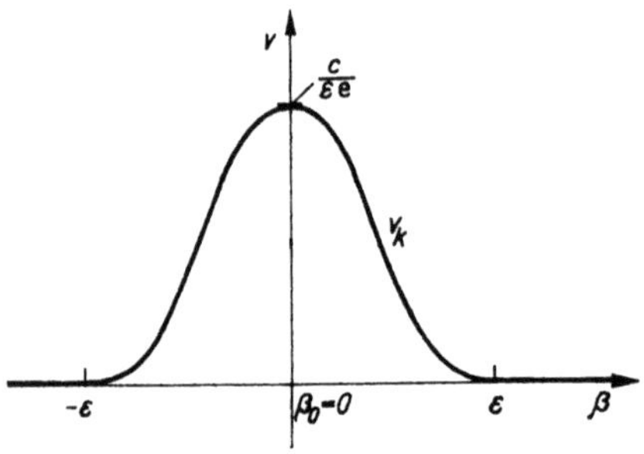

Fig. 5.10

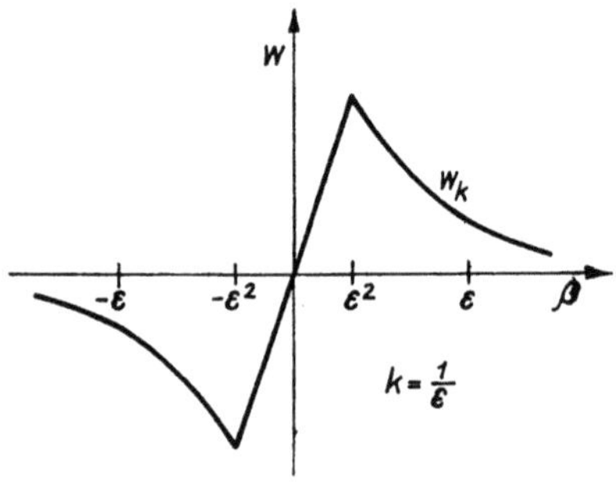

Fig. 5.11

Theorem 5.6. *Let $\Omega = B(x^0, r) \subset \mathbb{R}^2$ and $g_0 : \partial\Omega \to \mathbb{R}$ be the non-negative function defined in (11). If g_0 has a root at $y^0 = (R, \beta_0) \in \partial B(x^0, r)$ and $g_0(\beta) = (\beta - \beta_0)^2\, \tilde{R}_2(\beta, \beta_0)$, $\tilde{R}_2(\beta_0, \beta_0) > 0$, then there exists a sequence of functions $u_k \in H_G(\bar{\Omega})$ such that*

$$\int_0^{2\pi} g_0(\beta)\, |u_k(y)|\, \mathrm{d}\beta \to 0 \tag{27}$$

and

$$u_k(x) \to d_1 P(x, y^0) + d_2\, \frac{\partial P(x, y^0)}{\partial \beta}. \tag{28}$$

Proof. a) We have $g_0(\beta) = g_0(y)\, R$. From $b_0(\beta_0) = 0$ and $g_0(\beta) \geqq 0$ it follows that

$$\frac{\partial g_0(\beta_0)}{\partial \beta} = 0 \quad \text{and} \quad g_0(\beta) = (\beta - \beta_0)^{2k}\, \tilde{R}_{2k}(\beta, \beta_0), \tag{29}$$

where

$$\tilde{R}_{2k}(\beta, \beta_0) = \frac{1}{k!}\, \frac{\partial^{2k} g_0(\beta^*)}{\partial \beta^{2k}} \quad \text{and} \quad |\tilde{R}_{2k}(\beta, \beta_0)| \leqq M.$$

In the examples of Sect. 5.2 we have $k = 1$. Let $\tilde{u}_{k'}$ be the solution of the Dirichlet problem relative to $d_1 v_{k'}$, where the $v_{k'}$ were constructed in (21)–(26). They can be uniquely approximated by $u_{k'} \in H_G(\bar{\Omega})$ satisfying (3).

b) In order to construct a sequence of functions $u_{k''}$ satisfying (4), the following remarks can be applied. If $(\beta - \beta_0)\, w_{k''} = d_2 v_{k''}$, where the function v_k was defined in (21), then the second formula in (4) is satisfied. The functions

$$w_{k''} = \frac{v_{k''}}{\beta - \beta_0}$$

are not integrable. We define (see Fig. 5.11)

$$w_k(\beta) := \begin{cases} \dfrac{v_k(\beta)}{\beta - \beta_0} & \text{if} \quad |\beta - \beta_0| > \dfrac{1}{k^2}, \\[2ex] k^4(\beta - \beta_0)\, v_k(\beta) & \text{if} \quad |\beta - \beta_0| \leqq \dfrac{1}{k^2}. \end{cases} \tag{30}$$

The functions $w_k \colon [0, 2\pi] \mapsto \mathbb{R}$ are continuous. Then

$$(\beta - \beta_0)\, w_k(\beta) = \begin{cases} v_k(\beta) & \text{if} \quad |\beta - \beta_0| > \dfrac{1}{k^2}, \\[2ex] k^4(\beta - \beta_0)^2\, v_k(\beta) & \text{if} \quad |\beta - \beta_0| \leqq \dfrac{1}{k^2}. \end{cases} \tag{31}$$

From $w_k(\beta_0 - \beta^*) = -w_k(\beta_0 + \beta^*)$, $\beta^* = \beta - \beta_0$, it follows that

$$\int_0^{2\pi} w_k(\beta)\, d\beta = 0. \tag{32}$$

Further, the relation holds:

$$\int_{\beta_0-\pi}^{\beta_0+\pi} (\beta - \beta_0)\, w_k(\beta)\, d\beta = \int_{\beta_0-\pi}^{\beta_0+\pi} v_k(\beta)\, d\beta - \int_{\beta_0-1/k^2}^{\beta_0+1/k^2} v_k(\beta)\, d\beta + \int_{\beta_0-1/k^2}^{\beta_0+1/k^2} k^4(\beta - \beta_0)^3\, v_k(\beta)\, d\beta.$$

We obtain the following inequalities:

$$\left| \int_{\beta_0-1/k^2}^{\beta_0+1/k^2} k^4(\beta - \beta_0)^2\, v_k(\beta)\, d\beta \right| \leqq \int_{\beta_0-1/k^2}^{\beta_0+1/k^2} v_k(\beta)\, d\beta \leqq \frac{c}{e} \int_{\beta_0-1/k^2}^{\beta_0+1/k^2} k\, d\beta = \frac{2c}{k\,e} \to 0$$

and

$$\int_{\beta_0-1/k^2}^{\beta_0+1/k^2} v_k(\beta)\, d\beta \leqq \frac{c}{e} \int_{\beta_0-1/k^2}^{\beta_0+1/k^2} k\, d\beta = \frac{2c}{k\,e} \to 0.$$

Finally, we get

$$\int_{\beta_0-\pi}^{\beta_0+\pi} (\beta - \beta_0)\, w_k(\beta)\, \mathrm{d}\beta = \int_{\beta_0-\pi}^{\beta_0+\pi} v_k(\beta)\, \mathrm{d}\beta - \int_{\beta_0-1/k^2}^{\beta_0+1/k^2} v_k(\beta)\, \mathrm{d}\beta + \int_{\beta_0-1/k^2}^{\beta_0+1/k^2} k^4(\beta - \beta_0)^2\, v_k(\beta)\, \mathrm{d}\beta \to 1 \quad \text{for} \quad k \to \infty. \tag{33}$$

c) In order to prove

$$\int_{\beta_0-\pi}^{\beta_0+\pi} (\beta - \beta_0)^2\, w_k(\beta)\, \mathrm{d}\beta \to 0 \quad \text{for} \quad k \to \infty \tag{34}$$

we consider the relation

$$\int_{\beta_0-\pi}^{\beta_0+\pi} (\beta - \beta_0)^2\, w_k(\beta)\, \mathrm{d}\beta = \int_{\beta_0-1/k}^{\beta_0+1/k} (\beta - \beta_0)\, v_k(\beta)\, \mathrm{d}\beta - \int_{\beta_0-1/k^2}^{\beta_0+1/k^2} (\beta - \beta_0)\, v_k(\beta)\, \mathrm{d}\beta + \int_{\beta_0-1/k^2}^{\beta_0+1/k^2} k^4(\beta - \beta_0)^2\, v_k(\beta)\, \mathrm{d}\beta. \tag{35}$$

The following inequalities hold:

$$\left| \int_{\beta_0-1/k}^{\beta_0+1/k} (\beta - \beta_0)\, v_k(\beta)\, \mathrm{d}\beta \right| \leqq \int_{\beta_0-1/k}^{\beta_0+1/k} |\beta - \beta_0|\, v_k(\beta)\, \mathrm{d}\beta \leqq \int_{\beta_0-1/k}^{\beta_0+1/k} \frac{c}{k\,\mathrm{e}}\, v_k(\beta)\, \mathrm{d}\beta = \frac{2c}{k\,\mathrm{e}} \to 0,$$

$$\left| \int_{\beta_0-1/k^2}^{\beta_0+1/k^2} (\beta - \beta_0)\, v_k(\beta)\, \mathrm{d}\beta \right| \leqq \frac{c}{\mathrm{e}} \int_{\beta_0-1/k^2}^{\beta_0+1/k^2} \frac{1}{k^2}\, k\, \mathrm{d}\beta = \frac{2c}{k^3\,\mathrm{e}} \to 0,$$

$$\left| \int_{\beta_0-1/k^2}^{\beta_0+1/k^2} k^4(\beta - \beta_0)^2\, v_k(\beta)\, \mathrm{d}\beta \right| \leqq \int_{\beta_0-1/k^2}^{\beta_0+1/k^2} \frac{1}{k^4}\, k^4 v_k(\beta)\, \mathrm{d}\beta = \frac{2c}{k\,\mathrm{e}} \to 0.$$

Using these inequalities, relation (34) follows from (35).

d) Similar to $u_{k'}$ in c) we can construct functions $d_2 w_{k''}$, $\tilde{u}_{k''}$ and $u_{k''} \in H_G(\bar{\Omega})$ satisfying (4). From Taylor's theorem we obtain

$$P(x, y) = P(x, y^0) = (\beta - \beta_0)\, \frac{\partial P(x, y^0)}{\partial \beta} + (\beta - \beta_0)^2\, \frac{1}{2!}\, \frac{\partial^2 P(x, y^*)}{\partial \beta^2}. \tag{36}$$

Integrating (36) with respect to $u_k(y)\, R$ we obtain formula (5), where

$$R(x, u_k) := \int_0^{2\pi} (\beta - \beta_0)^2\, \frac{\partial^2 P(x, y^*)}{\partial \beta^2}\, u_k(y)\, R\, \mathrm{d}\beta \tag{37}$$

and hence

$$|R(x, u_k)| \leqq M \int_0^{2\pi} (\beta - \beta_0)^2 |u_k(y)| R \, d\beta.$$

Let $u_k := d_1 v_{k'} + d_2 w_{k''}$. Then

$$|R(x, u_k)| \leqq M \left(\int_0^{2\pi} (\beta - \beta_0)^2 |v_{k'}(\beta)| \, d\beta + \int_0^{2\pi} (\beta - \beta_0)^2 |w_{k''}(\beta)| \, d\beta \right).$$

Using (34) and (26) we obtain

$$R(x, u_k) \to 0 \quad \text{for} \quad k \to \infty. \tag{38}$$

Finally, from (5), (21)–(26), and (30)–(34) it follows that

$$u_k(x) \to d_1 P(x, y^0) + d_2 \frac{\partial P(x, y^0)}{\partial \beta}.$$

e) In Theorem 4.6 we proved that the set of all functions

$$f(x) = \sum_{k=1}^{m} c_k G_{\mathrm{L}}(x, x^k), \qquad x \in \partial \Omega,$$

is dense in $\mathbb{C}(\partial \Omega)$, $\Omega = B(x^0, R)$. From this fact it follows that there exists a sequence of functions $f_{k'} \in H_G(\bar{\Omega})$ satisfying (3) and further a second sequence of functions $f_{k''} \in H_G(\bar{\Omega})$ satisfying (4). Then the sequence of functions $u_k := f_{k'} + f_{k''}$ satisfies

$$u_k(x) \to d_1 P(x, y^0) + d_2 \frac{\partial P(x, y^0)}{\partial \beta}. \ \blacksquare \tag{39}$$

Corollary 5.1. *In the case of m roots at* $y^1 = (R, \beta_1), \ldots, y^m = y(R, \beta_m)$ *there exists a sequence of functions*

$$u_k = f_{k'}^{(1)} + f_{k''}^{(1)} + \cdots + f_{k'}^{(m)} + f_{k''}^{(m)}$$

such that

$$u_k(x) \to d_{11} P(x, y^{(1)}) + d_{12} \frac{\partial P(x, y^{(1)})}{\partial \beta} + \cdots + d_{m1} P(x, y^m) + d_{m2} \frac{\partial P(x, y^m)}{\partial \beta}. \tag{40}$$

The following condition is a necessary condition for $\nu = \nu_K + \gamma \in \operatorname{ex} \mathfrak{B}(\delta_{x^0})$. In Proposition 5.1 through 5.4 we proved that this condition is also sufficient. It is an open problem that in general the geometrical condition of Theorem 5.7 is also sufficient.

Theorem 5.7. *Let* $\nu = \nu_K + \gamma \in \mathfrak{B}(\delta_{x^0})$ *be an extremal element,* $\nu_K \neq 0$ *and* $g_0(\beta) = 0$ *at* $\beta_1, \ldots, \beta_m$. *Further, let* g_0 *satisfy at* β_j *the relations* $g_0(\beta) = (\beta - \beta_j)^2 \tilde{R}_2(\beta, \beta_j)$, $\tilde{R}_2(\beta_j, \beta_j) \neq 0$, $j = 1, \ldots, m$. *Then* supp ν_K *consists at most of* $2m$ *points.*

Proof. a) Following Theorem 5.3 the measure $\nu = \nu_K + \gamma$, $\nu_K \neq 0$, is an extremal element of $\mathfrak{B}(\delta_{x^0})$ if and only if $H_G(\bar{\Omega})$ is dense in $\mathbf{L}^1(\nu)$, i.e., for each $f \in \mathbf{L}^1(\nu)$ there exists a sequence of functions $u_k \in H_G(\bar{\Omega})$ satisfying $\left(g_0(y)\, R = g_0(\beta)\right)$

$$\int |f - u_k| \,\mathrm{d}\nu = \int |f(x) - u_k(x)| \,\mathrm{d}\nu_K(x) + \int_0^{2\pi} |f(y) - u_k(y)|\, g_0(\beta) \,\mathrm{d}\beta < \varepsilon \tag{41}$$

for $k > k_0(\varepsilon)$, $y = (R, \beta)$. If $\nu_K \neq 0$ from Theorem 5.5 and the analyticity of g_0 it follows that $g_0(\beta) = 0$ at finitely many points $\beta_1, \ldots, \beta_m$. We prove the theorem in the case $m = 1$. The same proof holds in the case of m roots.

Let us assume that K consists of more than 2 points. Then there exist three subsets K_1, K_2, K_3 such that $K = K_1 \cup K_2 \cup K_3$ and $\nu_K(K_j) \neq 0$, $j = 1, 2, 3$. Let ν_{K_j} be the restriction of ν_K on K_j. Then also $\nu_{K_j}(K_j) \neq 0$, $j = 1, 2, 3$. From (41) it follows that

$$\int |f(x) - u_k(x)| \,\mathrm{d}\nu_{K_j}(x) \leqq \int |f(x) - u_k(x)| \,\mathrm{d}\nu_K(x) < \varepsilon \tag{42}$$

for $k > k_0(\varepsilon)$. From this inequality we get

$$\left|\int \left(f(x) - u_k(x)\right) \mathrm{d}\nu_{K_j}(x)\right| \leqq \int |f(x) - u_k(x)| \,\mathrm{d}\nu_{K_j} < \varepsilon \tag{43}$$

for $k > k_0(\varepsilon)$.

b) If $f(y) = 0$ on $\partial B(x^0, R)$ then

$$\int_0^{2\pi} |u_k(y)|\, g_0(\beta) \,\mathrm{d}\beta < \varepsilon \quad \text{for} \quad k > k_0(\varepsilon). \tag{44}$$

We assumed that in a neighbourhood $U(\beta_1)$ the relation

$$g_0(\beta) = \frac{(\beta - \beta_1)^2}{2!}\, g_0''(\tilde{\beta}), \qquad g_0''(\tilde{\beta}) \geqq c_0 > 0, \tag{45}$$

holds. In the complement $U' := B(x^0, R) \setminus U(\beta_1)$ we get $g_0(\beta) \geqq c_1 > 0$. From (44) it follows that

$$c_1 \int_{U'} |u_k(y)| \,\mathrm{d}\beta \leqq \int_{U'} |u_k(y)|\, g_0(\beta) \,\mathrm{d}\beta, \tag{46}$$

i.e.,

$$\int_{U'} |u_k(y)| \,\mathrm{d}\beta \to 0 \quad \text{for} \quad k \to \infty.$$

On $U(\beta_1)$ the following inequality holds:

$$\begin{aligned}\int_{U(\beta_1)} (\beta - \beta_1)^2 |u_k(y)| \,\mathrm{d}\beta &= \int_{U(\beta_1)} (\beta - \beta_1)^2 \frac{g_0''(\tilde{\beta})}{g_0''(\tilde{\beta})} |u_k(y)| \,\mathrm{d}\beta \\ &\leqq c_2 \int_{U(\beta_1)} (\beta - \beta_1)^2\, g_0''(\tilde{\beta})\, |u_k(y)| \,\mathrm{d}\beta \\ &= c_2 \int_{U(\beta_1)} g_0(\beta)\, |u_k(y)| \,\mathrm{d}\beta. \end{aligned} \tag{47}$$

Using (46) and (47) we get from (37)

$$|R(x, u_k)| \leqq M \left(\int_{U'} (\beta - \beta_1)^2 \, |u_k(y)| \, \mathrm{d}\beta + c_2 \int_{U(\beta_1)} g_0(\beta) \, |u_k(y)| \, \mathrm{d}\beta \right) < \varepsilon \tag{48}$$

for $k > k_1(\varepsilon)$.

c) Integrating (5) with respect to $\mathrm{d}\nu_{K_j}$ we get

$$\int u_k(x) \, \mathrm{d}\nu_{K_j}(x) = \left(\int P(x, y^0) \, \mathrm{d}\nu_{K_j}(x) \right) \left(\int_0^{2\pi} u_k(y) \, R \, \mathrm{d}\beta \right)$$
$$+ \left(\int \frac{\partial P(x, y^0)}{\partial \beta} \, \mathrm{d}\nu_{K_j}(x) \right) \left(\int_0^{2\pi} (\beta - \beta_1) \, u_k(y) \, R \, \mathrm{d}\beta \right)$$
$$+ \int R(x, u_k) \, \mathrm{d}\nu_{K_j}(x).$$

Let

$$X_{kj} = a_{1j} U_k + a_{2j} V_k + R_{kj}, \qquad j = 1, 2, 3, \tag{49}$$

where

$$a_{1j} := \int P(x, y^0) \, \mathrm{d}\nu_{K_j}(x) > 0, \qquad a_{2j} := \int \frac{\partial P(x, y^0)}{\partial \beta} \, \mathrm{d}\nu_{K_j}(x),$$

$$U_k := \int_0^{2\pi} u_k(y) \, R \, \mathrm{d}\beta, \qquad V_k := \int_0^{2\pi} (\beta - \beta_1) \, u_k(y) \, R \, \mathrm{d}\beta, \tag{50}$$

$$X_{kj} := \int u_k(x) \, \mathrm{d}\nu_{K_j}(x), \qquad R_{kj} := \int R(x, u_k) \, \mathrm{d}\nu_{K_j}(x), \qquad j = 1, 2, 3.$$

From (43) it follows that X_{kj} can approach each arbitrary value F_j which only depends on the value $\int f \, \mathrm{d}\nu_{Kj}$, $f \in \mathbf{L}^1(\nu)$. Using (46) and (48) we get

$$|R_{kj}| \leqq \varepsilon_2 \quad \text{for} \quad k \geqq k_2(\varepsilon). \tag{51}$$

Let $a_{2j} \neq 0$, $j = 1, 2$, and

$$F_{kj} := X_{kj} - R_{kj}, \qquad j = 1, 2. \tag{52}$$

Because

$$F_j = \lim_{k \to \infty} F_{kj} = \lim_{k \to \infty} X_{kj}, \qquad j = 1, 2,$$

can be an arbitrary value we can construct two sequences (u_k) and $(\tilde{u}_k)$ such that

$$F_{kj} = a_{1j} U_k + a_{2j} V_k, \qquad j = 1, 2, \qquad k \geqq k^{(1)}(\varepsilon),$$

and

$$\tilde{F}_{kj} = a_{1j} \tilde{U}_k + a_{2j} \tilde{V}_k, \qquad j = 1, 2, \qquad k \geqq k^{(2)}(\varepsilon)$$

define two linearly independent vectors $(F_{k1}, F_{k2}) = F$ and $(\tilde{F}_{k1}, \tilde{F}_{k2}) = \tilde{F}$ in $\mathbb{R}^2$. Then

$$c_1 F + c_2 \tilde{F} \neq 0 \quad \text{for} \quad |c_1| + |c_2| \neq 0.$$

From this fact it follows that the determinant of the matrix

$$A = \begin{pmatrix} a_{11} & a_{12} \\ a_{21} & a_{22} \end{pmatrix}$$

is different from zero. In the case $a_{2j} = 0$, $j = 1, 2$, only F_{k1} can be prescribed arbitrarily.

d) Because $\det A \neq 0$ we can calculate U_k and V_k,

$$U_k = b_{11}F_{k1} + b_{21}F_{k2},$$

$$V_k = b_{21}F_{k1} + b_{22}F_{k2}.$$

By assumption the limit

$$F_j = \lim_{k\to\infty} F_{kj}, \qquad j = 1, 2,$$

exists and depends only on f on K_j. From this it follows that the limits of

$$U_k = \int_0^{2\pi} u_k(y)\, R\, \mathrm{d}\beta \quad \text{and} \quad V_k = \int_0^{2\pi} (\beta - \beta_1)\, u_k(y)\, R\, \mathrm{d}\beta$$

exist and depend only on f on $K_1 \cup K_2$. Let $j = 3$. Then

$$\int f\, \mathrm{d}\nu_{K_3} = \lim_{k\to\infty} \int u_k\, \mathrm{d}\nu_{K_3} = a_{13} \lim_{k\to\infty} U_k + a_{23} \lim_{k\to\infty} V_k.$$

The left-hand side can be chosen arbitrarily (consequence of $\overline{H_G(\bar{\Omega})} = \mathrm{L}^1(\nu)$), the right-hand side is determined by the limits of U_k and V_k. This is a contradiction. Therefore the support of ν_K consists at most of 2 points.

e) The same proof holds in the case of m roots of g_0 at $\beta_1, \ldots, \beta_m$. We remark that the function $g_0\colon [0, 2\pi] \mapsto \mathbb{R}$ is analytic. From this it follows that only finitely many roots exist. ∎

5.5. The case of a harmonic density. Analytic continuation of special potentials

In the preceding sections we have seen that the set $\mathfrak{B}(\mu)$ contains more than one element. In general this set is non-denumerable. Since $\mathfrak{B}(\mu)$ is convex there exist extremal elements of this set uniquely determined. Special extremal elements have been constructed in Proposition 5.1 through Proposition 5.4. Geometric conditions for more general extremal elements were proved in Sect. 5.4.

Here we shall study the case of a harmonic density ϱ (Anger and Kleine [1], Kleine [1], Lorenzi and Pagani [1], Novikov [1], Weck [1]). Then the relation between ϱ and the boundary values of $G_{\mathrm{L}}\mu_\varrho$ on $\partial\Omega$ is one-to-one, where

$$G_{\mathrm{L}}\mu_\varrho(x) = \int_\Omega G_{\mathrm{L}}(x, y)\, \varrho(y)\, \mathrm{d}y. \tag{1}$$

In that case ϱ is uniquely determined by $\mathrm{rest}_{\partial\Omega}\,\varrho$. Up to now the extremal elements of $\mathfrak{B}(\mu_\varrho)$ have not been studied, but they are of special interest. Of interest is also the density $\varrho_P(x) := P_R(x, t)$, where P_R is the Poisson kernel relative to a ball $B(x^0, R) \subset \mathbb{R}^n$ and $t \in \partial B(x^0, R)$ fixed. The potential $G_{\mathrm{L}}\mu_P$, where $\mu_P := \mu_{\varrho_P}$, can be analytically continued up to a straight line segment joining x^0 and t (Theorem 5.10). In the case $B(x^0, R) \subset \mathbb{R}^2$ there exist concrete formulas for the problem under consideration (Theorem 5.12). One can find some results in $\mathbb{R}^3$ in Sansò et al. [1].

Further, the integral

$$F(u) := \int_\Omega \varrho^2(x)\,\mathrm{d}x, \quad \text{where} \quad G_{\mathrm{L}}\mu_\varrho|_{\partial\Omega} = g \quad \text{is fixed,} \tag{2}$$

attains its minimum if ϱ is harmonic, which will be proved in Theorem 5.8. Let ϱ be a sufficiently smooth function. Then

$$-\Delta G_{\mathrm{L}}\mu_\varrho = \varrho \tag{3}$$

holds in the classical sense. If $\eta = -\Delta\varphi$, $\varphi \in \mathrm{C}_0^\infty(\mathbb{R}^n)$, $\mathrm{supp}\,\varphi \subset \Omega$, we get for $\varrho_1 := \varrho + \eta$ and $x \notin \Omega$

$$\begin{aligned} G_{\mathrm{L}}\mu_{\varrho_1}(x) &= \int_\Omega G_{\mathrm{L}}(x, y)\,\big(\varrho(y) - \Delta\varphi(y)\big)\,\mathrm{d}y = \int_\Omega G_{\mathrm{L}}(x, y)\,\varrho(y)\,\mathrm{d}y + \varphi(x) \\ &= \int G_{\mathrm{L}}(x, y)\,\varrho(y)\,\mathrm{d}y. \end{aligned} \tag{4}$$

The special density η has no influence for the potential $G_{\mathrm{L}}\mu_\varrho$ at $x \notin \Omega$. In the case of the Helmholtz equation $\Delta u + \varkappa^2 u = 0$ such a density is called *non-radiating* (Bleistein [1], Colton [1]). Let ϱ be a density harmonic in Ω. If $u = G_{\mathrm{L}}\mu_\varrho$ then from (3) it follows that

$$\Delta^2 u = \Delta\Delta u = -\Delta\varrho = 0, \tag{5}$$

i.e., u is a solution of the biharmonic equation.

Theorem 5.8. *Let M be the set of all functions $u \in \mathrm{C}^4(\Omega) \cap \mathrm{C}(\bar{\Omega})$ satisfying $u = G_{\mathrm{L}}\mu_\varrho$ and $G_{\mathrm{L}}\mu_\varrho|_{\partial\Omega} = g$, where g is fixed,*

$$F(u) := \int_\Omega (\Delta u)^2\,\mathrm{d}x, \qquad u \in M. \tag{6}$$

If there exists a function $u_0 = G_{\mathrm{L}}\mu_{\varrho^*} \in M$ satisfying

$$F(u_0) \leqq F(u) \quad \textit{for all} \quad u \in M \tag{7}$$

then

$$\Delta^2 u_0 = -\Delta\varrho^* = 0 \quad \textit{in } \Omega. \tag{8}$$

Proof. Let $\eta \in \mathrm{C}_0^\infty(\Omega)$ be an arbitrary function. Then $u = u_0 + \varepsilon\eta \in M$ and $F(u_0) \leqq F(u_0 + \varepsilon\eta)$. We get

$$F(u_0 + \varepsilon\eta) = F(u_0) + 2\varepsilon \int_\Omega (\Delta u_0)\,(\Delta\eta)\,\mathrm{d}x + \varepsilon^2 F(\eta).$$

From

$$\left.\frac{\mathrm{d}F(u_0 + \varepsilon\eta)}{\mathrm{d}\varepsilon}\right|_{\varepsilon=0} = 2\int_\Omega (\Delta u_0)\,(\Delta\eta)\,\mathrm{d}x = 0 \quad \text{for all} \quad \eta \in C_0^\infty(\Omega) \tag{9}$$

it follows that

$$\int_\Omega (\Delta u_0)\,(\Delta\eta)\,\mathrm{d}x = \int_\Omega (\Delta^2 u_0)\,\eta\,\mathrm{d}x = 0 \quad \text{for all} \quad \eta \in C_0^\infty(\Omega). \tag{10}$$

Since $\Delta^2 u_0$ is continuous we obtain

$$\Delta^2 u_0 = -\Delta\varrho^* = 0 \quad \text{in } \Omega. \tag{11}$$

Using Hilbert space methods WECK [1] proved the existence of a uniquely determined harmonic density ϱ. ∎

In the following we consider a ball $\Omega = B(0, R) \subset \mathbb{R}^2$. The Poisson kernel P_R of $B(0, R)$ was defined in (3.4.43),

$$P_R(x, y) = -\frac{\partial G_{\mathrm{L}}^\Omega(x, y)}{\partial n_y} = \frac{1}{2\pi R}\,\frac{R^2 - |x|^2}{|x - y|^2},$$

where G_{L}^Ω is the Green function of Ω. The swept-out measure of μ_ϱ uniquely determined has the form (see (4.2.21))

$$\begin{aligned}\Pi_{\partial\Omega}\mu_\varrho(f) &= \int_\Omega \left(\int_{\partial\Omega} P_R(x, y)\,f(y)\,\mathrm{d}S(y)\right)\varrho(x)\,\mathrm{d}x \\ &= \int_{\partial\Omega}\left(\int_\Omega P_R(x, y)\,\varrho(x)\,\mathrm{d}x\right) f(y)\,\mathrm{d}S(y).\end{aligned} \tag{12}$$

Since ϱ is harmonic the representation

$$\varrho(x) = \int_{\partial\Omega} \varrho_0(t)\,P_R(x, t)\,\mathrm{d}S(t), \qquad \varrho_0 = \mathrm{rest}_{\partial\Omega}\,\varrho, \tag{13}$$

holds. From (12) and (13) it follows that

$$\begin{aligned}\Pi_{\partial\Omega}\mu_\varrho(f) &= \int_{\partial\Omega}\left(\int_{\partial\Omega}\left(\int_\Omega P_R(x, y)\,P_R(x, t)\,\mathrm{d}x\right)\varrho_0(t)\,\mathrm{d}S(t)\right) f(y)\,\mathrm{d}S(y) \\ &= \int_{\partial\Omega}\left(\int_{\partial\Omega} Q(y, t)\,\varrho_0(t)\,\mathrm{d}S(t)\right) f(y)\,\mathrm{d}S(y),\end{aligned} \tag{14}$$

where

$$Q(y, t) := \int_\Omega P_R(x, y)\,P_R(x, t)\,\mathrm{d}x. \tag{15}$$

The swept-out measure $\Pi_{\partial\Omega}\mu_\varrho$ concentrated on $\partial\Omega$ has the form μ_{σ_0},

$$\mu_{\sigma_0}(f) = \int_{\partial\Omega} f(y)\,\sigma_0(y)\,\mathrm{d}S(y).$$

Therefore we get

$$\sigma_0(y) = \int_{\partial\Omega} Q(y, t)\, \varrho_0(t)\, \mathrm{d}S(t). \tag{16}$$

Outside Ω we have

$$G_L\mu_{\sigma_0}(z) = G_L\mu_\varrho(z), \qquad z \notin \Omega. \tag{17}$$

Since $\overline{D(\partial\Omega; G_L)} = \mathbf{C}(\partial\Omega)$ the measure μ_{σ_0} is uniquely determined. In Example 2.21 we have got the same formulas if $V_0(x) = 0$, $u_0 = 1$ and $\Delta V_1 = 0$ (see formula (2.3.31)). Formula (16) gives the relation between the harmonic density ϱ (its boundary values ϱ_0) and the density σ_0, where $\mu_{\sigma_0} \in \mathfrak{B}(\mu_\varrho)$. These considerations hold relative to a domain $\Omega \subset \mathbb{R}^2$ with smooth boundary $\partial\Omega$. Then $P(x, y)$ is defined by the normal derivative of the Green function $G_L^\Omega(x, y)$ at $y \in \partial\Omega$. In this direction further investigations are necessary. But further considerations give concrete formulas only for a ball $B(x^0, R)$.

Now some remarks concerning the analytic continuation of special Newtonian potentials. In Example 3.1 we have seen that the potential $G_L m_{x^0,R}$ of a ball $B(x^0, R)$ with constant density can be analytically continued up to the point x^0. Then $G_L\delta_{x^0}$ and $G_L m_{x^0,R}$ give the same potential outside $B(x^0, R)$. Similar facts hold relative to a domain $\Omega \subset \mathbb{R}^2$ bounded by an ellipse. In Example 3.3 we have proved that for constant density ϱ_0 the potential $G_L\mu_{\varrho_0}$ can be analytically continued up to the straight line segment joining the two foci (see also Herglotz [2]). In the following we shall prove that a similar continuation is possible relative to $G_L\mu_P$, where

$$\mu_P(f) := \int_\Omega f(x)\, P_R(x, t)\, \mathrm{d}x, \qquad t \in \partial\Omega \quad \text{fixed}. \tag{18}$$

The density of the volume distribution μ_ϱ is $\varrho(x) = P(x, t)$. The potential $G_L\mu_P$ can be analytically continued up to the straight line segment joining the origin and $t \in \partial\Omega$. Let

$$\delta_{\overline{0t}}(f) := \frac{1}{R} \int_0^R f\left(s\frac{t}{R}\right) \mathrm{d}s. \tag{19}$$

Then $\int \mathrm{d}\delta_{\overline{0t}} = 1$. Our results hold relative to a ball $B(0, R)$. In the general case $B(x^0, R)$ the point x must be replaced by $x - x^0$ and y by $y - y^0$.

Theorem 5.9. *Let $B(0, R) \subset \mathbb{R}^2$ be a ball and $t \in \partial B(0, R)$ a fixed point. The measure μ_P defined in* (18) *can be replaced by the measure*

$$\nu_0(f) := \frac{R}{2} \int f(\zeta)\, \mathrm{d}\delta_{\overline{0t}}(\zeta) = \frac{1}{2} \int_0^R f\left(s\frac{t}{R}\right) \mathrm{d}s \tag{20}$$

concentrated on the line segment $\overline{0t}$ joining the points 0 *and t such that $\frac{R}{2}\delta_{\overline{0t}} \in \mathfrak{B}(\mu_P)$, i.e., the potential $G_L\mu_P$ can be analytically continued up to the straight* *line segment* $\overline{0t}$. *Further, the relation*

$$Q(y, t) = \frac{R}{2} \int P_R(\zeta, y) \, \mathrm{d}\delta_{0t}(\zeta) = \frac{1}{2} \int\limits_{s=0}^{R} P_R\left(s \frac{t}{R}, y\right) \mathrm{d}s \tag{21}$$

holds.

Proof. In polar coordinates the kernel Q has the form

$$Q(y, t) = \int\limits_{\Omega} P_R(x, y) \, P_R(x, t) \, \mathrm{d}x = \int\limits_{r=0}^{R} \left(\int\limits_{\partial B(0,r)} P_R(x, y) \, P_R(x, t) \, \mathrm{d}s(x) \right) \mathrm{d}r, \tag{22}$$

where $x \in \partial B(0, r)$. Further,

$$P_R(x, t) = \frac{1}{2\pi R} \frac{R^2 - |x|^2}{|x - t|^2} = \frac{1}{2\pi R} \frac{R^2 - r^2}{|x - t|^2}. \tag{23}$$

We now fix $r < R$ and apply the Kelvin transformation (Helms [1], Kellogg [1])

$$t \mapsto t^* = \frac{r^2}{R^2} t \tag{24}$$

relative to the circle $\partial B(0, r)$. The following relation holds for $x \in \partial B(0, r)$:

$$\frac{r}{R} \frac{|x - t|}{\left| x - \dfrac{r^2}{R^2} t \right|} = 1. \tag{25}$$

Hence it follows that

$$\frac{1}{2\pi R} \frac{R^2 - r^2}{|x - t|^2} = \frac{r}{R} \frac{1}{2\pi r} \frac{r^2 - \left(\dfrac{r^2}{R}\right)^2}{\left| x - \dfrac{r^2}{R^2} t \right|^2}.$$

This means that

$$P_R(x, t) = \frac{r}{R} P_r(t^*, x), \tag{26}$$

where P_r denotes the Poisson kernel with respect to $B(0, r)$. Using (26) one obtains

$$\int\limits_{\partial B(0,r)} P_R(x, y) \, P_R(x, t) \, \mathrm{d}s(x) = \int\limits_{\partial B(0,r)} \frac{r}{R} P_R(x, y) \, P_r(t^*, x) \, \mathrm{d}s(x)$$

$$= \frac{r}{R} P_R(t^*, y). \tag{27}$$

In the last relation we used the fact that $x \mapsto P_R(x, y)$ is harmonic in $B(0, r)$. From (22) it follows that

$$Q(y, t) = \int_{r=0}^{R} \frac{r}{R} P_R\left(\frac{r^2}{R^2} t, y\right) \mathrm{d}r = \frac{1}{2} \int_{r=0}^{R} P_R\left(\frac{r^2}{R^2} t, y\right) \frac{2r}{R} \mathrm{d}r$$

$$= \frac{1}{2} \int_{s=0}^{R} P_R\left(s \frac{t}{R}, y\right) \mathrm{d}s. \tag{28}$$

Therefore from (19) and (28) it follows that

$$Q(y, t) = \frac{1}{2} \int_{s=0}^{R} P_R\left(s \frac{t}{R}, y\right) \mathrm{d}s = \frac{R}{2} \int P_R(\zeta, y) \, \mathrm{d}\delta_{\overline{0t}}(\zeta), \qquad \zeta = s \frac{t}{R}.$$

The density $x \mapsto P_R(x, t)$ and the measure $\frac{R}{2} \delta_{\overline{0t}}$ yield the kernel $Q(x, t)$, i.e., the potential $G_L \mu_P$ can be analytically continued up to $\overline{0t}$. ▮

Remark 5.5. In the case of m points $t_1, \ldots, t_m \in \partial B(0, R)$,

$$\nu := a_1 \delta_{t_1} + \cdots + a_m \delta_{t_m}, \qquad a_j \geqq 0, \qquad \sum_{j=1}^{m} a_j = 1,$$

and

$$P\nu(x) := a_1 P_R(x, t_1) + \cdots + a_m P_R(x, t_m)$$

one can replace $\mu_{P\nu}$ by the measure

$$\nu_0 := \frac{R}{2} (a_1 \delta_{\overline{0t_1}} + \cdots + a_m \delta_{\overline{0t_m}}) \in \mathfrak{B}(\nu_P), \quad \text{i.e.,}$$

the potential $G_L \mu_{P\nu}$ can be analytically continued up to the straight line segments $\overline{0t_1}, \ldots, \overline{0t_m}$.

Similar considerations are possible in $\mathbb{R}^3$. Then

$$P_R(x, t) = \frac{1}{4\pi R} \frac{R^2 - |x|^2}{|x - t|^3}.$$

Further, formulas (12) through (18) are valid in $\mathbb{R}^n$. We define

$$\delta_{\overline{0t}}(f) := \frac{3}{2R} \int_0^R f\left(s \frac{t}{R}\right) \sqrt{\frac{s}{R}} \, \mathrm{d}s, \quad \text{where} \quad \int \mathrm{d}\delta_{\overline{0t}} = 1. \tag{29}$$

Theorem 5.10. *Let $B(0, R) \subset \mathbb{R}^3$ be a ball and $t \in \partial B(0, R)$ a fixed point. The measure μ_P defined in (18) can be replaced by the measure*

$$\nu_0(f) := \frac{R}{3} \int f(\zeta) \, \mathrm{d}\delta_{\overline{0t}}(\zeta) = \int f\left(s \frac{t}{R}\right) \sqrt{\frac{s}{R}} \frac{\mathrm{d}s}{2}$$

concentrated on the line segment $\overline{0t}$ *joining the points* 0 *and* t *such that* $\frac{R}{3}\,\delta_{\overline{0t}} \in \mathfrak{B}(\mu_P)$, i.e.,

$$Q(y,t) = \frac{3}{R}\int P_R(\zeta, y)\,\mathrm{d}\delta_{\overline{0t}}(\zeta) = \frac{3}{R}\int_{s=0}^{R} P_R\left(s\,\frac{t}{R}, y\right)\sqrt{\frac{s}{R}}\,\frac{\mathrm{d}s}{2}. \tag{30}$$

The potential $G_L\mu_P$ *can be analytically continued up to the straight line segment* $\overline{0t}$.

Proof. We apply the Kelvin transformation

$$t \mapsto t^* = \frac{r^2}{R^2}\,t,$$

for which the relation

$$\frac{r}{R}\,\frac{|x-t|}{|x-t^*|} = 1$$

holds. From this it follows that

$$\frac{1}{4\pi R}\,\frac{R^2-r^2}{|x-t|^3} = \frac{r^2}{R^2}\,\frac{1}{4\pi r}\,\frac{r^2-\left(\frac{r^2}{R}\right)^2}{|x-t^*|^3},$$

that means

$$P_R(x,t) = \left(\frac{r}{R}\right)^2 P_r(t^*, x).$$

Like in the proof of Theorem 5.9 the following relations hold:

$$\begin{aligned}
Q(y,t) &= \int_{r=0}^{R}\left(\int_{\partial B(0,r)} P_R(x,y)\,P_R(x,t)\,\mathrm{d}S(x)\right)\mathrm{d}r \\
&= \int_{r=0}^{R}\left(\int_{\partial B(0,r)}\left(\frac{r}{R}\right)^2 P_R(x,y)\,P_r(t^*,x)\,\mathrm{d}S(x)\right)\mathrm{d}r \\
&= \int_{r=0}^{R}\left(\frac{r}{R}\right)^2 P_R(t^*,y)\,\mathrm{d}r = \int_{r=0}^{R} P_R\left(s\,\frac{t}{R}, y\right)\sqrt{\frac{s}{R}}\,\frac{\mathrm{d}s}{2}.
\end{aligned} \tag{31}$$

In the last relation we substituted $s = r^2/R^2$. Using (29) formula (31) can be written in the form

$$Q(y,t) = \frac{3}{R}\int_{s=0}^{R} P_R(\zeta,y)\,\mathrm{d}\delta_{\overline{0t}}(\zeta).\ \blacksquare$$

In the following let $\Omega = B(0, R) \subset \mathbb{R}^2$ be a ball and

$$t = R\,e^{i\beta}, \qquad x = r\,e^{i\varphi}, \qquad y = R\,e^{i\alpha}. \tag{32}$$

From (21) it follows that in this case

$$Q_1(\alpha, \beta) := Q(y, t)$$

has the form

$$Q_1(\alpha, \beta) = \frac{1}{2} \int\limits_{r=0}^{R} \frac{(R^2 - r^2)\,dr}{2\pi R\big(R^2 + r^2 - 2Rr\cos(\alpha - \beta)\big)}, \tag{33}$$

where $Q_1(\alpha, \beta)$ depends only on the difference $l := \alpha - \beta$. Let

$$\tilde{Q}(l) := Q_1(\alpha - \beta, 0). \tag{34}$$

The kernel $l \mapsto \tilde{Q}(l)$ is 2π-periodic. The following identity holds:

$$\frac{R^2 - r^2}{(R^2 + r^2 - 2Rr\cos l)} = -1 + \frac{2R^2\sin^2 l}{(r - R\cos l)^2 + R^2\sin^2 l} - \frac{2(r - R\cos l)\,R\cos l}{(r - R\cos l)^2 + R^2\sin^2 l}.$$

From this identity it follows that

$$\tilde{Q}(l) = -\frac{1}{4\pi}\left[1 + (\cos l)\ln\big(2(1 - \cos l)\big) + (\sin l)\,(l - \operatorname{sgn} l)\right], \tag{35}$$

where $\operatorname{sgn} l = 1$ for $l > 0$ and $\operatorname{sgn} l = -1$ for $l < 0$. At the point $l = 0$ the kernel Q has a logarithmic singularity,

$$\tilde{Q}(l) \sim \ln\big(2(1 - \cos l)\big) \sim \ln l^2 = 2\ln l. \tag{36}$$

On $\mathbf{L}^2\big(\partial B(0, R)\big)$ resp. $\mathbf{C}\big(\partial B(0, R)\big)$ we define the following mapping T by

$$b(y) = Tk(y) := \int\limits_{\partial B(0,R)} Q(y, t)\,k(t)\,dS(t). \tag{37}$$

In (16) we used for b the symbol σ_0 and for k the symbol ϱ_0. Let

$$b_1(\alpha) := b(R\,e^{i\alpha}), \qquad k_1(\beta) := k(R\,e^{i\beta}).$$

From (34) and (37) we obtain

$$b_1(\alpha) = R\int\limits_{-\pi}^{+\pi} Q(\alpha - \beta)\,k_1(\beta)\,d\beta = R\int\limits_{-\pi}^{+\pi} Q(l)\,k_1(\alpha - l)\,dl. \tag{38}$$

In order to prove this equality we substitute $l := \alpha - \beta$ in the first integral and obtain for $\alpha > 0$

$$\int\limits_{\alpha-\pi}^{\alpha+\pi} Q(l)\,k_1(\alpha - l)\,dl = \int\limits_{\alpha-\pi}^{\pi} + \int\limits_{\pi}^{\alpha+\pi} = \int\limits_{-\pi}^{\alpha-\pi} + \int\limits_{\alpha-\pi}^{\pi} = \int\limits_{-\pi}^{+\pi},$$

where for a 2π-periodic function $f(l) = f(l - 2\pi)$ the following relation holds

$$\int_{\pi}^{\alpha+\pi} f(l)\,\mathrm{d}l = \int_{-\pi}^{\alpha-\pi} f(l - 2\pi)\,\mathrm{d}l = \int_{-\pi}^{\alpha-\pi} f(l)\,\mathrm{d}l. \tag{39}$$

A similar proof holds for $\alpha < 0$. The following theorems were proved by KLEINE [1].

Theorem 5.11. *The mapping* $b_1 = Tk_1$ *defined in* (38) *is a compact, linear one-to-one mapping on* $\mathbf{L}^2(\partial B(0, R))$.

Proof. a) The linearity follows from (38).

b) Let

$$h(x) = \int_{\partial B(0,R)} P(x, t)\, k(t)\, \mathrm{d}S(t) \tag{40}$$

and

$$b(y) = \int_{B(0,R)} P(x, y)\, h(x)\, \mathrm{d}x, \tag{41}$$

where $b(y)$ is the density of the swept-out measure $b(y)\,\mathrm{d}S(y)$ (see (16)). From (37) and (21) it follows that

$$b(y) = \int_{\partial B(0,R)} Q(y, t)\, k(t)\, \mathrm{d}S(t) = \int_{\partial B(0,R)} \left(\frac{1}{2} \int_{s'=0}^{R} P_R\left(\frac{s'}{R}\, y, t\right) \mathrm{d}s'\right) k(t)\, \mathrm{d}S(t)$$

$$= \frac{1}{2} \int_{s'=0}^{R} \left(\int_{\partial B(0,R)} P_R\left(\frac{s'}{R}\, y, t\right) k(t)\, \mathrm{d}S(t)\right) \mathrm{d}s' = \frac{1}{2} \int_{s'=0}^{R} h\left(\frac{s'}{R}\, y\right) \mathrm{d}s'. \tag{42}$$

Here $t = R\,\mathrm{e}^{\mathrm{i}\beta}$, $x = \dfrac{s'}{R}\, y = s'\,\mathrm{e}^{\mathrm{i}\alpha}$, $\mathrm{d}s' = \mathrm{d}r$. On $\partial B(0, R)$ we consider the system of functions

$$k_1 = 1, \cos\beta, \sin\beta, \cos 2\beta, \sin 2\beta, \cos 3\beta, \ldots \tag{43}$$

By means of (40) and (41) we obtain the table

$k_1(\beta)$	$h(x)$	$b_1(\alpha) = Tk_1(\alpha)$
1	1	$\dfrac{R}{2}$
$\cos n\beta$	$\dfrac{r^n}{R^n} \cos n\beta$	$\dfrac{R}{2}\,\dfrac{1}{n+1} \cos n\alpha$
$\sin n\beta$	$\dfrac{r^n}{R^n} \sin n\beta$	$\dfrac{R}{2}\,\dfrac{1}{n+1} \sin n\alpha$

(44)

In $\mathbf{L}^2([-\pi, \pi])$ the system (43) is complete, i.e., the set of all linear combinations of (43) is dense in $\mathbf{L}^2([-\pi, \pi])$. The functions of (43) are the eigenfunctions of the mapping $b = Tk$ defined in (42). One can find their eigenvalues in (44). The mapping $b_1 = Tk_1$ is one-to-one, since the Fourier coefficients of b_1 and k_1 differ only by the factor $R/2(n + 1)$. The kernel Q is quadratically integrable. From this it follows that the mapping $b_1 = Tk_1$ is a compact mapping (see DUNFORD and SCHWARTZ [1]). On page 16 we proved that the inverse T^{-1} of a compact one-to-one operator is discontinuous. In our case T^{-1} contains derivatives (see also the Abel integral equation and Sect. 1.3). ▮

Theorem 5.12. *Let b_1 be a 2π-periodic function for which b_1'' exists and is integrable. Then there exists a uniquely determined 2π-periodic function k_1 of the integral equation*

$$b_1(\alpha) = Tk_1(\alpha) = -\frac{R}{4\pi} \int_{-\pi}^{+\pi} \Big[1 + (\cos l) \ln\big(2(1 - \cos l)\big) + (l - \operatorname{sgn} l) \sin l\Big]\, k_1(\alpha - l)\, \mathrm{d}l, \tag{45}$$

which has the form

$$k_1(\beta) = T^{-1}b_1(\beta) = \frac{2}{R}\left[b_1(\beta) + \frac{1}{2\pi}\int_{-\pi}^{+\pi} b_1''(\beta - l) \ln\big(2(1 - \cos l)\big)\, \mathrm{d}l\right]. \tag{46}$$

In order to prove (46) we have to apply the Fourier transform. For $f \in \mathbf{L}^1(\mathbb{R})$ it is defined by

$$\hat{f}(\xi) := \int_{-\infty}^{+\infty} f(\alpha)\, \mathrm{e}^{-\mathrm{i}\alpha\xi}\, \mathrm{d}\alpha. \tag{47}$$

For a measure μ on $\mathbb{R}$ satisfying $\|\mu\| < \infty$ we define

$$\hat{\mu}(\xi) := \int \mathrm{e}^{-\mathrm{i}\alpha\xi}\, \mathrm{d}\mu(\alpha). \tag{48}$$

The inverse Fourier transform is defined by

$$f(\alpha) := \frac{1}{2\pi} \int_{-\infty}^{+\infty} \mathrm{e}^{\mathrm{i}\alpha\xi} \hat{f}(\xi)\, \mathrm{d}\xi. \tag{49}$$

The following properties hold for all $f \in \mathbf{L}^1(\mathbb{R})$ (see BAUER [2], HÖRMANDER [2]):

The mapping $f \mapsto \hat{f}$ is one-to-one,

$$\widehat{(c_1 f_1 + c_2 f_2)} = c_1 \hat{f}_1 + c_2 \hat{f}_2, \qquad c_1, c_2 \in R,$$

$$\widehat{f_1 * f_2} = \hat{f}_1 \cdot \hat{f}_2, \quad \text{where} \quad (f_1 * f_2)(\alpha) := \int_{-\infty}^{+\infty} f_1(\beta) f_2(\alpha - \beta)\, \mathrm{d}\beta.$$

Further, we define

$$\delta_x(f) := f(x),\ \delta_0(f) := f(0),\ \delta_0'(f) := f'(0),\ \delta_0''(f) := f''(0), \ldots \tag{50}$$

Then

$$\hat{\delta}_x(\xi) = \mathrm{e}^{-\mathrm{i}\xi x},\ \hat{\delta}_0 = 1,\ \hat{\delta}_0'(\xi) = -\mathrm{i}\xi,\ \hat{\delta}_0''(\xi) = -\xi^2, \ldots \tag{51}$$

Let $\chi_0 \colon \mathbb{R} \mapsto \mathbb{R}$ be the characteristic function of $[-\pi, \pi]$. We define

$$b_0 := \chi_0 b_1, \qquad k_0 := \chi_0 k_1, \qquad Q_0 := \chi_0 \tilde{Q}. \tag{52}$$

These functions are elements of $\mathbf{L}^1(\mathbb{R})$, for which the Fourier transforms exist. Then

$$\hat{b}_0(\xi) := \int_{-\infty}^{+\infty} \chi_0(\alpha)\, b_1(\alpha)\, \mathrm{e}^{-\mathrm{i}\alpha\xi}\, \mathrm{d}\xi = \int_{-\pi}^{+\pi} b_1(\alpha)\, \mathrm{e}^{-\mathrm{i}\alpha\xi}\, \mathrm{d}\alpha,$$

$$\hat{k}_0(\xi) := \int_{-\pi}^{+\pi} k_1(\alpha)\, \mathrm{e}^{-\mathrm{i}\alpha\xi}\, \mathrm{d}\alpha, \tag{53}$$

$$\hat{Q}_0(\xi) := \int_{-\pi}^{+\pi} \tilde{Q}(\alpha)\, \mathrm{e}^{-\mathrm{i}\alpha\xi}\, \mathrm{d}\alpha.$$

Proof of Theorem 5.12. a) The following relations hold:

$$\begin{aligned} \hat{b}_0(\xi) &= \int_{-\pi}^{+\pi} b_1(\alpha)\, \mathrm{e}^{-\mathrm{i}\alpha\xi}\, \mathrm{d}\alpha = \int_{-\pi}^{+\pi} \left(R \int_{-\pi}^{+\pi} \tilde{Q}(\alpha - \beta)\, k_1(\beta)\, \mathrm{d}\beta \right) \mathrm{e}^{-\mathrm{i}\alpha\xi}\, \mathrm{d}\xi \\ &= \int_{-\pi}^{+\pi} \left(R \int_{-\pi}^{+\pi} Q(\alpha - \beta)\, \mathrm{e}^{-\mathrm{i}\alpha\xi}\, \mathrm{d}\xi \right) k_1(\beta)\, \mathrm{d}\beta \\ &= \int_{-\pi}^{+\pi} \left(R \int_{-\pi-\beta}^{\pi-\beta} Q(l)\, \mathrm{e}^{-\mathrm{i}(l+\beta)}\, \mathrm{d}l \right) k_1(\beta)\, \mathrm{d}\beta \\ &= \int_{-\pi}^{\pi} \left(R \int_{-\pi}^{\pi} Q(l)\, \mathrm{e}^{-\mathrm{i}\xi l}\, \mathrm{d}l \right) k_1(\beta)\, \mathrm{e}^{-\mathrm{i}\xi\beta}\, \mathrm{d}\beta = R\hat{Q}_0(\xi) \cdot \hat{k}_0(\xi). \end{aligned} \tag{54}$$

In (46) we used the substitution $\alpha - \beta = l$ and transformations of the integrals like in (39). To determine Q_0 one needs the Fourier transforms $\hat{k}_0$ and $\hat{b}_0 = \widehat{Tk_0}$. We change (44) into a table of complex functions

$$\begin{array}{ll} k_1(\beta) & b_1(\alpha) = Tk_1(\alpha) \\ \hline \exp(\mathrm{i}n\beta) & \dfrac{R}{2} \dfrac{1}{n+1} \exp(\mathrm{i}n\alpha) \end{array} \tag{55}$$

for all integers n. On the right-hand side we have to write $|n| + 1$, because for negative $-|n|$ the following relations hold:

$$\cos(-|n|\,\beta) = \cos(|n|\,\beta), \qquad \sin(-|n|\,\beta) = -\sin(|n|\,\beta).$$

For these negative $-|n|$ we obtain from (43)

$$h(x) = \frac{r^{|n|}}{R^{|n|}} \cos(|n| \beta), \qquad h(x) = -\frac{r^{|n|}}{R^{|n|}} \sin(-|n| \beta).$$

Then

$$b_1(\alpha) = \frac{R}{2(|n|+1)} \cos(-|n| \alpha) \text{ resp. } b_1(\alpha) = \frac{R}{2(|n|+1)} \sin(-|n| \alpha).$$

In order to compute $\hat{b}_0$ and $\hat{k}_0$ we calculate

$$\widehat{(\chi_0(\alpha)\, e^{in\alpha})}(\xi) = \int_{-\pi}^{+\pi} e^{in\alpha}\, e^{-i\alpha\xi}\, d\alpha$$

$$= \begin{cases} 2\,\dfrac{\sin(\pi(n-\xi))}{n-\xi} & \text{if} \quad \xi \neq n, \\ 2 & \text{if} \quad \xi = n. \end{cases} \tag{56}$$

Then we obtain

$\hat{k}_0(\xi)$	$\hat{b}_0(\xi)$		
$2\,\dfrac{\sin(\pi(\xi-n))}{\xi-n}$	$\dfrac{R}{2}\,\dfrac{1}{\lvert n\rvert+1}\,2\,\dfrac{\sin(\pi(\xi-n))}{\xi-n},$	$\xi \neq n$	(57)
2	$\dfrac{R}{2}\,\dfrac{1}{\lvert n\rvert+1}\,2\pi,$	$\xi = n$	

If we transform a convergent Fourier series

$$k_0(\beta) = \sum c_k\, e^{in\beta} \tag{58}$$

we get

$$\hat{k}_0 = (2\pi) \sum c_n \delta_n. \tag{59}$$

In (54) we have proved the relation

$$\hat{b}_0(\xi) = R\hat{Q}_0(\xi)\, \hat{k}_0(\xi). \tag{60}$$

In order to determine the coefficients c_n we need the functions of (60) at the points $\xi = n$. From (57) and (60) it follows that

$$\hat{Q}_0(n) = \frac{1}{2}\,\frac{1}{|n|+1}. \tag{61}$$

b) In Theorem 5.11 we proved that the mapping $k \mapsto TK$ is one-to-one. Therefore the inverse $T^{-1} =: S$ exists and is also linear. From

$$b_0 = Tk_0 = R\tilde{Q}_0 * k_0$$

it follows that

$$\hat{b}_0(\xi) = R\hat{Q}_0(\xi)\,\hat{k}_0(\xi)$$

and

$$\hat{k}_0(\xi) = \big(1/R\hat{Q}_0(\xi)\big)\,\hat{b}_0(\xi)\,. \tag{62}$$

If $1/R\hat{Q}_0(\xi) =: \hat{S}_0(\xi)$ is known we get, using the inverse Fourier transform,

$$k_0(\beta) = (S_0 * b_0)\,(\beta)\,, \tag{63}$$

where S_0 is the inverse Fourier transform of $\hat{S}_0$. From $\hat{S}_0(\xi) = 1/\hat{Q}_0(\xi)$ and (61) it follows that

$$\hat{S}_0(n) = \frac{2}{R}\,(|n| + 1) \tag{64}$$

for all integers n. In order to compute S_0 by the inverse Fourier transform we need S_0 for all $\xi \in \mathbb{R}$. Taking into account (64) we use a representation of the form

$$S_0 = \frac{2}{R}\,(S_1 + S_2 * S_3) \tag{65}$$

such that

$$\hat{S}_0 = \frac{2}{R}\,(\hat{S}_1 + \hat{S}_2 \cdot \hat{S}_3)\,. \tag{66}$$

From (64) and (65) it follows that

$$\hat{S}_1(\xi) = 1 \quad \text{and} \quad S_1 = \delta_0\,. \tag{67}$$

Further, if

$$\hat{S}_2(\xi) = \xi^2 \quad \text{and} \quad \hat{S}_3(\xi) = \sum_{\substack{n=-\infty\\ n\neq 0}}^{+\infty} \frac{1}{|n|}\,\frac{\sin \pi(\xi - n)}{\xi - n}$$

we get for $\xi = n$

$$\hat{S}_2(n) \cdot \hat{S}_3(n) = n^2 \cdot \frac{1}{|n|} = |n|\,. \tag{68}$$

The inverse Fourier transform of S_2 is

$$S_2 = -\delta_0''\,. \tag{69}$$

S_3 is to be computed by

$$S_3(\alpha) = \frac{1}{2\pi} \int\limits_{-\infty}^{+\infty} \hat{S}_3(\xi)\, e^{i\alpha\xi}\, d\xi = \frac{1}{2\pi} \sum_{\substack{n=-\infty\\ n\neq 0}}^{+\infty} \frac{1}{|n|} \int\limits_{-\infty}^{+\infty} \frac{\sin\big(\pi(\xi - n)\big)}{(\xi - n)}\, e^{i\alpha\xi}\, d\xi$$

$$= \frac{1}{2\pi} \sum_{\substack{n=-\infty\\ n\neq 0}}^{+\infty} \frac{1}{|n|}\, \chi_0(\alpha)\, e^{in\alpha} = \frac{1}{\pi}\, \chi_0(\alpha) \sum_{n=1}^{\infty} \frac{1}{n} \cos n\alpha\,. \tag{70}$$

Now let us study the convergence of the series

$$S_4(\alpha) = \sum_{n=1}^{\infty} \frac{1}{n} \cos n\alpha .$$

Introducing the notation $z_0 = e^{i\alpha}$ we get

$$S_4(\alpha) = \operatorname{Re} \sum_{n=1}^{\infty} \frac{1}{n} z_0^n .$$

The complex power series

$$S_5(z) = \sum_{n=1}^{\infty} \frac{1}{n} z^n$$

is convergent inside the unit circle, i.e., for $|z| < 1$. For these z we obtain

$$S_5(z) = \log \frac{1}{1-z} .$$

In complex analysis it is proved that this series is convergent for all z satisfying $|z| = 1$, $z \neq 1$. Therefore

$$S_4(\alpha) = \operatorname{Re} \log \frac{1}{1 - e^{i\alpha}} = \log \frac{1}{|1 - e^{i\alpha}|} = \frac{1}{2} \log \frac{1}{2(1 - \cos \alpha)}, \qquad \alpha \neq 2k\pi . \tag{71}$$

From (70) and (71) it follows that

$$S_3(\alpha) = -\frac{1}{2\pi} \chi_0(\alpha) \ln \big(2(1 - \cos \alpha)\big) . \tag{72}$$

From (65), (67), (69), and (72) we get formula (46). ∎

Finally, some additional remarks on the inverse source problem of Example 2.1 and the inverse source problem in gravimetry of Example 2.2. At first let $\Omega = B(x^0, R) \subset \mathbb{R}^n$ be a ball and μ a measure on $\bar{\Omega}$, such that the swept-out measure has the form μ_σ, where σ is a function defined on $\partial\Omega$ (see (4.1.10)). In the case of a ball in $\mathbb{R}^2$ we assume that σ has continuous derivatives of second order relative to the angle β. These conditions are fulfilled if supp $\mu \subset \Omega$. Then the boundary values ϱ_0 of the density ϱ are given by formula (45), where $k_1 = \varrho_0$ and $b_1 = \sigma$. In that case there exists to μ a uniquely determined harmonic density ϱ producing outside Ω the same potential as μ. For more general measures see Kapanadze [1] and for more general domains Sanso et al. [1]. If we have to study the inverse source problem in gravimetry, a measure μ concentrated on $\bar{\Omega}$ is to be determined such that

$$\operatorname{grad} G_L\mu|_{\partial\Omega} = g \tag{73}$$

holds. For such measures μ the first partial differential derivatives of $G_L\mu$ on $\partial\Omega$ have to exist and should be continuous. The classical mass distributions

μ_ϱ satisfy these conditions. If for a further measure ν concentrated on $\bar{\Omega}$ the first partial derivatives of $G_L\nu$ exist on $\partial\Omega$ and are continuous, then from

$$\operatorname{grad} G_L\nu(x) = \operatorname{grad} G_L\mu(x), \qquad x \notin \Omega, \tag{74}$$

it follows that (73) is also fulfilled. One can construct such measures ν by applying the balayage principle relative to a ball $B(x^0, r) \subset \Omega \subset \mathbb{R}^n$. Special mass distributions μ satisfying (73) are constructed in Škorvanek [1] and Zidarov [1, 2, 3]. In the publications of Zidarov one can find an extensive bibliography and historical remarks.

Remark 5.6. The inverse source problem in gravimetry cannot be uniquely solved. As a first step one has to study all measures $\nu \in \mathfrak{B}(\mu)$ for which the gradient of $G_L\nu$ exists on $\partial\Omega$ and which are continuous. Additional conditions are necessary to uniquely determine a $\nu_0 \in \mathfrak{B}(\mu)$.

5.6. An inverse source problem for the heat equation

This section deals with the treatment of an inverse source problem for the heat equation relative to the closed infinite strip $\bar{\Omega}_{(0,T)}$ (see Fig. 5.12), where

$$\Omega_{(0,T)} := \{(x, t) \in \mathbb{R}^{n+1} : 0 < t < T\}. \tag{1}$$

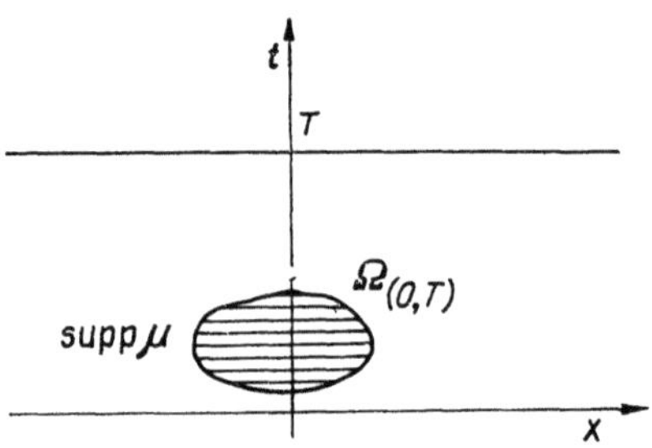

Fig. 5.12

Given a positive measure μ on this strip, we consider the convex and weakly compact set $\mathfrak{B}_H(\mu)$ of all positive measures on this strip producing the same potential for $t > T$ as the given measure μ (see Example 2.15). In the case considered the potentials $G_H\nu$ are analytic for $t > T$. This property is often used in our considerations. The extremal elements of $\mathfrak{B}_H(\mu)$, which are characterized by an approximation theory (Theorem 5.13), are of special interest. The set $\mathfrak{B}_H(\delta_0)$, where δ_0 is the Dirac measure at $(x', t') = (0, 0)$, is studied from a systematic point of view (Anger [10], Anger and Czerner [1]). There are two classes of extremal measures. The first class consists of the harmonic measures $\tilde{\mu}_{(x',t'),\tau}$, which solve the Cauchy problem relative to Ω_τ^* at $(x', t') = (0, 0)$ (Anger [12]), the second class consists of measures whose support has two parts (see also Sect. 5.2 and Sect. 5.4). The support of the first measure is contained in the strip $\Omega_{(0,T)}$, whereas the support of the second measure is concentrated on the boundary $\partial\Omega_T^*$ of the strip. These special measures must have a physical interpretation, especially in diffusion. It is an open problem to study the inverse source problem considered relative to a cylinder $\Omega = \Omega_0 \times (0, T)$, or more generally, an open set $\Omega \subset \mathbb{R}^{n+1}$ (DOOB [1], BLIEDTNER and HANSEN [1]). Here $\Omega_0 \subset \mathbb{R}^n$ is a given domain. In this case the potentials $G_{\mathrm{H}}\mu$ are not analytic. A further open problem is the following: Do relations exist between the extremal elements of the heat equation and the Laplace equation, whose solutions are time-independent solutions of the heat equation? Using time-depending equations we get more information on the extremal elements (ANGER [13, 16]).

In this section we also apply the ideas developed for the Laplace equation. Instead of the heat potential

$$G_{\mathrm{H}}\mu(\tilde{x}) = \int G_{\mathrm{H}}(\tilde{x}, \tilde{z})\, \mathrm{d}\mu(\tilde{z}), \qquad \tilde{x} = (x, t), \qquad \tilde{z} = (z, \tau), \tag{2}$$

we consider the linear form

$$\mu(G_{\mathrm{H}}^*\lambda) = \int G_{\mathrm{H}}^*\lambda\, \mathrm{d}\mu = \int G_{\mathrm{H}}\mu\, \mathrm{d}\lambda. \tag{3}$$

Here

$$G_{\mathrm{H}}^*\lambda(\tilde{z}) = \int G_{\mathrm{H}}^*(\tilde{z}, \tilde{x})\, \mathrm{d}\lambda(\tilde{x}) = \int G_{\mathrm{H}}(\tilde{x}, \tilde{z})\, \mathrm{d}\lambda(\tilde{x}) \tag{4}$$

is the adjoint heat potential, which is a solution of the adjoint heat equation if $\tilde{z} \notin \operatorname{supp} \lambda$.

Following Theorem 3.2 the solution of the initial-value problem (Cauchy problem) relative to Ω_τ is given by

$$u(x, t) = \int_{\mathbb{R}^n} G_{\mathrm{H}}(x, t; y, \tau)\, \Phi(y)\, \mathrm{d}y_1 \dots \mathrm{d}y_n, \qquad t > \tau, \tag{5}$$

and for the adjoint heat equation relative to Ω_τ^* by

$$u(x, t) = \int_{\mathbb{R}^n} G_{\mathrm{H}}(y, \tau; x, t)\, \Phi(y)\, \mathrm{d}y_1 \dots \mathrm{d}y_n, \qquad t < \tau. \tag{6}$$

We introduce the following notations:

$$\mathrm{d}\mu_{(x,t),\tau}(\tilde{y}) := G_{\mathrm{H}}(x, t; y, \tau)\, \mathrm{d}y_1 \dots \mathrm{d}y_n, \qquad t > \tau, \tag{7}$$

and

$$\mathrm{d}\tilde{\mu}_{(x,t),\tau}(\tilde{y}) := G_{\mathrm{H}}(y, \tau; x, t)\, \mathrm{d}y_1 \dots \mathrm{d}y_n, \qquad t < \tau. \tag{8}$$

Then (5) can be written in the form

$$u(x, t) = \int \Phi(y)\, \mathrm{d}\mu_{(x,t),\tau}(\tilde{y}) \tag{5'}$$

and (6) in the form

$$u(x, t) = \int \Phi(y)\, \mathrm{d}\tilde{\mu}_{(x,t),\tau}(\tilde{y}). \tag{6'}$$

The measures $\mu_{(x,t),\tau}$ and $\tilde{\mu}_{(x,t),\tau}$ correspond to the harmonic measures μ_x of the Laplace equation.

For solutions of the adjoint heat equation we define

$$\begin{aligned} Af(z, \tau) &:= \int f(y, T) \, \mathrm{d}\tilde{\mu}_{(z,\tau),T}(\tilde{y}), \qquad \tau < T, \\ Af(z, \tau) &:= f(z, \tau), \qquad \tau = T. \end{aligned} \tag{9}$$

Let $\mu \geqq 0$ be a measure satisfying $\operatorname{supp} \mu \subset \bar{\Omega}_{(0,T)}$ and $\int \mathrm{d}\mu = \|\mu\| < \infty$. In the following we consider measures $\nu \geqq 0$ satisfying

$$\operatorname{supp} \nu \subset \bar{\Omega}_{(0,T)}, \qquad G_{\mathrm{H}}\nu(x, t) = G_{\mathrm{H}}\mu(x, t), \qquad t > T. \tag{10}$$

Further, let λ be a measure on $\partial\Omega_{T_1}$, $T_1 > T$. We integrate (10) with respect to $\mathrm{d}\lambda$ and obtain

$$\int G_{\mathrm{H}}\mu \, \mathrm{d}\lambda = \int G_{\mathrm{H}}^{*}\lambda \, \mathrm{d}\mu = \int G_{\mathrm{H}}\nu \, \mathrm{d}\lambda = \int G_{\mathrm{H}}^{*}\lambda \, \mathrm{d}\nu. \tag{11}$$

Special measures λ have the form

$$\mathrm{d}\lambda_{\Phi}(x) = \Phi(x, T_1) \, \mathrm{d}x_1 \ldots \mathrm{d}x_n, \tag{12}$$

where $x \mapsto \Phi(x, T_1)$ is a bounded and continuous function. Then

$$G_{\mathrm{H}}^{*}\lambda(\tilde{z}) \to \Phi(x, T_1) \quad \text{for} \quad \tilde{z} = (z, \tau) \to (x, T_1).$$

We denote by $\mathbf{C}_{\omega}(\bar{\Omega}_{(0,T)})$ the space of all functions continuous on $\bar{\Omega}_{(0,T)}$ tending to zero at infinity. Further, let $D(\bar{\Omega}_{(0,T)})$ be the space of all functions $u \in \mathbf{C}_{\omega}(\bar{\Omega}_{(0,T)})$, which are solutions of the adjoint heat equation. The dual space $\mathbf{C}_{\omega}^{*}(\Omega_{(0,T)})$ consists of all measures μ satisfying $\operatorname{supp} \mu \subset \bar{\Omega}_{(0,T)}$ and $\|\mu\| < \infty$ (see also (3.3.33)–(3.3.35)).

Proposition 5.7. *Let μ be a measure with compact support. Then $G_{\mathrm{H}}\lambda(\tilde{x}) \to 0$ for $\tilde{x} = (x, t) \to \omega$, where ω is the point at infinity. The same property holds relative to the adjoint potential $G_{\mathrm{H}}^{*}\lambda$.*

Proof. Let $\tilde{x} = (x, t) \in \mathbb{R}^{n+1}$. The fundamental solution has the form

$$E(x, t) := G_{\mathrm{H}}(x, t; 0, 0) = \left(\frac{1}{2\sqrt{\pi t}}\right)^{n} \exp\left(-\frac{\sum_{j=1}^{n} x_j^2}{4t}\right), \qquad t > 0,$$

$$E(x, t) = 0 \quad \text{otherwise.}$$

We consider E on the special set

$$\partial\Omega_{[0,T]}^{*} := \{(x, t) : t = T, |x| \leqq T\} \cup \{(x, t) : |x| = T, 0 \leqq t \leqq T\}.$$

On the first set we have $|x| = rT$, $0 \leqq r \leqq 1$. Then

$$E(x, T) = \left(\frac{1}{2\sqrt{\pi T}}\right)^{n} \exp\left(-\frac{r^2 T}{4}\right) \leqq \left(\frac{1}{2\sqrt{\pi T}}\right)^{n}.$$

On the second set we have

$$E(x, t) = \left(\frac{1}{2\sqrt{\pi t}}\right)^{n} \exp\left(-\frac{T^2}{4t}\right), \qquad t > 0.$$

Its maximum is attained for $t_0 = T^2/2n = T(T/2n) > T$ if $T > 2n$. For these values we obtain on $\partial\Omega^*_{[0,T]}$

$$E(x, t) \leqq \left(\frac{1}{2\sqrt{\pi T}}\right)^n. \tag{13}$$

From the last inequality it follows that

$$|G_H\lambda(\tilde{x})| \leqq |G_H\lambda^+(\tilde{x})| + |G_H\lambda^-(x)| = \int G_H(\tilde{x}, \tilde{y})\,\mathrm{d}(\lambda^+ + \lambda^-)(\tilde{y})$$
$$\leqq \int \left(\frac{1}{2\sqrt{\pi(t-\tau)}}\right)^n \mathrm{d}(\lambda^+ + \lambda^-)(\tilde{y}) \quad \text{if} \quad |\tilde{x}| > R_0, \qquad \tilde{y} = (y, \tau),$$

hence

$$G_H\lambda(\tilde{x}) \to 0 \quad \text{for} \quad \tilde{x} = (x, t) \to \omega. \; \blacksquare$$

For a positive measure μ satisfying supp $\mu \subset \bar{\Omega}_{(0,T)}$ and $\|\mu\| < \infty$ we define

$$\begin{aligned} \mathfrak{B}_H(\mu) := \{\nu \geqq 0 \colon &\operatorname{supp} \nu \subset \bar{\Omega}_{(0,T)} \quad \text{compact}, \\ &\int u \,\mathrm{d}\nu = \int u \,\mathrm{d}\mu \quad \text{for all} \quad u \in D(\bar{\Omega}_{(0,T)})\}. \end{aligned} \tag{14}$$

If $u(\tilde{z}) = G_H(x, \tilde{z})$ from (14) follows (10).

Proposition 5.8. *Let* $\nu \in \mathfrak{B}_H(\mu)$. *Then* $\int \mathrm{d}\nu = \int \mathrm{d}\mu$.

Proof. Integrating the relation $G_H\nu(\tilde{x}) = G_H\mu(\tilde{x})$ for $t > T$ with respect to $\mathrm{d}\lambda_\Phi$, $\Phi(x, T_1) \equiv 1$, $T_1 > T$, defined in (12) we obtain from (11)

$$\int G^*_H\lambda_\Phi \,\mathrm{d}\nu = \int G^*_H\lambda_\Phi \,\mathrm{d}\mu.$$

Using the maximum principle we get $G^*_H\lambda_\Phi(z) = 1$ if $\tau < T_1$. From this it follows that

$$\int \mathrm{d}\nu = \int \mathrm{d}\mu. \; \blacksquare$$

Proposition 5.9. *The set* $\mathfrak{B}_H(\mu)$ *is convex and weakly compact.*

Proof. From

$$\nu = r\nu_1 + (1 - r)\,\nu_2, \qquad \nu_1, \nu_2 \in \mathfrak{B}_H(\mu), \qquad 0 \leqq r \leqq 1,$$

it follows that

$$G_H\nu(\tilde{x}) = rG_H\nu_1(\tilde{x}) + (1 - r)\,G_H\nu_2(\tilde{x}) = rG_H\mu(\tilde{x}) + (1 - r)\,G_H\mu(\tilde{x}) = G_H\mu(\tilde{x})$$

if $t > T$. Since $\mathfrak{B}_H(\mu) \subset \mathbf{C}^*_\omega(\bar{\Omega}_{(0,T)})$ we can introduce on $\mathfrak{B}_H(\mu)$ the weak topology like in Definition 4.1. The further proof is similar to the proof in Theorem 5.2. $\blacksquare$

Let

$$H_G(\bar{\Omega}_{(0,T)}) := \left\{f \in \mathbf{C}_\omega(\bar{\Omega}_{(0,T)}) \colon f(x) = \sum_{k=1}^{N} c_k G_H(x^k, x),\, x^k \notin \bar{\Omega}_{(0,T)}\right\}. \tag{15}$$

Similar to Theorem 5.3 we can prove

Theorem 5.13. *A measure $\nu \in \mathfrak{B}_{\mathrm{H}}(\mu)$ is extremal if and only if $H_G(\bar{\Omega}_{(0,T)})$ is dense in $\mathbf{L}^1(\nu)$ relative to the norm* (5.3.11).

Theorem 5.14. *Let $(x', t') = (0, 0)$ be the origin in $\mathbb{R}^{n+1}$. The measures $\tilde{\mu}_{(x',t'),\tau}$, $\tau < T$, defined in* (8) *are extremal elements of $\mathfrak{B}_{\mathrm{H}}(\delta_0)$.*

Proof. If the measure $\nu := \tilde{\mu}_{(x',t'),\tau}$ is not extremal, then

$$\nu = r\nu_1 + (1 - r)\,\nu_2, \qquad \nu_1, \nu_2 \in \mathfrak{B}_{\mathrm{H}}(\delta_0), \qquad \nu_1 \neq \nu_2, \qquad 0 < r < 1.$$

From

$$\nu \geqq r\nu_1 \quad \text{and} \quad \nu \geqq (1 - r)\,\nu_2$$

it follows that $\operatorname{supp} \nu_1 \subset \partial\Omega_T^*$ and $\operatorname{supp} \nu_2 \subset \partial\Omega_T^*$. Since

$$G_{\mathrm{H}}\nu_1(\tilde{x}) = G_{\mathrm{H}}\nu_2(\tilde{x}) = G_{\mathrm{H}}\delta_0(\tilde{x}) = G_{\mathrm{H}}(\tilde{x}, 0), \qquad t > T, \tag{16}$$

we can analytically continue the potentials $G_{\mathrm{H}}\nu_j$ up to the hyperplane $t = \tau$. For $t < \tau$ these potentials are identically zero. The set $\partial\Omega_T^*$ has Lebesgue measure zero. Therefore we can integrate

$$G_{\mathrm{H}}\nu_1(\tilde{x}) = G_{\mathrm{H}}\nu_2(\tilde{x}), \qquad t \neq T,$$

with respect to

$$P^*(\partial)\,\varphi(\tilde{x})\,\mathrm{d}x := \big(-\partial_t\varphi(\tilde{x}) - \Delta\varphi(\tilde{x})\big)\,\mathrm{d}\tilde{x}, \qquad \varphi \in C_0^\infty(\mathbb{R}^{n+1}),$$

and get

$$\int G_{\mathrm{H}}\nu_1(\tilde{x})\,P^*(\partial)\,\varphi(\tilde{x})\,\mathrm{d}\tilde{x} = \int G_{\mathrm{H}}\nu_2(\tilde{x})\,P^*(\partial)\,\varphi(\tilde{x})\,\mathrm{d}\tilde{x}.$$

From

$$\begin{aligned}\int G_{\mathrm{H}}\nu_j(\tilde{x})\,P^*(\partial)\,\varphi(\tilde{x})\,\mathrm{d}\tilde{x} &= \int \Big(\int G_{\mathrm{H}}(\tilde{x}, \tilde{z})\,\mathrm{d}\nu_j(\tilde{z})\Big)\,P^*(\partial)\,\varphi(\tilde{x})\,\mathrm{d}\tilde{x} \\ &= \int \Big(\int G_{\mathrm{H}}(\tilde{x}, \tilde{z})\,P^*(\partial)\,\varphi(\tilde{x})\,\mathrm{d}\tilde{x}\Big)\,\mathrm{d}\nu_j(\tilde{z}) = \int \varphi(\tilde{z})\,\mathrm{d}\nu_j(\tilde{z})\end{aligned}$$

it follows that

$$\int \varphi(\tilde{z})\,\mathrm{d}\nu_1(\tilde{z}) = \int \varphi(\tilde{z})\,\mathrm{d}\nu_2(\tilde{z}) \quad \text{for all} \quad \varphi \in \mathbf{C}_0^\infty(\mathbb{R}^{n+1}). \tag{17}$$

By Theorem 3.13 this relation holds for each $\varphi \in \mathbf{C}_0(\mathbb{R}^{n+1})$, i.e., $\nu_1 = \nu_2$. ∎

Let

$$\langle f, \mu\rangle := \int f\,\mathrm{d}\mu, \qquad f \in \mathbf{C}_\omega(\bar{\Omega}_{(0,T)}), \qquad \mu \in \mathbf{C}_\omega^*(\bar{\Omega}_{(0,T)}). \tag{18}$$

The measures in (18) satisfy $\|\mu\| < \infty$. Definition (18) also holds for each bounded $f \in \mathbf{C}(\bar{\Omega}_{(0,T)})$. The mapping A defined in (9) is continuous, i.e.,

$$\|Af\|_{\mathbf{C}} \leqq M\,\|f\|_{\mathbf{C}}. \tag{19}$$

From

$$|\langle Af, \mu\rangle| \leqq \|\mu\|\,\|Af\|_{\mathbf{C}} \leqq \|\mu\|\,\|A\|\,\|f\|_{\mathbf{C}}$$

follows the existence of the adjoint mapping

$$A^*: \mathbf{C}^*_\omega(\bar{\Omega}_{(0,T)}) \mapsto \mathbf{C}^*_\omega(\bar{\Omega}_{(0,T)}), \quad \text{which satisfies}$$
$$\langle Af, \mu \rangle = \langle f, A^*\mu \rangle. \tag{20}$$

Let

$$J(\partial\Omega^*_T) := \{f \in \mathbf{C}(\bar{\Omega}_{(0,T)}) : f(y) = 0 \text{ on } \partial\Omega^*_T\}. \tag{21}$$

We get

$$Af_2 = 0 \quad \text{for all} \quad f_2 \in J(\partial\Omega^*_T), \tag{22}$$

hence

$$\operatorname{supp} A^*\mu \subset \partial\Omega^*_T. \tag{23}$$

From (19) it follows that

$$\langle Af, \nu \rangle = \int Af \, d\nu = \int \left(\int G_H(\tilde{x}, \tilde{z}) \, f(x, T) \, dx \, d\nu(\tilde{z}) \right.$$
$$= \int \left(\int G_H(\tilde{x}, \tilde{z}) \, d\nu(\tilde{z}) \right) f(x, T) \, dx = \int f(x, T) \, dA^*\nu(\tilde{x}), \tag{24}$$

where

$$dA^*\nu(\tilde{x}) := \left(\int G_H(\tilde{x}, \tilde{z}) \, d\nu(\tilde{z}) \right) dx \tag{25}$$

is the so-called *swept-out measure* of ν. If

$$\nu = \nu_1 + \nu_2, \qquad \operatorname{supp} \nu_1 \subset \bar{\Omega}_{(0,T)}, \qquad \operatorname{supp} \nu_2 \subset \partial\Omega^*_T,$$

then

$$A^*\nu(f) = \int \left(\int G_H(\tilde{x}, \tilde{z}) \, d\nu(\tilde{z}) \right) f(x, T) \, dx + \int f(\tilde{x}) \, d\nu_2(\tilde{x}). \tag{26}$$

The following considerations are similar to Sect. 5.2. We consider the measure

$$\nu = \nu_K + \gamma \in \mathfrak{B}_H(\delta_0), \tag{27}$$

where $\operatorname{supp} \nu_K \subset K \subset \Omega_{(0,T)}$, K compact. Since

$$G_H\delta_0(\tilde{x}) = G_H\mu_{(0,0),T}(\tilde{x}) = (G_H A^*\nu_K)(\tilde{x}) + G_H\gamma(\tilde{x}), \qquad t > T, \tag{28}$$

we get

$$G_H\gamma(\tilde{x}) = G_H(\tilde{\mu}_{(0,0),T} - A^*\nu_K)(\tilde{x}), \qquad t > T. \tag{29}$$

The measure

$$\gamma(f) = \int f(x, T) \left(G_H(\tilde{x}, 0) - \int G_H(\tilde{x}, \tilde{z}) \, d\nu_K(\tilde{z}) \right) dx = \int g_0(x, T) \, dx \tag{30}$$

is uniquely determined. Here

$$g_0(x, T) := G_H(\tilde{x}, 0) - \int G_H(\tilde{x}, \tilde{z}) \, d\nu_K(\tilde{z}), \qquad \tilde{x} = (x, T). \tag{31}$$

Afterwards we shall consider the two examples

$$\nu_K = a\delta_{\tilde{z}}, \qquad \tilde{z} = (z, \tau), \tag{32}$$

$$\nu_K = a_1\delta_{\tilde{z}^1} + a_2\delta_{\tilde{z}^2}, \qquad z^1 = (|z|, \tau), \qquad z^2 = (-|z|, \tau). \tag{33}$$

Proposition 5.10. *Let $g_0 : \partial\Omega_T^* \mapsto \mathbb{R}$ defined in* (31) *be strictly positive. Then there exists a constant $b > 0$ such that*

$$g_0(x, T) \geqq b \int G_{\mathrm{H}}(\tilde{x}, \tilde{z}) \, \mathrm{d}\nu_K(\tilde{z}) \quad \text{on } \partial\Omega_T^*. \tag{34}$$

Proof. Let

$$\tilde{g}_a(x, T) := G_{\mathrm{H}}(\tilde{x}, 0) - a \int G_{\mathrm{H}}(\tilde{x}, \tilde{z}) \, \mathrm{d}\nu_K(\tilde{z}) = G_{\mathrm{H}}(\tilde{x}, 0) \left(1 - ah(x)\right), \qquad a > 0, \tag{35}$$

where

$$h(x) = \int \left(G_{\mathrm{H}}(\tilde{x}, \tilde{z})/G_{\mathrm{H}}(\tilde{x}, 0)\right) \mathrm{d}\nu_K(\tilde{z}). \tag{36}$$

If $a = 1$, then $\tilde{g}_1 = g_0$ on $\partial\Omega_T^*$; if $a < 1$, then $\tilde{g}_a > g_0$. From

$$G_{\mathrm{H}}(\tilde{x}, \tilde{z})/G_{\mathrm{H}}(\tilde{x}, 0) = C(T, \tau) \exp \left(-\tau \sum_{j=1}^n x_j^2 + 2T \sum_{j=1}^n z_j x_j - T \sum_{j=1}^n z_j^2\right) \frac{1}{4(T - \tau)\, T}$$

it follows that the function h tends to zero for $|x| \to \infty$. From $g_0 > 0$ we obtain the inequality $1 > h(x) > 0$. There exists at least one point $x^0 \in \partial\Omega_T^*$, such that

$$\sup \{h(x), x \in \partial\Omega_T^*\} = h(x^0).$$

The function h is analytic for $\operatorname{supp} \nu_K \subset \Omega_{(0,T)}$. In the case $\Omega_{(0,T)} \subset \mathbb{R}^2$ the function h is an analytic function of the real variable x vanishing at infinity. Therefore the relation $h(x) = h(x^0)$ holds only for finite many points $x^1, \ldots, x^m$. If $a_0 = 1/h(x^0)$, then $\tilde{g}_{a_0}(x^0) = 0$. From $\tilde{g}_1 > 0$ it follows that $a_0 = 1 + b, b > 0$, and $\tilde{g}_{a_0} \geqq 0$. Finally we obtain

$$\begin{aligned} g_0(x, T) &= \tilde{g}_1(x, T) = G_{\mathrm{H}}(x, 0) \left(1 - a_0 h(x) + bh(x)\right) \\ &= \tilde{g}_{a_0}(x, T) + b \int G_{\mathrm{H}}(\tilde{x}, \tilde{z}) \, \mathrm{d}\nu_K(\tilde{z}) \geqq b \int G_{\mathrm{H}}(\tilde{x}, \tilde{z}) \, \mathrm{d}\nu_K(\tilde{z}). \ \blacksquare \end{aligned}$$

Proposition 5.11. *Let $g_0 : \partial\Omega_T^* \mapsto \mathbb{R}$ be strictly positive and $u_k \in H_G(\bar{\Omega}_{(0,T)})$ satisfying*

$$\int |u_k(\tilde{x})| \, \mathrm{d}\gamma(\tilde{x}) = \int |u_k(\tilde{x})| \, g_0(x, T) \, \mathrm{d}x \to 0 \quad \textit{for} \quad k \to 0. \tag{37}$$

Then

$$\int |u_k(\tilde{z})| \, \mathrm{d}\nu_K(\tilde{z}) \to 0 \quad \textit{and} \quad u_k(0) \to 0 \quad \text{for} \quad k \to \infty. \tag{38}$$

Proof. From Proposition 5.10 it follows that

$$\begin{aligned} \int |u_k(\tilde{z})| \, \mathrm{d}\nu_K(\tilde{z}) &= \int \left| \int G_{\mathrm{H}}(\tilde{x}, \tilde{z}) \, u_k(\tilde{x}) \, \mathrm{d}x \right| \mathrm{d}\nu_K(\tilde{z}) \leqq \int \left(\int G_{\mathrm{H}}(\tilde{x}, \tilde{z}) \, |u_k(\tilde{x})| \, \mathrm{d}x \right) \mathrm{d}\nu_K(\tilde{z}) \\ &= \int \left(\int G_{\mathrm{H}}(\tilde{x}, \tilde{z}) \, \mathrm{d}\nu_K(\tilde{z}) \right) |u_k(\tilde{x})| \, \mathrm{d}x \leqq \frac{1}{b} \int g_0(x, T) \, |u_k(\tilde{x})| \, \mathrm{d}x. \end{aligned}$$

Therefore from

$$\int g_0(x, T) \, |u_k(\tilde{x})| \, \mathrm{d}x \to 0 \quad \text{we obtain} \quad \int |u_k(\tilde{z})| \, \mathrm{d}\nu_K(\tilde{z}) \to 0 \quad \text{for} \quad k \to \infty.$$

Furthermore, the following inequalities hold:

$$\begin{aligned}|u_k(0)| &= \left|\int G_H(\tilde{x}, 0)\, u_k(\tilde{x})\, dx\right| = \left|\int \left(g_0(x, T) + \int G_H(\tilde{x}, \tilde{z})\, d\nu_K(\tilde{z})\right) u_k(\tilde{x})\, dx\right| \\ &= \left|\int g_0(x, T)\, u_k(\tilde{x})\, dx + \int u_k(\tilde{z})\, d\nu_K(\tilde{z})\right| \\ &\leqq \int g_0(x, T)\, |u_k(\tilde{x})|\, dx + \int |u_k(\tilde{z})|\, d\nu_K(\tilde{z}).\end{aligned}$$

From this relation we obtain $u_k(0) \to 0$ for $k \to \infty$. ▮

Theorem 5.15. *Let $g_0 \geqq 0$ and $\nu = \nu_K + \gamma \in \mathfrak{B}(\delta_0)$ be an extremal element. If $g_0 > 0$, then $\nu_K = 0$, if $\nu_K \neq 0$, then g_0 must have at least one root.*

Proof. The measure ν is an extremal element of $\mathfrak{B}(\delta_0)$ if and only if $H_G(\bar{\Omega}_{(0,T)})$ is dense in $\mathbf{L}^1(\nu)$. We consider the special function $f \in \mathbf{L}^1(\nu)$,

$$f(\tilde{x}) = 0 \quad \text{on } \partial\Omega_T^* \quad \text{and} \quad f(\tilde{z}) = 1 \quad \text{on supp } \nu_K.$$

Then there exists a sequence of functions $u_k \in H_G(\bar{\Omega}_{(0,T)})$ such that

$$\int |0 - u_k(\tilde{x})|\, g_0(x, T)\, dx \to 0 \quad \text{and} \quad \int |1 - u_k(\tilde{z})|\, d\nu_K(\tilde{z}) \to 0.$$

Let $g_0 > 0$. From Proposition 5.11 it follows that

$$\int u_k(\tilde{z})\, d\nu_K(\tilde{z}) \to 0 \quad \text{while} \quad \int u_k(\tilde{z})\, d\nu_K(\tilde{z}) \to \int d\nu_K(\tilde{z}) \neq 0.$$

This is a contradiction. In the case $g_0 > 0$ and $\nu \in \text{ex}\, \mathfrak{B}(\delta_0)$, the measure ν_K must be zero. If $\nu_K \neq 0$ the function g_0 must have at least one root. ▮

Similar to Theorem 5.6 one can prove the following theorem (Anger and Czerner [1]) in $\mathbb{R}^2$:

Theorem 5.16. *Let $g_0 : \partial\Omega_T^* \mapsto \mathbb{R}$ be the nonnegative function defined in (31). If g_0 has a root at $\tilde{x}^0 \in \partial\Omega_T^*$ and $g_0(x, T) = (x - x^0)^2\, \tilde{R}_2(x, x^0)$, $\tilde{R}_2(x^0, x^0) \neq 0$, then there exists a sequence of functions $u_k \in H_G(\bar{\Omega}_{(0,T)})$ such that*

$$\int g_0(x, T)\, |u_k(\tilde{x})|\, dx \to 0 \quad \textit{for} \quad k \to \infty \tag{39}$$

and

$$u_k(\tilde{z}) \to d_1 G_H(\tilde{x}^0, \tilde{z}) + d_2 \frac{\partial G_H(\tilde{x}^0, \tilde{z})}{\partial x}. \tag{40}$$

Now we shall study the special measures defined in (32) and (33). Let $\tilde{z} = (z, \tau) \in \Omega_{(0,T)} \subset \mathbb{R}^2$. We consider measures ν of the form (Anger [12])

$$\nu_a := a\delta_{\tilde{z}} + \gamma \in \mathfrak{B}(\delta_0), \tag{41}$$

here $a \geqq 0$ and $d\gamma(\tilde{x}) = \tilde{g}_a(x, T)\, dx$, where $\tilde{g}_a$ was defined in (35) in this special case having the form

$$\tilde{g}_a(x, T) := G_H(\tilde{x}, 0) - aG_H(\tilde{x}, \tilde{z}) = G_H(\tilde{x}, 0)\left(1 - ah(x)\right) \geqq 0, \tag{42}$$

$$h(x) = \frac{1}{(1 - \tau/T)^{1/2}} \exp\left(-\frac{\tau x^2 - 2Txz + Tz^2}{4(T - \tau)\, T}\right).$$

From $\tilde{g}_a(x_0, T) = 0$ it follows that $h(x_0) = 1/a_0$ and

$$\exp\left(\frac{\tau x_0^2 - 2Tx_0 z + Tz^2}{4(T-\tau)\,T}\right) = a_0/(1 - \tau/T)^{1/2}.$$

This is equivalent to

$$\tau x_0^2 - 2Tx_0 z + Tz^2 = 4T(T-\tau)\log a_0/(1-\tau/T)^{1/2}. \tag{43}$$

Further, from $\tilde{g}_{a_0}(x_0, T) = 0$ and $\tilde{g}_{a_0}(x, T) \geqq 0$ we obtain

$$h(x) \geqq \frac{1}{a_0} \quad \text{and} \quad h(x_0) = \frac{1}{a_0}.$$

Therefore $h'(x_0) = 0$, which is equivalent to

$$2\tau x_0 - 2Tz = 0, \qquad x_0 = \frac{Tz}{\tau}. \tag{44}$$

From (42) and (43) it follows that

$$-\frac{z^2}{4\tau} = \log a_0/(1-\tau/T)^{1/2}$$

and

$$a_0 = (1-\tau/T)^{1/2}\exp\left(-\frac{z^2}{4\tau}\right). \tag{45}$$

Theorem 5.17. *Let $\nu_a = a\delta_{\tilde{z}} + \gamma \in \mathfrak{B}_{\mathrm{H}}(\delta_0)$ be the measure defined in* (41), (42). *If $a_0 = (1-\tau/T)^{1/2}\exp(-z^2/4\tau)$, then $\tilde{g}_{a_0} \geqq 0$ has a root at $x_0 = Tz/\tau$. The measure ν_{a_0} is an extremal element of $\mathfrak{B}_{\mathrm{H}}(\delta_0)$.*

Proof. Following formula (30) the measure γ has the form

$$\mathrm{d}\gamma(\tilde{x}) = \tilde{g}_a(x, T)\,\mathrm{d}x, \qquad \tilde{g}_a(x, T) = G_{\mathrm{H}}(\tilde{x}, 0) - aG_{\mathrm{H}}(\tilde{x}, \tilde{z}) \geqq 0.$$

If ν_{a_0} is not an extremal element then

$$\nu_{a_0} = r\nu_1 + (1-r)\,\nu_2, \qquad \nu_1, \nu_2 \in \mathfrak{B}_{\mathrm{H}}(\delta_0), \qquad \nu_1 \neq \nu_2, \qquad 0 < r < 1.$$

From

$$\nu_{a_0} \geqq r\nu_1 \quad \text{and} \quad \nu_{a_0} \geqq (1-r)\,\nu_2$$

it follows that

$$\nu_1 = a_1\delta_{\tilde{z}} + \gamma_1, \qquad a_1 > 0, \qquad \tilde{g}_{a_1}(x, T) = G_{\mathrm{H}}(\tilde{x}, 0) - a_1 G_{\mathrm{H}}(x, \tilde{z}) \geqq 0,$$

$$\nu_2 = a_2\delta_{\tilde{z}} + \gamma_2, \qquad a_2 > 0, \qquad \tilde{g}_{a_2}(x, T) = G_{\mathrm{H}}(\tilde{x}, 0) - a_2 G_{\mathrm{H}}(\tilde{x}, \tilde{z}) \geqq 0,$$

where

$$a_0 = ra_1 + (1-r)\,a_2 \tag{46}$$

and

$$\tilde{g}_{a_0}(x, T) = r\tilde{g}_{a_1}(x, T) + (1 - r)\,\tilde{g}_{a_2}(x, T) \geqq 0 .$$

Since $\tilde{g}_{a_0}(x_0, T) = 0$ we get

$$\tilde{g}_{a_1}(x_0, T) = \tilde{g}_{a_2}(x_0, T) = 0 . \tag{47}$$

From (46) we obtain

$$a_1 < a_0 < a_2 \quad \text{or} \quad a_2 < a_0 < a_1 .$$

Then

$$\tilde{g}_{a_2}(x_0, T) > 0 \quad \text{or} \quad \tilde{g}_{a_1}(x_0, T) > 0 ,$$

which contradicts $\tilde{g}_{a_1}(x_0, T) = \tilde{g}_{a_2}(x_0, T) = 0$. ∎

Now we consider the following measures (ANGER and CZERNER [1]):

$$\nu_{a_1,a_2} := a_1\delta_{\tilde{z}^1} + a_2\delta_{\tilde{z}^2} + \gamma \in \mathfrak{B}_{\mathrm{H}}(\delta_0) , \qquad a_1, a_2 > 0 . \tag{48}$$

Here $z^1 = (|z|, \tau) \in \mathbb{R}^2$, $z^2 = (-|z|, \tau) \in \mathbb{R}^2$ (see Fig. 5.12). From formula (30) it follows that $\mathrm{d}\gamma(\tilde{x}) = \tilde{g}_{a_1,a_2}(x, T)\,\mathrm{d}x$, where

$$\tilde{g}_{a_1,a_2}(x, T) = \frac{1}{2\sqrt{\pi}}\left[\frac{1}{T^{1/2}}\exp\left(-\frac{x^2}{4T}\right) - \frac{a_1}{(T-\tau)^{1/2}}\exp\left(-\frac{(x-|z|)^2}{4(T-\tau)}\right)\right.$$
$$\left. - \frac{a_2}{(T-\tau)^{1/2}}\exp\left(-\frac{(x+|z|)^2}{4(T-\tau)}\right)\right] \geqq 0 . \tag{49}$$

In the further considerations we assume $a_1 = a_2 = a$. In this case $\tilde{g}_{a_1,a_2}(-x, T) = \tilde{g}_{a_1,a_2}(x, T)$. We determine a root x_0 satisfying

$$\tilde{g}_{a,a}(x_0, T) = 0 , \qquad \tilde{g}_{a,a}(x, T) \geqq 0 . \tag{50}$$

From (50) we obtain $\tilde{g}'_{a,a}(x_0, T) = 0$. The derivative has the form

$$-4\sqrt{\pi}\,\tilde{g}'_{a,a}(x, T) = \frac{x}{T^{3/2}}\exp\left(-\frac{x^2}{4T}\right) - a\,\frac{(x-|z|)}{(T-\tau)^{3/2}}\exp\left(-\frac{(x-|z|)^2}{4(T-\tau)}\right)$$
$$- a\,\frac{x+|z|}{(T-\tau)^{3/2}}\exp\left(-\frac{(x+|z|)}{4(T-\tau)}\right), \tag{51}$$

which is zero if $x_0 = 0$. Further, $\tilde{g}''_{a,a}(x_0, T) > 0$ is sufficient for a minimum. The second derivative has the form

$$4\sqrt{\pi}\,\tilde{g}''_{a,a}(0, T) = -\frac{1}{T^{3/2}} + \frac{a}{(T-\tau)^{3/2}}\left(2 - \frac{|z|^2}{(T-\tau)}\exp\left(-\frac{|z|^2}{4(T-\tau)}\right)\right)$$
$$= \frac{1}{T^{1/2}(T-\tau)}\left(\frac{\tau}{T} - \frac{|z|^2}{2(T-\tau)}\right) > 0 \tag{52}$$

if

$$a_0 = \frac{1}{2}(1 - \tau/T)^{1/2} \exp\left(\frac{|z|^2}{4(T-\tau)}\right) \tag{53}$$

and

$$|z| < \left(2\tau(1 - \tau/T)\right)^{1/2}. \tag{54}$$

In this case

$$\tilde{g}_{a_0,a_0}(0, T) = 0.$$

Theorem 5.18. *Let* $\nu_{a_1,a_2} = a_1\delta_{\tilde{z}^1} + a_2\delta_{\tilde{z}^2} + \gamma \in \mathfrak{B}_H(\delta_0)$ *defined in* (48). *If* $|z| < \left(2\tau(1 - \tau/T)\right)^{1/2}$ *and*

$$a_1 = a_2 = a_0 = \frac{1}{2}(1 - \tau/T)^{1/2} \exp\left(\frac{|z|^2}{4(T-\tau)}\right),$$

then ν_{a_0,a_0} *is an extremal element of* $\mathfrak{B}_H(\delta_0)$.

Proof. In (49) through (54) we proved that

$$\tilde{g}_{a_0,a_0}(0, T) = \tilde{g}'_{a_0,a_0}(0, T) = 0, \qquad \tilde{g}_{a_0,a_0}(x, T) \geqq 0, \qquad \tilde{g}''_{a_0,a_0}(0, T) > 0.$$

If ν_{a_0,a_0} is not an extremal element, then there exist two measures ν_1, ν_2 $\in \mathfrak{B}_H(\delta_0)$, $\nu_1 \neq \nu_2$, satisfying

$$\nu_{a_0,a_0} = r\nu_1 + (1 - r)\,\nu_2, \qquad 0 < r < 1.$$

From $\nu_{a_0,a_0} \geqq r\nu_1$ and $\nu_{a_0,a_0} \geqq (1 - r)\,\nu_2$ it follows that

$$\nu_1 = c_1\delta_{\tilde{z}^1} + d_1\delta_{\tilde{z}^2} + \gamma_1, \tag{55}$$

$$\nu_2 = c_2\delta_{\tilde{z}^1} + d_2\delta_{\tilde{z}^2} + \gamma_2. \tag{56}$$

We consider functions f identically zero outside a neighbourhood of $\tilde{z}^1$, $f(\tilde{z}_1) = 1$, and functions f with the same properties at $\tilde{z}^2$. From (55) and (56) it follows that for $a = a_0$

$$a = rc_1 + (1 - r)\,c_2, \tag{57}$$

$$a = rd_1 + (1 - r)\,d_2. \tag{58}$$

From (57) we get $c_1 = c_2 = a$ or $c_1 < a < c_2$ or $c_2 < a < c_1$. From (58) we get $d_1 = d_2 = a$ or $d_1 < a < d_2$ or $d_2 < a < d_1$. If $c_1 = c_2 = a$, then necessarily $d_1 = d_2 = a$, since from $a < d_1$ or $a < d_2$ it follows that condition (50) is not fulfilled. The same considerations hold in the other case. Therefore

$$c_1 < a < c_2 \quad \text{and} \quad d_2 < a < d_1$$

or

$$c_2 < a < c_1 \quad \text{and} \quad d_1 < a < d_2.$$

From

$$\tilde{g}_{a_0,a_0}(x, T) = r g_{c_1,d_1}(x, T) + (1 - r)\, g_{c_2,d_2}(x, T) \geqq 0$$

and

$$g_{c_1,d_1}(x, T) \geqq 0, \qquad g_{c_2,d_2}(x, T) \geqq 0$$

we obtain

$$g_{c_1,d_1}(0, T) = g_{c_2,d_2}(0, T) = 0. \tag{59}$$

From (49) and (55) it follows that

$$c_j + d_j = (1 - \tau/T)^{1/2} \exp\left(-\frac{|z|^2}{4(T - \tau)}\right) = 2a \tag{60}$$

and from (51)

$$4\sqrt{\pi}\, \tilde{g}'_{c_j,d_j}(0, T) = (c_j - d_j)\, b(T - \tau)^{3/2} \exp\left(-\frac{|z|^2}{4(T - \tau)}\right) \neq 0.$$

This is a contradiction to $\tilde{g}_{c_j,d_j}(0, T) = 0$ and $\tilde{g}_{c_j,d_j}(x, T) \geqq 0$, from which follows $\tilde{g}'_{c_j,d_j}(0, T) = 0$. ∎

As a third example we consider measures of the form (Anger and Czerner [1])

$$\nu = a_1 \delta_{\tilde{z}^1} + a_2 \delta_{\tilde{z}^2} + \gamma \in \mathfrak{B}_{\mathrm{H}}(\delta_0), \tag{61}$$

where $\tilde{z}^1 = (z_1, \tau_1)$, $\tilde{z}^2 = (z_2, \tau_2)$, $a_1 > 0$, $a_2 > 0$ (see Fig. 5.13).

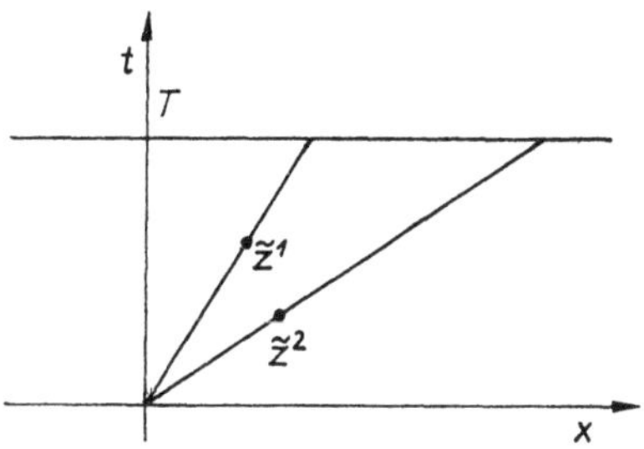

Fig. 5.13

Here $d\gamma(\tilde{x}) = \tilde{g}_{a_1,a_2}(x, T)\, dx$ and

$$\tilde{g}_{a_1,a_2}(x, T) = \frac{1}{2\sqrt{\pi}} \left(\frac{1}{T^{1/2}} \exp\left(-\frac{x^2}{4T}\right) - \frac{a_1}{(T - \tau_1)^{1/2}} \exp\left(-\frac{(x - z_1)^2}{4(T - \tau_1)}\right) - \frac{a_2}{(T - \tau_2)^{1/2}} \exp\left(-\frac{(x - z_2)^2}{4(T - \tau_2)}\right)\right) \geqq 0. \tag{62}$$

From $\tilde{g}_{a_1,a_2}(x_0, T) = \tilde{g}'_{a_1,a_2}(x_0, T) = 0$ it follows that

$$a_1 = (1 - \tau_1/T)^{1/2} \exp\left(-\frac{x_0^2}{4T}\right) \exp\left(\frac{(x_0 - z_1)^2}{4(T - \tau_1)}\right) A_1, \tag{63}$$

$$a_2 = (1 - \tau_2/T)^{1/2} \exp\left(-\frac{x_0^2}{4T}\right) \exp\left(\frac{(x_0 - z_2)^2}{4(T - \tau_2)}\right) A_2, \tag{64}$$

where

$$A_1 = \left(\frac{x_0 - z_2}{T - \tau_2} - \frac{x_0}{T}\right) \Big/ \left(\frac{x_0 - z_2}{T - \tau_2} - \frac{x_0 - z_1}{t - \tau_1}\right), \tag{65}$$

$$A_2 = \left(\frac{x_0}{T} - \frac{x_0 - z_1}{T - \tau_1}\right) \Big/ \left(\frac{x_0 - z_2}{T - \tau_2} - \frac{x_0 - z_1}{T - \tau_1}\right). \tag{66}$$

From $a_1 > 0$, $a_2 > 0$ we obtain $A_1 > 0$, $A_2 > 0$. Then

$$\left(\frac{x_0 - z_2}{T - \tau_2} - \frac{x_0}{T}\right) \Big/ \left(\frac{x_0}{T} - \frac{x_0 - z_1}{T - \tau_1}\right) = \frac{A_1}{A_2} > 0. \tag{67}$$

From this inequality it follows that the root x_0 must be an element of the interval

$$\left]\frac{z_2 T}{\tau_2}, \frac{z_1 T}{\tau_1}\right[\quad \text{respectively} \quad \left]\frac{z_1 T}{\tau_1}, \frac{z_2 T}{\tau_2}\right[. \tag{68}$$

The following condition is sufficient for $\tilde{g}_{a_1,a_2}(x_0, T) = \tilde{g}'_{a_1,a_2}(x_0, T) = 0$, $\tilde{g}''_{a_1,a_2}(x_0) > 0$:

$$\frac{T(z_1\tau_2 - z_2\tau_1)^2}{8\tau_1\tau_2(T - \tau_1)(T - \tau_2)} < \frac{2\tau_1\tau_2(z_1\tau_2 - z_2\tau_1)}{(\tau_2 - \tau_1)(z_1\tau_2 + z_2\tau_1) + 2\tau_1\tau_2\big((z_1 - z_2)T + z_2\tau_1 - z_1\tau_2\big)} \tag{69}$$

Remark 5.7. If $\tau_1 = \tau_2 = \tau$, $z_1 = |z|$, $z_2 = -|z|$, we obtain inequality (54).

Theorem 5.19. *The measure* $\nu_{a_1,a_2} = a_1\delta_{\tilde{z}^1} + a_2\delta_{\tilde{z}^2} + \gamma \in \mathfrak{B}_{\mathrm{H}}(\delta_0)$ *defined in* (61) *is an extremal element if* (63), (64), (68), *and* (69) *are fulfilled.*

Proof. This proof is similar to the proof of Theorem 5.18. ▮

The fourth example is an example in $\mathbb{R}^2$. In $\mathbb{R}$ the set of roots of g has dimension zero. We shall show that in $\mathbb{R}^2$ the set of roots of g can have di-

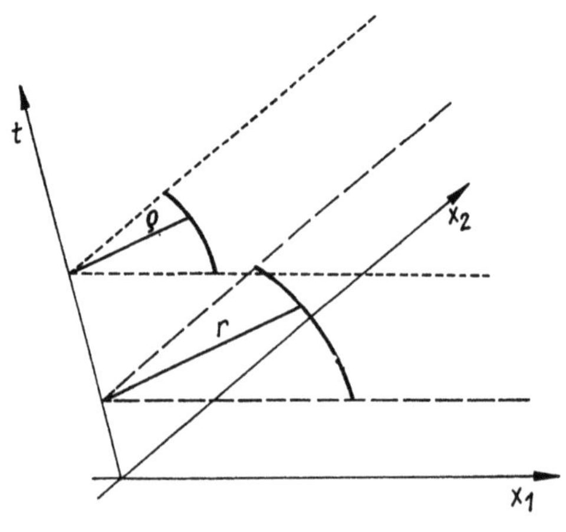

Fig. 5.14

mension 1. In $\mathbb{R}^3$ the set of roots of g can have dimension 2. This fact may be important in physics, because the space $\mathbb{R}^3$ is the space in which the physical laws hold. Let

$$K := \{(z_1, z_2, \tau) \in \mathbb{R}^3 : z_1 = r\cos\alpha,\ z_2 = r\sin\alpha,\ \tau = \text{const}\} \tag{70}$$

be a circle having the centre at $(0, 0, \tau)$, $0 < \tau < T$. We shall construct a measure ν_K on K such that the corresponding function g_0 defined in (31) is zero on a circle whose centre is at $(0, 0, T)$ (see Fig. 5.14). In this case the function g_0 has the form

$$\begin{aligned} g_0(x_1, x_2, T) &= \frac{1}{4\pi}\left[\frac{1}{T}\exp\left(-\frac{x_1^2 + x_2^2}{4T}\right) - \int_0^{2\pi} \frac{a(\alpha)}{T-\tau}\right. \\ &\quad \left.\times \exp\left(-\frac{(x_1 - r\cos\alpha)^2 + (x_2 - r\sin\alpha)^2}{4(T-\tau)}\right) r\,\mathrm{d}\alpha\right] \\ &= \frac{1}{4\pi}\left[\frac{1}{T}\exp\left(-\frac{x_1^2 + x_2^2}{4T}\right) - \frac{r}{T-\tau}\right. \\ &\quad \left.\times \exp\left(-\frac{x_1^2 + x_2^2 + r^2}{4(T-\tau)}\right) A(x_1, x_2)\right]. \end{aligned} \tag{71}$$

Let $x_1 = \varrho\cos\beta$, $x_2 = \varrho\sin\beta$, $a(\alpha) = a = \text{const}$. Then

$$g_0(\varrho, T) = \frac{1}{4\pi}\left[\frac{1}{T}\exp\left(-\frac{\varrho^2}{4T}\right) - \frac{r}{T-\tau}\exp\left(-\frac{\varrho^2 + r^2}{4(T-\tau)}\right) A_0(\varrho, r)\right], \tag{72}$$

where

$$A_0(\varrho, r) = a\int_0^{2\pi} \exp\left(\frac{\varrho r\cos\vartheta}{2(T-\tau)}\right) \mathrm{d}\vartheta = 2\pi a I_0\left(\frac{\varrho r}{2(T-\tau)}\right). \tag{73}$$

Here I_n, $n = 0, 1, 2, \ldots$, is the *modified Bessel function* of order n, which can be represented in the form

$$I_n(\xi) = \sum_{k=0}^{\infty} \frac{1}{k!\,\Gamma(n+k+1)}\left(\frac{\xi}{2}\right)^{2k+n}. \tag{74}$$

From $\xi \geqq 0$ it follows that $I_n(\xi) \geqq 0$. The function g_0 defined in (72) is constant for $\varrho = \text{const}$. From $g_0(\varrho_0) = 0$ and $g_0(\varrho) \geqq 0$ it follows that $g_0'(\varrho_0) = 0$. Therefore we have to solve the following equations, $\xi = \varrho r/2(T-\tau)$:

$$g_0(\varrho) = \frac{1}{4\pi}\left[\frac{1}{T}\exp\left(-\frac{\varrho^2}{4T}\right) - \frac{2\pi r a}{T-\tau} I_0(\xi)\exp\left(-\frac{\varrho^2 + r^2}{4(T-\tau)}\right)\right] = 0, \tag{75}$$

$$\begin{aligned} g_0'(\varrho) &= \frac{1}{4\pi}\left[-\frac{\varrho}{2T^2}\exp\left(-\frac{\varrho^2}{4T}\right)\right. \\ &\quad \left. + \frac{ra}{(T-\tau)^2}\left(\varrho I_0(\xi) - r I_0'(\xi)\right)\exp\left(-\frac{\varrho^2 + r^2}{4(T-\tau)}\right)\right] = 0. \end{aligned} \tag{76}$$

Further, the second derivative has the form

$$g_0''(\varrho) = \frac{1}{4\pi}\left[-\frac{1}{2T^2}\left(1 - \frac{\varrho^2}{2T}\right)\exp\left(-\frac{\varrho^2}{4T}\right)\right]$$
$$+ \frac{ra}{4(T-\tau)^2}\left[-\frac{\varrho^2 I_0(\xi)}{2(T-\tau)} - \frac{r\varrho I_0'(\xi)}{2(T-\tau)} + I_0(\xi)\right.$$
$$\left. + \frac{r\varrho I_0'(\xi)}{2(T-\tau)} - \frac{r^2 I_0''(\xi)}{2(T-\tau)}\right]\exp\left(-\frac{\varrho^2+r^2}{4(T-\tau)}\right). \tag{77}$$

From (75) it follows that

$$a = \frac{(T-\tau)}{2\pi r I_0(\xi)}\exp\left(-\frac{\varrho^2}{4T} + \frac{\varrho^2+r^2}{4(T-\tau)}\right). \tag{78}$$

Since $I_0' = I_1$ we obtain from (75) and (76)

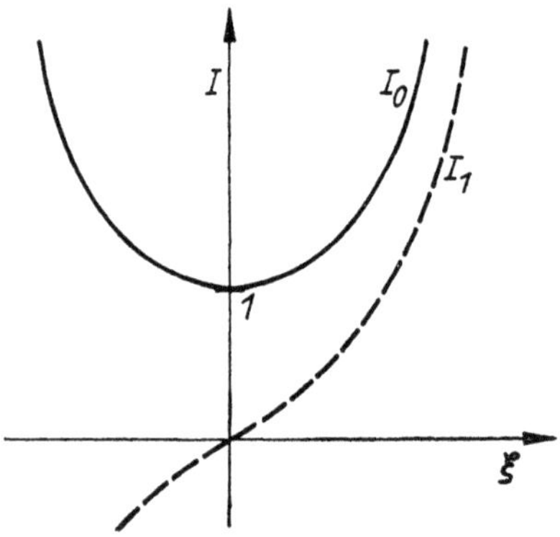

Fig. 5.15

$$I_1(\xi) = \frac{\varrho\tau}{rT} I_0(\xi),$$

which can be written in the form

$$I_1(\xi) = k\xi I_0(\xi), \qquad k = \frac{2\tau(T-\tau)}{r^2 T}. \tag{79}$$

Let $\xi \neq 0$. Using the relation $I_0 - (2/\xi) I_1 = I_2$ we get from (79)

$$I_2(\xi) = 2\left(\frac{1}{2} - k\right) I_0(\xi). \tag{80}$$

Since $I_1(\xi) < I_0(\xi)$ for all $\xi \in \mathbb{R}$ it follows that

$$\frac{\varrho\tau}{rT} < 1 \quad \text{or} \quad \varrho < \frac{rT}{\tau}. \tag{81}$$

From $I_2 > 0$ and $I_0 > 0$ for all $\xi > 0$ we obtain

$$k < \frac{1}{2} \quad \text{or} \quad 4\tau(T-\tau) < r^2 T. \tag{82}$$

If $k > 1/2$ equation (80) has no solution. In this case $\xi = 0$ is the only root of (79). Now let $k < 1/2$, $f_1(\xi) := I_1(\xi)/I_0(\xi)$, $f_2(\xi) := k\xi$. Then $f_1'(0) = 1/2$, $f_1(\xi) < 1$, $f_2'(0) = k < 1/2$. If $\xi > 1/k$ we get $f_2(\xi) > 1$. There exists a uniquely determined ξ_0 satisfying (79) or (80).

Let us calculate $g''(\varrho)$ if ϱ is a solution of (75), (76) or (79). From $I_0' = I_1$, $I_0'' + (1/\xi)\, I_0' - I_0 = 0$, and (79) we get

$$I_0'' = I_0(1 - k).$$

Then

$$\begin{aligned} g''(\varrho) &= \frac{arI_0(\xi)}{4T(T-\tau)}\left[-\left(1 - \frac{\varrho^2}{2T}\right) - \frac{\varrho^2}{2(T-\tau)^2}\right. \\ &\quad \left. + \frac{\varrho^2}{(T-\tau)^2} + \frac{T}{T-\tau} - \frac{r^2T(1-k)}{2(T-\tau)^2}\right] \\ &= \frac{arI_0(\xi)}{4T(T-\tau)}\left[\varrho^2\frac{t^2}{2T(T-\tau)^2} - \frac{r^2T}{2(T-\tau)^2} + \frac{2\tau}{T-\tau}\right] \\ &= \frac{arI_0(\xi)}{8T^2(T-\tau)^3}\left[\varrho^2\tau^2 - r^2T^2\left(1 - \frac{4\tau(T-\tau)}{r^2T}\right)\right] \\ &= \frac{arI_0(\xi)}{8T^2(T-\tau)^3}\left[\varrho^2\tau^2 - r^2T(1-2k)\right]. \end{aligned} \tag{83}$$

If $k > 1/2$ and $\varrho = 0$ then $g''(0) > 0$. In this case we have a minimum at this point. If $k < 1/2$ and $\varrho = 0$ we get $g''(0) < 0$, i.e., at $\varrho = 0$ there is a maximum. If in the case considered ϱ is a solution of (79), $r^2 > 4\tau(T-\tau)/T$, and

$$\varrho^2\tau^2 > r^2T(1 - 2k)$$

then $g''(\varrho) > 0$, i.e., at this point we have a minimum. We get the final result: if $k < 1/2$ and

$$0 < \frac{T^2}{2}\left(r^2 - \frac{4\tau(T-\tau)}{T}\right) < \varrho^2 < \frac{r^2T^2}{\tau^2} \tag{84}$$

then $g_0(\varrho) = 0$, $g_0'(\varrho) \geqq 0$. At this point g_0 has a minimum.

6. Some one-dimensional inverse problems in differential equations

6.1. Explicitly solvable inverse problems in heat conduction

At first we shall study the problems formulated in Examples 2.11 through 2.13. These are inverse problems which can be explicitly solved (CANNON [2], ROSSI COSTA [1]). The heat equation

$$\partial_t u = a(t)\, \partial_x^2 u, \qquad a(t) > 0 \quad \text{continuous,} \tag{1}$$

can be transformed into the equation

$$\partial_\tau v = \partial_x^2 v \tag{2}$$

by means of the transformation

$$\tau := \theta(t) = \int_0^t a(t^*)\, \mathrm{d}t^*. \tag{3}$$

Then

$$\theta'(t) = a(t) > 0. \tag{4}$$

From (3) and (4) follows the existence of the inverse function $\psi := \theta^{-1}$, where

$$t = \theta^{-1}(\tau) =: \psi(\tau) \tag{5}$$

with

$$\psi'(\tau) = 1/\theta'(t) = 1/a(t) = 1/a\big(\psi(\tau)\big). \tag{6}$$

From

$$v(x, \tau) := u\big(x, \psi(\tau)\big) \tag{7}$$

it follows that

$$\begin{aligned} \partial_\tau v(x, \tau) &= \partial_t u\big(x, \psi(\tau)\big)\, \psi'(\tau) = \partial_t u\big(x, \psi(\tau)\big)\, a^{-1}\big(\psi(\tau)\big) = \partial_x^2 u\big(x, \psi(\tau)\big) \\ &= \partial_x^2 v(x, \tau). \end{aligned} \tag{8}$$

By using the transformation (3) a solution v of (8) gives a solution u of (1), i.e.,

$$u(x, t) = v\big(x, \theta(t)\big). \tag{9}$$

Theorem 6.1. *The solution u of Example* 2.11 *is given by*

$$u(x, t) = (\sin \pi x) \exp\left(-\pi^2 \int_0^t a(t^*) \, dt^*\right). \tag{10}$$

From (2.2.67) *it follows that*

$$g(t) = \exp\left(-\pi^2 \int_0^t a(t^*) \, dt^*\right). \tag{11}$$

If g is a positive, continuously differentiable function with $g'(t) < 0$ *for all* $0 < t$ *and* $g(0) = 1$, *then*

$$a(t) = -\frac{g'(t)}{\pi^2 g(t)}, \qquad 0 < t. \tag{12}$$

The pair (a, u) *constitutes the unique solution to Example* 2.11.

Proof. From (2.2.14)–(2.2.18) and (1)–(9) it follows that

$$u(x, t) = \int_0^1 G\big(x, \theta(t); y, 0\big) \sin \pi y \, dy = \sin \pi x \exp\left(-\pi^2 \int_0^t a(t^*) \, dt^*\right).$$

For $x = 1/2$ we get

$$u\left(\frac{1}{2}, t\right) = g(t) = \exp\left(-\pi^2 \int_0^t a(t^*) \, dt^*\right).$$

Differentiating (11) we obtain

$$g'(t) = -\pi^2 a(t) \exp\left(-\int_0^t a(t^*) \, dt^*\right) = -\pi^2 a(t) \, g(t),$$

hence

$$a(t) = -\frac{g'(t)}{\pi^2 g(t)} > 0 \quad \text{for all} \quad 0 < t.$$

The solution (12) is uniquely determined by (11). ∎

Theorem 6.2. *The thermal conductivity* $a(t)$ *in Example* 2.12 *has the form*

$$a(t) = -\frac{g(t)}{\pi}\left(1 + \pi \int_0^t g(t^*) \, dt^*\right)^{-1}, \tag{13}$$

where $g(t) < 0$ *and* $-\int_0^t g(t^*) \, dt^* < 1/\pi$. *The pair* (a, u) *constitutes the unique solution to Example* 2.12.

Proof. From (2.2.14)–(2.2.18), (2.2.68)–(2.2.70), and (1)–(9) it follows that

$$u(x, t) = (\sin \pi x) \exp\left(-\pi^2 \int_0^t a(t^*)\, dt^*\right).$$

From (2.2.71) we get

$$-a(t)\, \pi \exp\left(-\pi^2 \int_0^t a(t^*)\, dt^*\right) = g(t). \tag{14}$$

The following function is an integral of (14):

$$\exp\left(-\pi^2 \int_0^t a(t^*)\, dt^*\right) = 1 + \pi \int_0^t g(t^*)\, dt^*, \tag{15}$$

which can be proved by differentiating (15). Substituting this in (14), we obtain

$$a(t) = -\frac{g(t)}{\pi}\left(1 + \pi \int_0^t g(t^*)\, dt^*\right)^{-1}. \tag{16}$$

This function is positive if

$$-\int_0^t g(t^*)\, dt^* < \frac{1}{\pi}.$$

The problem is uniquely solvable since (13) has the unique solution (15). ∎

Theorem 6.3. *The boundary-flux measurement for a semi-infinite conductor yields a thermal conductivity $a(t)$ which has the form*

$$a(t) = \frac{\pi}{2h_0^2}\, g(t) \int_0^t g(t^*)\, dt^*, \tag{17}$$

where $g(t) > 0$ and the constant h_0 is positive. The pair (a, u) constitutes the unique solution to Example 2.13.

Proof. From (2.2.9)–(2.2.13), (2.2.72)–(2.2.75) and (1)–(9) it follows that

$$u(x, t) = -2 \int_0^{\theta(t)} \partial_x G_{\mathrm{H}}\big(x, \theta(t); 0, \tau\big)\, h\big(\psi(\tau)\big)\, d\tau,$$

where

$$\tau = \theta(t) = \int_0^t a(t^*)\, dt^*, \qquad \psi(\tau) = \theta^{-1}(\tau).$$

From

$$\tau = \theta(t^*), \qquad t^* = \psi(\tau), \qquad \psi\big(\theta(t^*)\big) = t^*, \qquad d\tau = a(t^*)\, dt^*$$

we get

$$u(x, t) = -2 \int_0^t \partial_x G_{\mathrm{H}}\big(x, \theta(t); 0, \theta(t^*)\big)\, h(t^*)\, a(t^*)\, \mathrm{d}t^* .$$

Since

$$\partial_x^2 G_{\mathrm{H}}\big(x, \theta(t); 0, \theta(t^*)\big) = -\frac{1}{a(t^*)}\, \partial_{t^*} G_{\mathrm{H}}\big(x, \theta(t); 0, \theta(t^*)\big)$$

and

$$G_{\mathrm{H}}\big(x, \theta(t); 0, \theta(t^*)\big) \to 0 \quad \text{for} \quad t^* \to t, \qquad x \neq 0,$$

we obtain for $h = h_0 = \text{const}$

$$\partial_x u(x, t) = -\int_0^t \partial_x^2 G_{\mathrm{H}}\big(x, \theta(t); 0, \theta(t^*)\big)\, h_0 a(t^*)\, \mathrm{d}t^*$$

$$= 2 \int_0^t \partial_{t^*} G_{\mathrm{H}}\big(x, \theta(t); 0, \theta(t^*)\big)\, h_0\, \mathrm{d}t^* = -2 h_0 G_{\mathrm{H}}\big(x, \theta(t); 0, 0\big).$$

Therefore

$$\partial_x u(0, t) = \lim_{x \to 0} \partial_x u(x, t) = -2 h_0 G_{\mathrm{H}}\big(0, \theta(t); 0, 0\big).$$

From (2.2.75) it follows that

$$2 h_0 a(t)\, G_{\mathrm{H}}\big(0, \theta(t); 0, 0\big) = g(t)$$

and

$$a(t) = \frac{\sqrt{\pi}}{h_0}\, g(t) \left(\int_0^t a(t^*)\, \mathrm{d}t^* \right)^{1/2} . \tag{18}$$

This is equivalent to

$$\frac{\mathrm{d}}{\mathrm{d}t} \left(\int_0^t a(t^*)\, \mathrm{d}t^* \right)^{1/2} = \frac{\sqrt{\pi}}{2 h_0}\, g(t) .$$

Hence

$$\left(\int_0^t a(t^*)\, \mathrm{d}t^* \right)^{1/2} = \frac{\sqrt{\pi}}{2 h_0} \int_0^t g(t^*)\, \mathrm{d}t^*$$

and

$$\int_0^t a(t^*)\, \mathrm{d}t^* = \frac{\pi}{4 h_0^2} \left(\int_0^t g(t^*)\, \mathrm{d}t^* \right)^2 .$$

Differentiation yields

$$a(t) = \frac{\pi}{2h_0^2} g(t) \int_0^t g(t^*) \, dt^* . \tag{19}$$

Equation (18) has the unique solution (19). ▮

One can find further problems in which $a(t)$ is to be determined in CANNON [2] and ROSSI COSTA [1].

6.2. A nonlinear inverse problem in which the coefficients depend on the gradient of the solution

In the following we shall study a one-dimensional inverse problem relative to a compressible fluid flow. In $\mathbb{R}^2$ we assume that the density $\varrho(q)$ depending on the speed q of the flow is a positive, sufficiently smooth, and nonincreasing function in an interval $0 < q \leqq q_{\max} < +\infty$, where $\varrho(q)$ is the coefficient of the elliptic system of nonlinear differential equations (18). For simplicity the density $\varrho(q)$ is to be determined by two measurements of q on the one-dimensional set B_1 (see Fig. 6.1) and on $\tilde{B}_1$ (see Fig. 6.2). The statement is given by Theorem 6.4. Further, there are relations between the two measurements (Remark 6.1). Since $q(x) \leqq q_0 := \sup \{q(y), y \in B_1\}$ (see BERS [1], SCHIFFER [1]) and ϱ is a nonincreasing function we get an estimation of ϱ for all points x of the flow. Therefore the one-dimensional measurements of q are sufficient. This is a first example to determine the coefficient ϱ in the nonlinear system (18) by measurements on the boundary. It is necessary to study such special cases to get a feeling for the problems under consideration. May be that this example is more interesting in other areas.

The flow of a perfect fluid is described in the so-called Eulerian representation

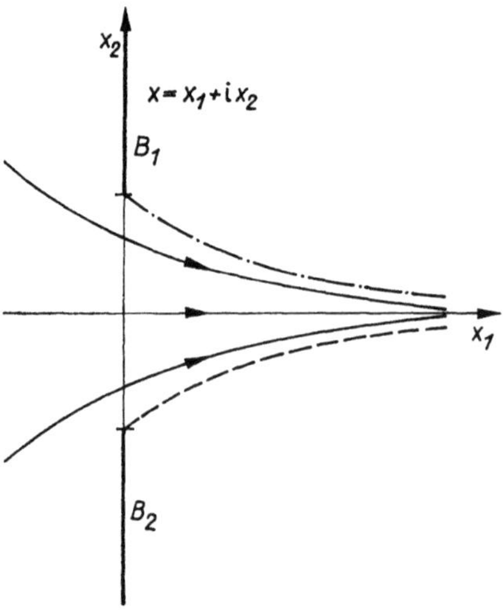

Fig. 6.1

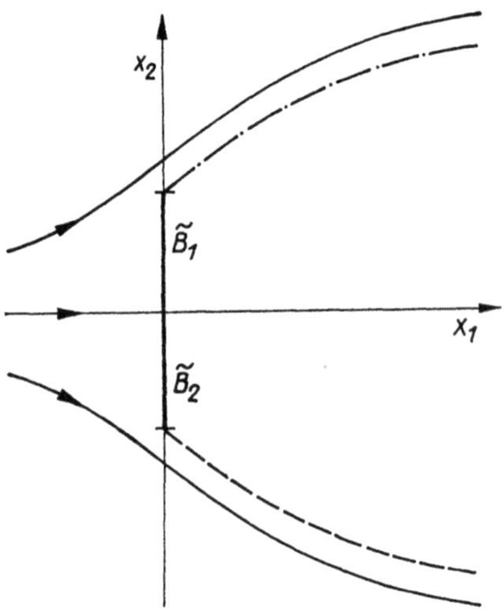

Fig. 6.2

by giving the density ϱ and the velocity components u_1, u_2 as functions of the Cartesian coordinates x_1, x_2 and time t. A complete description requires also the knowledge of two other thermodynamic variables, pressure p and temperature. We consider a general *barotropic fluid*, i.e., a fluid in which

$$p = p(\varrho), \tag{1}$$

$p(\varrho)$ being some sufficiently smooth function. For an ideal gas we have

$$p = a\varrho^\gamma, \qquad a = \text{const}, \tag{2}$$

where $\gamma > 1$ is a constant. The standard value for air is 1.4.

The velocity components and the density are connected by the continuity equations which express the law of conservation of mass

$$\partial_t \varrho + \sum_{j=1}^{2} \partial_{x_j}(\varrho u_j) = 0. \tag{3}$$

On the other hand, the velocity components must obey the *Euler equations of motion* which are the analogue of Newton's second law. If one neglects body forces, in particular gravity, these equations read

$$\partial_t u_j + \sum_{k=1}^{2} u_k\, \partial_{x_k} u - \frac{1}{\varrho}\, \partial_{x_j} p = 0, \qquad j = 1, 2. \tag{4}$$

Note that in view of relation (1) the pressure can be eliminated from the Euler equation.

We shall deal exclusively with *steady flows*, i.e., with solutions of the equations of motion which do not depend upon time. Furthermore, we shall assume that the flow is irrotational, i.e.,

$$\partial_{x_1} u_2 - \partial_{x_2} u_1 = 0. \tag{5}$$

From (4) it follows that (Bernoulli's law)

$$\frac{q^2}{2} + \int \frac{\mathrm{d}p}{\varrho} = \text{const}, \tag{6}$$

where $q^2 := u_1^2 + u_2^2$. By (1) and (6) we obtain

$$\varrho = \varrho(q), \tag{7}$$

where $\varrho(q)$ is a nonincreasing function defined implicitly by the relation

$$\frac{q^2}{2} + \int^{\varrho} \frac{p'(\xi)}{\xi}\, \mathrm{d}\xi = \text{const}. \tag{8}$$

Let

$$c^2 := \frac{\mathrm{d}p}{\mathrm{d}\varrho} \quad \text{and} \quad M := \frac{q}{c} \tag{9}$$

be the Mach number. From (8) it follows that

$$q \frac{dq}{d\varrho} + \frac{1}{\varrho}\frac{dp}{d\varrho} = 0 \quad \text{and} \quad \frac{dp}{d\varrho} = -\frac{\varrho q}{\dfrac{d\varrho}{dq}} = -\frac{q}{\varrho'(q)}, \qquad -\varrho'(q) > 0,$$

and

$$c^2 = -\frac{\varrho q}{\varrho'(q)}, \qquad M^2 = -\frac{q}{\varrho}\frac{d\varrho}{dq}, \qquad 1 - M^2 = \frac{1}{\varrho}\frac{d(\varrho q)}{dq}.$$

The flow is called *subsonic* if $M < 1$ and *supersonic* if $M > 1$. There exists a value q_{cr}, $0 < q_{cr} < q_{max}$, such that $M^2 < 1$ for $q < q_{cr}$.

From (2) and (7) it follows that

$$\varrho(q) = \left(1 - \frac{\gamma - 1}{a\gamma}\frac{q^2}{2}\right)^{1/(\gamma-1)}, \qquad c^2 = a\gamma - \frac{\gamma - 1}{2} q^2,$$
$$M^2 = \frac{q^2}{a\gamma - [(\gamma - 1)/2]\, q^2}. \tag{10}$$

Thus the flow is subsonic if the speed q is less than the critical speed

$$q_{cr} = \left(\frac{2a\gamma}{\gamma + 1}\right)^{1/2} \tag{11}$$

and since ϱ must be positive, only speeds below the maximum speed

$$q_{max} = \left(\frac{2a\gamma}{\gamma - 1}\right)^{1/2} \tag{12}$$

are possible. Using (12) relation (10) becomes

$$\varrho(q) = \left[\frac{\gamma - 1}{2\gamma}(q_{max}^2 - q^2)\right]^{1/(\gamma-1)} \tag{13}$$

We have seen that for an ideal gas the density ϱ can be determined as a function of the speed q. In the following we try to determine $\varrho(q)$, the coefficient in the elliptic system (18) which is nonlinear, by measurements of q in the general case. For a plane, the *continuity equation* reads

$$\partial_{x_1}(\varrho u_1) + \partial_{x_2}(\varrho u_2) = 0. \tag{14}$$

From (14) it follows that there exists a *stream function* $\Psi(x_1, x_2)$ such that

$$\varrho u_1 = \partial_{x_2}\Psi, \qquad \varrho u_2 = -\partial_{x_1}\Psi. \tag{15}$$

Further, from the equation (5) it follows that there exists a *velocity potential* Φ such that

$$u_1 = \partial_{x_1}\Phi, \qquad u_2 = \partial_{x_2}\Phi. \tag{16}$$

Then

$$q^2 = (\partial_{x_1}\Phi)^2 + (\partial_{x_2}\Phi)^2 =: |\nabla\Phi|^2. \tag{17}$$

Therefore we have to solve the nonlinear system

$$\begin{aligned} \varrho(|\nabla\phi|)\,\partial_{x_1}\Phi &= \partial_{x_2}\Psi, \\ \varrho(|\nabla\phi|)\,\partial_{x_2}\Phi &= -\partial_{x_1}\Psi. \end{aligned} \tag{18}$$

We shall study the following two examples for subsonic flows (KLEINE [3]).

In the first example (experiment 1) the domain is bounded by two rigid walls, which we take as lying on the x_2-axis, and by two symmetrically situated free boundaries extending to infinity (see Fig. 6.1). The speed $q(x)$ vanishes far upstream, has a constant value

$$q(x) = q_0 \quad \text{on the free streamlines,} \tag{19}$$

and tends to the limit q_0 downstream. On the rigid walls we have

$$\partial_{x_1}\Phi = u_1 = 0 \tag{20}$$

expressing the fact that the obstacle is a stream line. From (18) and (19) it follows that on the wall $B = B_1 \cup B_2$

$$\partial_{x_2}\Psi(0, x_2) = 0, \quad \text{i.e.,} \quad \Psi(0, x_2) = \text{const.} \tag{21}$$

To obtain a problem with unique solution, we can require that

$$\begin{aligned} \Psi(x) &= \frac{\pi}{2} \quad \text{on } B_1 \text{ and the upper free stream line,} \\ \Psi(x) &= -\frac{\pi}{2} \quad \text{on } B_2 \text{ and the lower free stream line.} \end{aligned} \tag{22}$$

This boundary value problem can be solved by mapping the flow domain bounded by the impermeable walls and the free stream boundaries into the *potential plane* $t = \Phi + i\Psi$, whereas its image is the strip

$$-\infty < \Phi < +\infty, \qquad -\frac{\pi}{2} < \Psi < \frac{\pi}{2}.$$

The precise shape of the free stream lines is, as always, part of the problem. For existence and uniqueness see BERS [1], SCHIFFER [1] and GILBARG [1].

The second example (experiment 2) is the classical Kirchhoff example of a symmetric flow of velocity past a vertical plate with stream lines detachment at its ends. The physical ($x = x_1 + ix_2$)-plane and its images in the hodographic ($u_1 + iu_2$)-plane, i.e., the velocity plane, are illustrated in Fig. 6.2 and Fig. 6.4, where in Fig. 6.3 one can find the corresponding considerations for the first example.

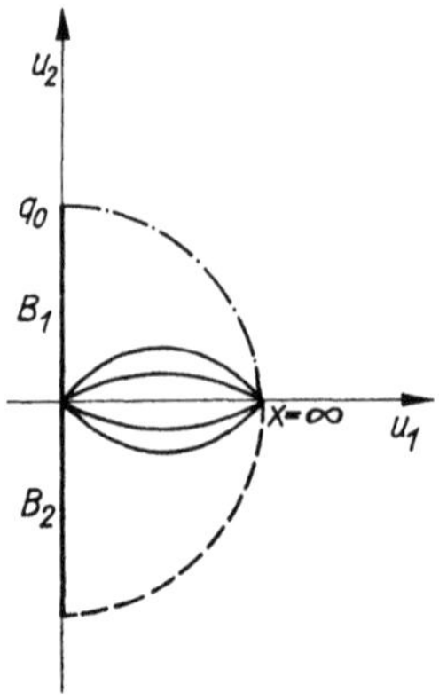

Fig. 6.3

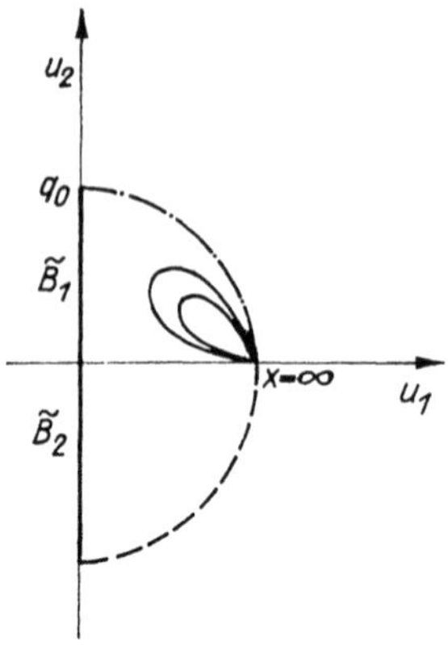

Fig. 6.4

On $\tilde{B} = \tilde{B}_1 \cup \tilde{B}_2$ we have

$$\partial_{x_1}\Phi = u_1 = 0 \tag{23}$$

and

$$q(x) = q_0 \quad \text{on the free stream lines.} \tag{24}$$

From (15) and (23) it follows that on B

$$\partial_{x_2}\Psi(0, x_2) = 0, \quad \text{i.e.,} \quad \Psi(0, x_2) = \text{const.} \tag{25}$$

We assume

$$\Psi = 0 \quad \text{on } B \text{ and the free stream lines.} \tag{26}$$

At the origin $q = 0$ holds. Let there $\Phi = 0$. For the corresponding flow domain we consider the mapping into the potential plane $t = \Phi + \mathrm{i}\Psi$ cut off along the non-negative Φ-axis (see Fig. 6.8). For existence and uniqueness see GILBARG [1]. In the following we shall study these mappings.

The treatment of the nonlinear system (18) is simplified considering as new variables the velocity components u_1, u_2 or some other coordinates in the *hodographic plane*, say, the polar coordinate q and ϑ (see Fig. 6.3),

$$q = (u_1^2 + u_2^2)^{1/2}, \qquad \vartheta = \arctan \frac{u_2}{u_1}. \tag{27}$$

Using the new variables (27) the solutions Φ and Ψ of (18) must satisfy the linear system of partial differential equations (BERS [1], SCHIFFER [1])

$$\partial_\vartheta\Phi = \frac{q}{\varrho}\,\partial_q\Psi, \qquad \partial_q\Phi = q\,\frac{\mathrm{d}}{\mathrm{d}q}\left(\frac{1}{q\varrho}\right)\partial_\vartheta\Psi. \tag{28}$$

Let us introduce some useful formal transformations of the hodographic equations (28):

$$r = r(q) := \exp\left(\int\limits_{q_0}^{q}\left(\frac{1}{\varrho}\,\frac{\mathrm{d}(\varrho q)}{\mathrm{d}q}\right)^{1/2}\frac{\mathrm{d}q}{q}\right), \qquad r(q_0) = 1, \tag{29}$$

$0 \leqq q_0 \leqq q_{\text{cr}}$. Its inverse is denoted by $q = q(r)$. From (27) and (29) we get with

$$\varrho^*(r) = \varrho \Bigg/ \left(\frac{1}{\varrho} \frac{\mathrm{d}(\varrho q)}{\mathrm{d}q} \right)^{1/2} \tag{30}$$

and $w^* = u_1^* - \mathrm{i} u_2^* = r\, \mathrm{e}^{-\mathrm{i}\vartheta}$ the system

$$\varrho^* \, \partial_\vartheta \Phi = r \, \partial_r \Psi, \qquad \varrho^* \, \partial_r \Psi = -\frac{1}{r} \, \partial_\vartheta \Psi, \tag{31}$$

which is invariant with respect to conformal transformations. The hodograph in the (r, ϑ)-coordinate system is a unit-semidisk. The boundary conditions for the system (31) are the following:

$$\begin{aligned} &\text{Example 1: } \Psi = \pi/2 \quad \text{on the upper boundary} \quad -\vartheta > 0, \\ &\qquad\qquad\; \Psi = -\pi/2 \quad \text{on the lower boundary} \quad \vartheta > 0. \end{aligned} \tag{32}$$

$$\text{Example 2: } \Psi = 0 \quad \text{on the whole boundary.} \tag{33}$$

The conformal transformation

$$\zeta = \frac{2}{\dfrac{1}{w^*} - w^*}$$

maps the unit-semidisk onto the right half plane $\operatorname{Re} \zeta > 0$. The point $w^* = 1$ corresponds to $+\infty$, $w^* = \mathrm{e}^{\mathrm{i}\vartheta}$ to

$$\zeta = \frac{2}{\mathrm{e}^{-\mathrm{i}\vartheta} - \mathrm{e}^{\mathrm{i}\vartheta}} = \frac{-\mathrm{i}}{\sin \vartheta}$$

and $w^* = \mathrm{i} u_2$ to $\zeta = \mathrm{i} u_2/(1 + u_2^2)$ (see Fig. 6.5).

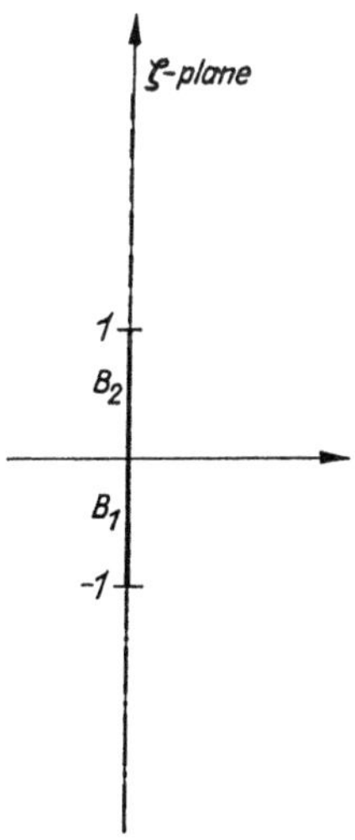

Fig. 6.5

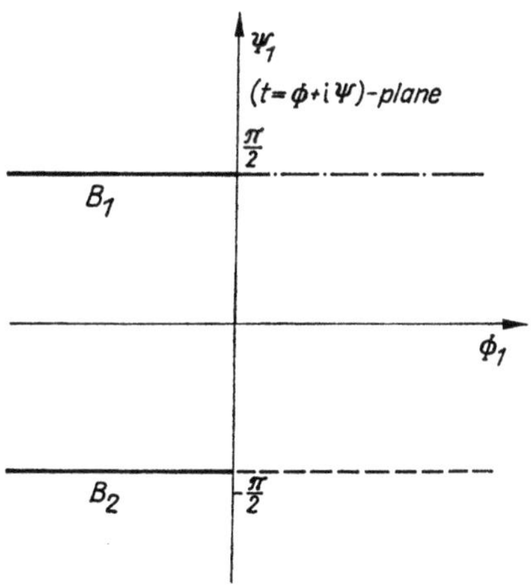

Fig. 6.6

Setting $\zeta = R\,\mathrm{e}^{-\mathrm{i}\alpha}$ system (31) takes on the form

$$\varrho^*\,\partial_R\Phi = -\frac{1}{R}\,\partial_\alpha\Psi, \qquad \varrho^*\,\partial_\alpha\Phi = R\,\partial_R\Psi. \tag{34}$$

By means of the transformation (experiment 1)

$$t = \Phi_1 + \mathrm{i}\Psi_1 = \log\zeta = \log R - \mathrm{i}\alpha \tag{35}$$

the half-space $\operatorname{Re}\zeta > 0$ is mapped onto the strip $-\pi/2 < \Psi < \pi/2$ (see Fig. 6.6). In the case of experiment 2 (see Fig. 6.7) we consider the mapping

$$t = \Phi_2 + \mathrm{i}\Psi_2 = -\zeta^2 = R^2(\cos 2\alpha + \mathrm{i}\sin 2\alpha), \tag{36}$$

which maps $\operatorname{Re}\zeta > 0$ into the whole t-plane cut off along the real axis (see Fig. 6.8)

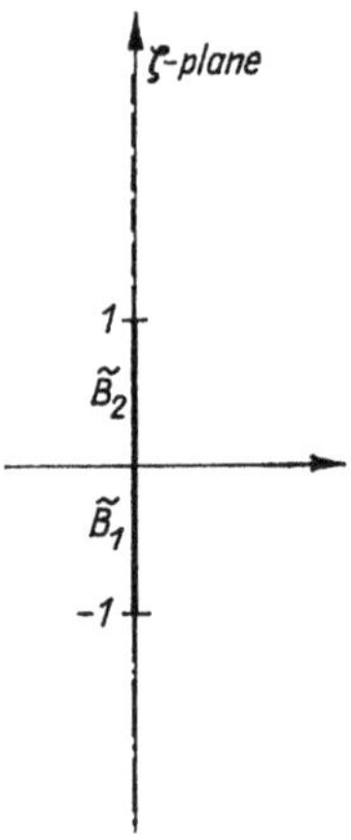

Fig. 6.7

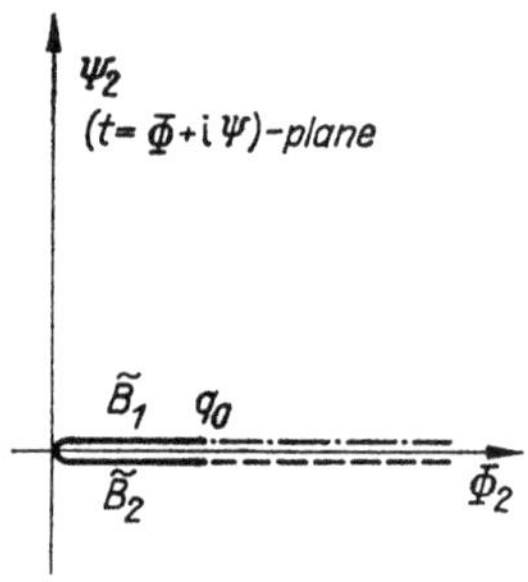

Fig. 6.8

Now we shall study the inverse problem. In order to uniquely determine $\varrho(q)$ we shall use two experiments, namely experiment 1 and experiment 2. In Remark 6.1 we shall prove that there exist relations between both experiments. May be that one experiment is sufficient to determine $\varrho(q)$. In experiment 1 the upper rigid wall is mapped onto the set $M := \{(u_1, u_2) : q < q_0, \vartheta = -\pi/2\}$. In experiment 2 that lower half of the obstacle is mapped onto the same set. We remark that the quantity $r = r(q)$ defined in (29) depends on the measured value q. At the boundaries considered we have $u_1 = 0$ and $q = -u_2 = -\partial_{x_2}\Phi > 0$. In (16) we defined a mapping T_1 between the physical plane $x = x_1 + \mathrm{i}x_2$ and the velocity plane $u_1 + \mathrm{i}u_2$, where $q = (u_1^2 + u_2^2)^{1/2}$. In (29) we defined a mapping T_2 between q and a new variable r. The mapping T_3 defines a relation between the polar coordinates (r, ϑ) and the ζ-plane. Finally, the transformation T_4 defines the relations between the ζ-plane and the potential plane $t = \Phi + \mathrm{i}\Psi$. The composed mapping

$$T := T_4T_3T_2T_1$$

is mapping from the physical plane into the potential plane $t = \Phi + i\Psi$. The mapping $x \mapsto Tx = t$ is one-to-one, except for isolated points. The mappings T_1, T_2, T_3 are equal for both experiments, while T_4 is different in experiment 1 and experiment 2. The different quantities q, (r, ϑ), (R, α) and so on, and its inverses can be calculated by means of the mappings T_j and further relations.

The following relations hold:

$$\partial_q \Phi = \frac{\partial \Phi}{\partial q} = \frac{\partial \Phi}{\partial x_2} \frac{dx_2}{dq} = -q \frac{dx_2}{dq}.$$

For experiment 1 from (34) and (35) it follows that

$$\frac{\partial \Phi_1}{\partial R} = \frac{1}{\varrho^* R} \quad \text{on } M. \tag{37}$$

For experiment 2 from (34) and (36) it follows that

$$\frac{\partial \Phi_2}{\partial R} = \frac{2R}{\varrho^*} \quad \text{on } M. \tag{38}$$

Since

$$\zeta = \frac{2}{\dfrac{1}{w^*} - w^*} = R\, e^{-i\alpha}$$

we obtain for $w^* = r \exp(+i\pi/2)$ the value $\zeta = 2ir/(1 + r^2)$. Therefore $R = |\zeta| = 2r/(1 + r^2) = 2/(1/r + r)$. From these considerations we get

$$\frac{\partial_q \Phi_1}{\partial_q \Phi_2} = \frac{\partial_R \Phi_1\, \partial_q R}{\partial_R \Phi_2\, \partial_q R} = \frac{1}{2R^2} = \frac{1}{8} \left(\frac{1}{r} + r \right)^2. \tag{39}$$

On M we define

$$f(q) := 2 \frac{\partial_q \Phi_1}{\partial_q \Phi_2} = 2 \frac{\dfrac{dx_2^{(1)}}{dq}}{\dfrac{dx_2^{(2)}}{dq}}. \tag{40}$$

Then

$$f(q) = \frac{1}{4} \left(\frac{1}{r(q)} + r(q) \right)^2 \tag{41}$$

and

$$r(q) = \sqrt{f(q)} - \sqrt{f(q) - 1}, \tag{42}$$

where $1 \leqq f(q)$ and $0 < r(q) \leqq 1$. From (42) it follows that

$$\frac{dr}{r} = -\frac{df}{2\sqrt{f}\sqrt{f - 1}}.$$

Using relation (29) we get

$$\frac{-\mathrm{d}f}{2\sqrt{f}\sqrt{f-1}} = \left(\frac{1}{\varrho}\frac{\mathrm{d}(\varrho q)}{\mathrm{d}q}\right)^{1/2}\frac{\mathrm{d}q}{q}$$

and

$$\left(\frac{\mathrm{d}f}{\mathrm{d}q}\right)^2 \frac{q^2}{4f(f-1)} = \frac{1}{\varrho}\frac{\mathrm{d}(\varrho q)}{\mathrm{d}q} = 1 + \frac{q}{\varrho}\frac{\mathrm{d}\varrho}{\mathrm{d}q},$$

from which follows

$$\left(\left(\frac{\mathrm{d}f}{\mathrm{d}q}\right)^2 \frac{q}{4f(f-1)} - \frac{1}{q}\right)\mathrm{d}q = \frac{\mathrm{d}\varrho}{\varrho}.$$

Integrating this relation we obtain

$$\varrho(q) = \frac{1}{q}\exp\left(\int^{q}\left(\frac{\mathrm{d}f}{\mathrm{d}q}\right)^2 \frac{q}{4f(f-1)}\,\mathrm{d}q\right). \tag{43}$$

We have proved

Theorem 6.4 (Determination of the density in a subsonic flow). *Let a compressible fluid flow be governed by the system* (18). *In a first experiment a solution is to be determined satisfying the boundary conditions* (19), (20), and (22). *If a solution of a second experiment satisfying the boundary conditions* (23), (24), *and* (26) *is determined and the speed* $q = (\partial x_1^2\Phi + \partial x_2^2\Phi)^{1/2}$ *(overspecified data) is measured on* B *resp.* $\tilde{B}$, *then* $\varrho(q)$ *is determined on* B *by formula* (43). *Since the maximal speed* q_0 *is attained on* B *resp.* $\tilde{B}$, *from this fact it follows that the inequality* $\varrho(q(x)) \leqq \varrho(q_0)$ *holds for all* x *of the flow.*

There are relations between the two quantities $\Phi_1(q)$ and $\Phi_2(q)$ meassured on M. They have to fulfill some compatibility conditions, which will be proved in the following. From (29) it follows that

$$\frac{\mathrm{d}r}{\mathrm{d}q} = r\left(\frac{1}{\varrho}\frac{\mathrm{d}(\varrho q)}{\mathrm{d}q}\right)^{1/2}\frac{1}{q}, \tag{44}$$

and from this relation

$$\frac{1}{\varrho}\frac{\mathrm{d}\varrho}{\mathrm{d}q} = \left(\frac{\mathrm{d}r}{\mathrm{d}q}\right)^2 \frac{q}{r^2} - \frac{1}{q} =: h_1. \tag{45}$$

The quantity q can be calculated by means of (40) and (42). Further, from (37) we get

$$\frac{\partial\Phi_1}{\partial q} = \frac{\partial\Phi_1}{\partial R}\frac{\mathrm{d}R}{\mathrm{d}r}\frac{\mathrm{d}r}{\mathrm{d}q} \doteq \frac{1}{\varrho^* R}\frac{\mathrm{d}R}{\mathrm{d}r}\frac{\mathrm{d}r}{\mathrm{d}q}.$$

Together with (30) and (44) we obtain

$$\frac{\partial \Phi_1}{\partial q} = \frac{1}{\varrho} \frac{q}{rR} \frac{\mathrm{d}R}{\mathrm{d}r} \left(\frac{\mathrm{d}r}{\mathrm{d}q}\right)^2 =: \frac{h_2}{\varrho}, \tag{46}$$

where h_2 can be calculated by means of (40) and (42). Differentiating the logarithm of (46) one gets

$$\frac{\mathrm{d}}{\mathrm{d}q} \log \frac{\partial \Phi_1}{\partial q} = \frac{1}{h_2} \frac{\mathrm{d}h}{\mathrm{d}q} - \frac{1}{\varrho} \frac{\mathrm{d}\varrho}{\mathrm{d}q} = \frac{1}{h_2} \frac{\mathrm{d}h_2}{\mathrm{d}q} - h_1. \tag{47}$$

Relation (47) is the compatibility condition for the measurements of experiment 1 and experiment 2. The right-hand side of (47) contains the function $f(q) = 2\, \partial_q \Phi_1 / \partial_q \Phi_2$. We have proved

Remark 6.1. For the measurements in experiment 1 and in experiment 2 there exists the compatibility condition (47).

7. Some inverse problems for nonlinear equations

7.1. A uniqueness theorem for determining conductivity from overspecified data

We consider the problem of finding conductivity of a region given measurement of either the steady state temperature u or the steady state flux $k(u)\,\partial_n u$ on part of the boundary and a measurement of both of these quantities on the remainder of the boundary.

We shall study solutions u of the nonlinear differential equation

$$\operatorname{div}(k(u)\operatorname{grad} u) = f(x), \qquad k(u) \geqq k_0 > 0, \tag{1}$$

in a bounded open subset $\Omega \subset \mathbb{R}^2$ with smooth boundary $\partial\Omega$ consisting of the union of the two arcs $\partial\Omega_1$ and $\partial\Omega_2$ (see Fig. 7.1) satisfying on $\partial\Omega_2$ both the conditions (Pilant and Rundell [3])

$$u = g, \tag{2}$$

$$k(u)\,\partial_n u = h_2, \tag{3}$$

while on the arc $\partial\Omega_1$

$$B(u) = h_1 \tag{4}$$

holds. The boundary operator $B(u)$ is either of Dirichlet or Neumann type. The function f is assumed to be a known function. The case where k is known and $f = f(u)$ is to be determined will be studied in Sect. 7.2. One could consider conditions (3) and (4) to be the proper data. Condition (2) represents the overspecified boundary data which is to be used to recover $k(u)$. In some papers (Alessandrini [1], Kohn and Vogelius [1, 2]) a related problem, that of

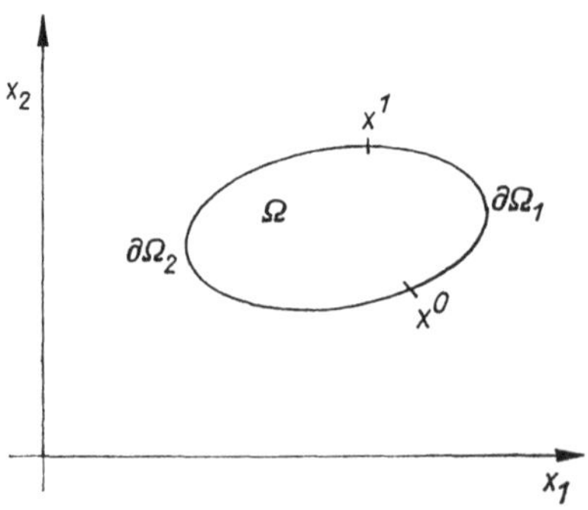

Fig. 7.1

determining the coefficient $\gamma(x)$ in the operator

$$P(x, \partial)\, u := \operatorname{div}\,(\gamma(x)\,\operatorname{grad}\, u), \qquad \gamma \in \mathbb{C}^\infty(\Omega), \tag{5}$$

from overspecified boundary data, is studied. Both inverse problems are fundamentally different. The unknown coefficient γ is not a function of the dependent variable, but rather of the independent variables. The information content of the last problem is greater than that of $k(u)$. In Sect. 1.3 and Sect. 1.4 we gave several counterexamples.

In (2.2.28) we introduced the transformation

$$v(x) = (Tk)\,(x) := \int\limits_{u(x^0)}^{u(x)} k(r)\,\mathrm{d}r, \qquad k \geqq k_0 > 0, \tag{6}$$

which is one-to-one. Let s_1 and s_2 be the arc lengths along $\partial\Omega_1$ and $\partial\Omega_2$ measured from (x_1^0, x_2^0). Then (1)–(4) becomes

$$\Delta v = f(x), \tag{7}$$

$$\partial_n v = h_2 \quad \text{on } \partial\Omega_2, \tag{8}$$

$$v(y) = \int\limits_{g(0)}^{g(s_2)} k(r)\,\mathrm{d}r \quad \text{on } \partial\Omega_2, \qquad y = \big(y_1(s_2), y_2(s_2)\big), \tag{9}$$

and either

$$\partial_n v = h_1 \quad \text{on } \partial\Omega_1 \tag{10}$$

or

$$v(y) = \int\limits_{h_1(0)}^{h_1(s_1)} k(r)\,\mathrm{d}r \quad \text{on } \partial\Omega_1, \qquad y = \big(y_1(s_1), y_2(s_1)\big). \tag{11}$$

Let us consider these different cases.

a) $B(u) = k(u)\,\partial_n u$. This boundary value problem leads to

$$\Delta v = f, \tag{12}$$

$$\partial_n v = h_2 \quad \text{on } \partial\Omega_2, \qquad \partial_n v = h_1 \quad \text{on } \partial\Omega_1, \tag{13}$$

with $v(x^0) = 0$. From (12) and (13) the existence of v follows (Gilberg and Trudinger [1], Miranda [1]). On $\partial\Omega_2$ v has the form (overspecified data)

$$v(s_2) = \int\limits_{g(0)}^{g(s_2)} k(r)\,\mathrm{d}r. \tag{14}$$

Equation (14) is an integral equation for the unknown function k. We remark once more that v is uniquely determined by (12) and (13). Let us consider the following example relative to the domain

$$\Omega := \{(x_1, x_2) \in \mathbb{R}^2 : 0 < x_1 < 1,\ 0 < x_2 < 1\}.$$

We shall study the differential equation

$$\partial_{x_1}\big(k(u)\,\partial_{x_1}u\big) + \partial_{x_2}\big(k(u)\,\partial_{x_2}u\big) = 0 \text{ in } \Omega, \tag{15}$$

where u satisfies the boundary conditions

$$\begin{aligned}
k\big(u(0, x_2)\big)\,\partial_{x_1}u(0, x_2) &= -1, \quad 0 < x_2 < 1,\\
k\big(u(1, x_2)\big)\,\partial_{x_1}u(1, x_2) &= -1, \quad 0 < x_2 < 1,\\
k\big(u(x_1, 0)\big)\,\partial_{x_2}u(x_1, 0) &= 0, \qquad 0 < x_1 < 1,\\
k\big(u(x_1, 1)\big)\,\partial_{x_2}u(x_1, 1) &= 0, \qquad 0 < x_1 < 1,
\end{aligned} \tag{16}$$

and the overspecified data

$$u(x_1, 0) = g(x_1), \qquad 0 < x_1 < 1. \tag{17}$$

Using transformation (6) we obtain

$$\Delta v(x) = 0 \quad \text{in } \Omega, \tag{18}$$

$$\begin{aligned}
\partial_{x_1}v(0, x_2) &= -1, \qquad \partial_{x_1}v(1, x_2) = -1,\\
\partial_{x_2}v(x_1, 0) &= 0, \qquad \partial_{x_2}v(x_1, 1) = 0.
\end{aligned} \tag{19}$$

The solution v of (18), (19) is

$$v(x_1, x_2) = -x_1. \tag{20}$$

From (6) and (17) it follows that

$$\int\limits_{g(0)}^{g(s_2)} k(r)\,\mathrm{d}r = v(x_1, 0) = -x_1. \tag{21}$$

By differentiating (20) we get

$$g'(x_1)\,k\big(g(x_1)\big) = -1. \tag{22}$$

Further, from (6) and (20) it follows that

$$0 = \partial_{x_2}v = k(u)\,\partial_{x_2}u, \quad \text{i.e.,} \quad \partial_{x_2}u = 0.$$

Therefore

$$u(x_1, x_2) = u(x_1, 0) = g(x_1). \tag{23}$$

Let $r := u(x_1, x_2) = g(x_1)$. Then

$$v(x_1, x_2) = \int\limits_{u(x^0)}^{u(x_1,x_2)} k(r)\,\mathrm{d}r = \int\limits_0^{x_1} k(r)\,\frac{\mathrm{d}r}{\mathrm{d}\xi}\,\mathrm{d}\xi = \int\limits_0^{x_1} -\frac{1}{g'(\xi)}\,g'(\xi)\,\mathrm{d}\xi = -x_1. \tag{24}$$

One immediately sees that (16) and (17) are fulfilled.

b) $Bu = u$, $h_1 = \text{const}$. In that case v satisfies

$$\partial_{\mathrm{n}} v = h_2 \quad \text{on } \partial\Omega_2,$$

$$v(y) = \int_{h_1}^{h_1} k(r)\, \mathrm{d}r = 0 \quad \text{if} \quad h_1 = \text{const on } \partial\Omega_1. \tag{25}$$

The overspecification on $\partial\Omega_1$ leads to equation (14).

c) Finally, we shall study the problem relative to $Bu = u$, $h_1 \neq \text{const}$. Let $G(x, y)$ be the Green function of Ω, where

$$G(x, y) = 0 \quad \text{if} \quad y \in \partial\Omega_1,$$
$$\frac{\partial G(x, y)}{\partial_{\mathrm{n}_y}} = 0 \quad \text{if} \quad y \in \partial\Omega_2. \tag{26}$$

For the existence of G see GILBARG and TRUDINGER [1] and STEPHAN [1]. From Green's first identity it follows that v can be represented in the form

$$v(x) = \int_{\Omega} G(x, y)\, f(y)\, \mathrm{d}y - \int_{\partial\Omega_2} G\big(x, y(s_2)\big)\, h_2(s_2)\, \mathrm{d}s_2$$
$$+ \int_{\partial\Omega_1} \partial_{\mathrm{n}_y} G\big(x, y(s_1)\big) \left(\int_{h_1(0)}^{h_1(s_1)} k(r)\, \mathrm{d}r \right) \mathrm{d}s_1. \tag{27}$$

Using on $\partial\Omega_2$ the overspecified boundary condition we obtain

$$\int_{g(0)}^{g(s_2)} k(r)\, \mathrm{d}r = \int_{\Omega} G\big(x(s_2), y\big)\, f(y)\, \mathrm{d}y - \int_{\partial\Omega} G\big(x(s_2), y(s_2)\big)\, h_2(s_2)\, \mathrm{d}s_2$$
$$+ \int_{\partial\Omega_1} \partial_{\mathrm{n}_y} G\big(x(s_2), y(s_1)\big) \left(\int_{h_1(0)}^{h_1(s_1)} k(r)\, \mathrm{d}r \right) \mathrm{d}s_1. \tag{28}$$

We define

$$K(u) := \int_{u(x^0)}^{u(x)} k(r)\, \mathrm{d}r \tag{29}$$

and

$$\Psi\big(x(s_2)\big) = \int_{\Omega} G\big(x(s_2), y\big)\, f(y)\, \mathrm{d}y - \int_{\partial\Omega_2} G\big(x(s_2), y(s_2)\big)\, h_2(s_2)\, \mathrm{d}s_2, \tag{30}$$

where Ψ is a known function defined on $\partial\Omega_2$. Further, let

$$T\psi\big(x(s_2)\big) := \int_{\partial\Omega_1} \partial_{\mathrm{n}_y} G\big(x(s_2), y(s_1)\big)\, \psi(s_1)\, \mathrm{d}s_1, \tag{31}$$

where $\psi = K(h_1)$. T has a singularity of the order $1/|x - y|$. Often it is called a *Hilbert transform operator*. The integral exists for $x \in \partial\Omega_1$ if ψ satisfies a Hölder condition at x. In our case the function $T\psi$ is defined on $\partial\Omega_2$. In GILBARG and TRUDINGER [1] the following maximum principle is proved.

Theorem 7.1. *Let $\Omega \subset \mathbb{R}^n$, $n \geqq 2$, be a bounded domain with smooth boundary $\partial\Omega$, v a solution of the Laplace equation $\Delta v = 0$ (more general, a solution of (2.1.41) with $f = b = 0$). If u is continuous at $y^0 \in \partial\Omega$ and $u(y^0) > u(x)$ for all $x \in \Omega$, then the outer normal derivative of u at y^0, if it exists, satisfies the strict inequality $\partial_n u(y^0) > 0$. A similar property holds relative to the minimum of u. Then $v := -u$ has a maximum at y^0.*

Corollary 7.1. *If $w \in \mathrm{C}(\bar{\Omega})$, $w \neq$ const, is a solution of the Laplace equation $\Delta w = 0$ satisfying $\partial_n w(y) = 0$ for all y which are elements of the interior of $\partial\Omega_2$, then w cannot attain its maximum or its minimum either in the interior of Ω or on $\partial\Omega_2$, where $\partial_n w(y) = 0$. Thus the maximum and the minimum values of w on $\bar{\Omega}$ must lie on $\partial\Omega_1$.*

Let $w := T\psi$ be the mapping defined in (31). From (27) and the properties of G it follows that

$$\partial_n w(x) = 0 \quad \text{for all} \quad x \in \partial\Omega_2, \quad x \neq x^0, x^1.$$

Let $\psi \in \mathrm{C}(\partial\Omega_1)$ such that $T\psi(x)$, $x \in \partial\Omega_2$, exists. From Corollary 7.1 we obtain

$$\inf\{T\psi(x), x \in \partial\Omega_2\} \geqq \inf\{\psi(y), y \in \partial\Omega_1\}$$

and

$$\sup\{T\psi(x), x \in \partial\Omega_2\} \leqq \sup\{\psi(y), y \in \partial\Omega_1\}.$$

From these two inequalities it follows that

$$-\|\psi\|_{\mathrm{C}(\partial\Omega_1)} \leqq \inf\{\psi(y), y \in \partial\Omega_1\} \leqq T\psi(x) \leqq \sup\{\psi(y), y \in \partial\Omega_1\} \leqq \|\psi\|_{\mathrm{C}(\partial\Omega_1)}$$

or

$$\|T\psi\|_{\mathrm{C}(\partial\Omega_2)} \leqq \|\psi\|_{\mathrm{C}(\partial\Omega_1)}. \tag{32}$$

Therefore $\|T\| \leqq 1$. Let $w_0 = \text{const}$. Then $\psi_0 = w_0$ on $\partial\Omega_1$ and $\partial_n w_0 = 0$ on $\partial\Omega_2$. Because w_0 is uniquely determined by ψ_0 on $\partial\Omega_1$ and $\partial_n w_0 = 0$ on $\partial\Omega_2$ we get

$$w_0 = Tw_0.$$

From this it follows that

$$\|T\| = 1. \tag{33}$$

The integral equation (28) can be formally written in the form

$$K(g(s)) = \Psi(s) + TK(h_1). \tag{34}$$

Let h_1 and g be monotone. Then from $\eta := h_1(s_1)$ on $\partial\Omega_1$ and $\xi := g(s_2)$ on $\partial\Omega_2$ the existence of its inverses

$$s_2 = g^{-1}(\xi), \qquad s_1 = h_1^{-1}(\eta)$$

follows. Then we obtain

$$K(\xi) = \Psi\big(g^{-1}(\xi)\big) + \int_{\partial\Omega_1} \partial_{n_y} G\big(g^{-1}(\xi), \eta\big)\, h_1'\big(h_1^{-1}(\eta)\big)\, K(\eta)\, \mathrm{d}\eta \tag{35}$$

which can be formally written as

$$K = \overline{\Psi} + \overline{T}(K). \tag{36}$$

Using the counterexamples of Sect. 2.2 we get the important

Remark 7.1. The eigenspace of the mapping T defined in (31) and (34) may range in dimension 1 (Example 2.7) to infinity (Example 2.6). In Example 2.6 the boundary is not smooth.

Theorem 7.2. *If the continuous functions g and h_1 are such that $g(x^0) = h_1(x^0)$, $g(x^1) = h_1(x^1)$ and* range $\{h_1\} \subset$ range $\{g\}$, *where the latter range is not a single point, and if the boundary value problem* (1), (3), (4) *has a continuous solution u on $\overline{\Omega}$, then there is at most one solution of* (1)–(4) *for given $f = f(x)$.*

Proof. Suppose (k_1, u_1) and (k_2, u_2) satisfy (1)–(4). Let $v_1 = S_{k_1}^{-1}(u_1)$, $v_2 = S_{k_2}^{-1}(u_2)$, $k = k_1 - k_2$, $w = v_1 - v_2$. From (1)–(4) it follows that w satisfies

$$\begin{aligned} &\Delta w = 0 \text{ in } \Omega, \qquad \partial_n w = 0 \text{ on } \partial\Omega_2, \\ &w = \int_{h_1(0)}^{h_1(s_1)} k(r)\,\mathrm{d}r \quad \text{on } \partial\Omega_1. \end{aligned} \tag{37}$$

Further, we have

$$w = \int_{g(0)}^{g(s_2)} k(r)\,\mathrm{d}r \quad \text{on } \partial\Omega_2. \tag{38}$$

By Corollary 7.1 w cannot attain its maximum or its minimum either in the interior of Ω or on the arc $\partial\Omega_2$ where $\partial_n w = 0$. Thus the maximum and the minimum values of w on $\overline{\Omega}$ must lie in the range of the function

$$w = \int_{h_1(0)}^{h_1(s_2)} k(r)\,\mathrm{d}r \quad \text{for} \quad s_1 \in \partial\Omega_1.$$

By assumption this implies that the range of w must lie in the range of the values of the function

$$w = \int_{g(0)}^{g(s_2)} k(r)\,\mathrm{d}r.$$

The continuity of $g(s_2)$ then demands that w must attain its maximum and minimum on $\partial\Omega_2$ which can only happen if w is constant. Since both of $h_1(s_1)$ and $g(s_2)$ cannot be constant we conclude that $w = 0$ and from (38) that the function $k(u)$ must be zero for u in the range of g. ∎

Corollary 7.2. *If h_1 and g are both monotone functions on the respective arcs $\partial\Omega_1$ and $\partial\Omega_2$ then there is at most one solution for the inverse problem* (1)—(4).
Proof. If g is increasing then h_1 must be decreasing. In this case we have range $\{h_1\}$ = range $\{g\}$. ∎

7.2. A nonlinear parabolic inverse coefficient problem

In the following we shall consider the reaction diffusion equation (see Example 2.4)

$$\partial_t u - \Delta u = f(u) + \gamma(x, t), \qquad x \in \Omega_0, \qquad 0 < t, \tag{1}$$

with boundary conditions

$$\partial_n u(x, t) = -h(x, t), \qquad x \in \partial\Omega_0, \qquad 0 < t, \tag{2}$$

and with initial condition

$$u(x, 0) = \Phi(x), \qquad x \in \Omega_0. \tag{3}$$

The overposed boundary condition is

$$u(x^0, t) = g(t) \quad \text{for some} \quad x^0 \in \partial\Omega. \tag{4}$$

The purpose is to show that conditions on the data can be given such that there is a unique solution to (1)–(4), i.e., the pair (u, f) is unique determined. Let

$$\Omega = \Omega_0 \times]0, T[.$$

We define the following norms and seminorms:

$$|u| := \sup \{|u(x, t)|, x \in \Omega_0\}, \tag{5}$$

$$\|u\| := \sup \{|u(x, t)|, (x, t) \in \Omega\}, \tag{6}$$

$$\|u(\cdot, t)\|_\alpha := \sup \left\{\frac{|u(x, t) - u(y, t)|}{|x - y|^\alpha}, x \neq y, x, y \in \Omega_0\right\}, \tag{7}$$

$$\|f\| := \sup \{|f(\xi)|, \xi \in \operatorname{dom}(f)\}, \tag{8}$$

$$|f|_\alpha := \sup \left\{\frac{|f(\xi) - f(\eta)|}{|\xi - \eta|^\alpha}, \xi \neq \eta, \xi, \eta \in \operatorname{dom}(f)\right\}, \tag{9}$$

$$\|f\|_\alpha := \|f\| + |f|_\alpha. \tag{10}$$

The corresponding function spaces are denoted by $\mathbf{C}$ and $\mathbf{C}^\alpha$. By Lip_1 we shall denote the space $\mathbf{C}^\alpha$ where $\alpha = 1$.

We shall impose the following assumptions on the data (Pilant and Run-

DELL [5]):

$$A_1: \Phi(x) = 0, \qquad x \in \Omega_0,$$

$$A_2: \gamma(x, t) \in \mathrm{Lip}_1(\Omega), \qquad \gamma > 0,$$

$$A_3: \partial_t h(x, t) \in \mathrm{Lip}_1(\partial\Omega_0 \times [0, \infty[), \qquad h(x, t) \leqq h(x^0, t),$$

$$\partial_t h(x, t) > 0, \qquad h(x, 0) = 0,$$

$$A_4: g'(t) \in \mathrm{Lip}_1[0, \infty[), \quad g(0) = 0, \quad g'(t) > 0, \quad g'(0) = \gamma(x^0, 0),$$

$$A_5: \|g'(t) - \partial_t\psi(x^0, t)\|_1 < c_1.$$

By a solution pair to (1)–(4) we mean that

1. $f \in \mathbf{C}^\alpha$ for some $0 < \alpha \leqq 1$,
2. $u(x, t) \in \mathbf{C}^{2+\alpha}(\bar{\Omega}) \cap \mathbf{C}^{1+\alpha}([0, \infty[)$,
3. (1)–(4) hold.

The condition $f \in \mathbf{C}^\alpha$ is sufficient to show existence of the direct problem (1)–(3) (FRIEDMAN [1]).

Let $\psi(x, t)$ satisfy

$$\begin{aligned} &\partial_t\psi - \Delta\psi = \gamma(x, t), && (x, t) \in \Omega_0 \times]0, T[, \\ &\partial_n\psi(x, t) = h(x, t), && (x, t) \in \partial\Omega_2 \times]0, T[, \\ &\psi(x, 0) = 0. \end{aligned} \tag{11}$$

Note that if u satisfies (1)–(3) and if $f \in \mathbf{C}^\alpha$ is a Lipschitz continuous function then $u(x, t)$ can be written as (FRIEDMAN [1])

$$u(x, t) = \psi(x, t) + \int_0^t \int_{\Omega_0} K(x, t; y, \tau)\, f\big(u(y, \tau)\big)\, dy\, d\tau, \tag{12}$$

where $K(x, t; y, \tau)$ is the Green function for the heat equation with homogeneous Neumann data.

For simplicity we shall study the case $\Omega_0 = \{x: 0 < x < \infty\} \subset \mathbb{R}$. One can find results for $\Omega_0 \subset \mathbb{R}^n$ in PILANT and RUNDELL [1]. By using a fixed point theorem we shall show existence and uniqueness to the inverse problem

$$\partial_t u - \partial_x^2 u = f(u) + \gamma(x, t), \qquad 0 < x, \qquad 0 < t, \tag{13}$$

$$\partial_x u(0, t) = h(t), \qquad 0 < t, \tag{14}$$

$$u(x, 0) = \Phi(x), \qquad 0 < x, \tag{15}$$

with overposed data

$$u(0, t) = g(t), \qquad 0 < t. \tag{16}$$

The comparison function ψ is a solution of

$$\partial_t \psi - \partial_x^2 \psi = \gamma(x, t), \qquad 0 < x, \qquad 0 < t, \tag{17}$$

$$\partial_x \psi(0, t) = h(t), \qquad 0 < t, \tag{18}$$

$$\psi(x, 0) = \Phi(x), \qquad 0 < x. \tag{19}$$

In this special case the Neumann function K has the form, $t > \tau$,

$$K(x, t; y, \tau) := \frac{1}{2\sqrt{\pi(t-\tau)}} \left[\exp\left(-\frac{(x-y)^2}{4(t-\tau)} \right) + \exp\left(-\frac{(x+y)^2}{4(t-\tau)} \right) \right]. \tag{20}$$

The difference $v := u - \psi$ satisfies

$$\partial_t v - \partial_x^2 v = f(u), \qquad \partial_x v(0, t) = 0, \qquad v(x, 0) = 0.$$

Therefore we have the representation

$$u(x, t) - \psi(x, t) = v(x, t) = \int_0^t \int_0^\infty K(x, t; y, \tau)\, f\big(u(y, \tau)\big)\, \mathrm{d}y\, \mathrm{d}\tau. \tag{21}$$

From (13)–(16) it follows that

$$g'(t) - \partial_x^2 u(0, t; f) = f\big(g(t)\big) + \gamma(0, t). \tag{22}$$

The quantities g' and γ are known, but $\partial_x^2 u$ and f are not. For a given $f = f^{(0)}$ we can solve (13)–(15) and get a solution $u^{(0)} := u(x, t; f^{(0)})$. The expression

$$g'(t) - \partial_x^2 u(0, t; f^{(0)}) - \gamma(0, t)$$

will in general not be equal to $f^{(0)}\big(g(t)\big)$. We define $f^{(1)}$ by

$$f^{(1)}\big(g(t)\big) := g'(t) - \partial_x^2 u(0, t; f^{(0)}) - \gamma(0, t) \tag{23}$$

and obtain a solution of (13)–(15) denoted by $u^{(1)}(x, t; f^{(1)})$ setting $f\big(u(x, t)\big) = f\big(g(t)\big)$ if $u(x, t) = g(t)$.
Further, let

$$f^{(k+1)}\big(g(t)\big) := g'(t) - \partial_x^2 u^{(k)}(0, t; f^{(k)}) - \gamma(0, t), \qquad k \geqq 0. \tag{24}$$

We define the boundary-coefficient operator $T_g[f]$ by

$$T_g[f]\,(t) := g'(t) - \partial_x^2 u(0, t; f) - \gamma(0, t). \tag{25}$$

Then (24) can be written as

$$f^{(k+1)}\big(g(t)\big) := T_g[f^{(k)}]\,(t). \tag{26}$$

We obtain

$$\begin{aligned} f^{(k+1)}\big(g(t)\big) &= T_g[f^{(k)}]\,(t) = g'(t) - \partial_t u^{(k)}(0, t; f^{(k)}) + f^{(k)}\big(u^{(k)}(0, t)\big) \\ &= f^{(k)}\big(g(t)\big) + \big\{f^{(k)}\big(u^{(k)}(0, t)\big) - f^{(k)}\big(g(t)\big) + g'(t) - \partial_t u^{(k)}(0, t; f^{(k)})\big\} \\ &= f^{(k)}\big(g(t)\big) + \mathscr{F}[g(t) - u^{(k)}(0, t; f^{(k)})]. \end{aligned} \tag{27}$$

In order to calculate the iterates $f^{(k)}$ at the point $x = 0$ only the difference $g(t) - u^{(k)}(0, t; f^{(k)})$ is available. Therefore (26) is written in the form

$$f^{(k+1)}\big(g(t)\big) = f^{(k)}\big(g(t)\big) + \mathscr{F}[g(t) - u^{(k)}(0, t; f^{(k)})]. \tag{28}$$

Definition 7.1. A function f will be called a *g-fixed point of* $f \mapsto T_g[f]\,(t)$ if for given $g = g(t)$

$$f\big(g(t)\big) = T_g[f]\,(t). \tag{29}$$

Lemma 7.1. *If f is uniformely Lipschitz, then the pair (u, f) is a solution of* (13)–(16) *if and only if f is a g-fixed point of $f \to T_g[f]$ defined in* (29).
Proof. a) Suppose, (u, f) is a solution of (13)–(16). Then

$$\begin{aligned} f\big(g(t)\big) &= f\big(u(0, t)\big) = \partial_t u(0, t) - \partial_x^2 u(0, t) - \gamma(0, t) \\ &= g'(t) - \partial_x^2 u(0, t) - \gamma(0, t) = T_g[f]\,(t), \end{aligned} \tag{30}$$

i.e., a g-fixed point of $T_g[f]$.

b) Conversely, suppose f is a g-fixed point and u satisfies (13)–(15). We have to show that (16) is satisfied. For any pair (u, f) satisfying (13)–(15) we have

$$\gamma(0, t) - \partial_x^2 u(0, t; f) = \partial_t u(0, t) - f\big(u(0, t)\big).$$

If f is a g-fixed point we have

$$\begin{aligned} f\big(g(t)\big) = T_g[f]\,(t) &= g'(t) - \big(\partial_x^2 u(0, t) + \gamma(0, t)\big) \\ &= g'(t) - \partial_t u(0, t) + f\big(u(0, t)\big). \end{aligned}$$

Hence

$$g'(t) - \partial_t u(0, t) = f\big(g(t)\big) - f\big(u(0, t)\big).$$

Let $\alpha(t) := g(t) - u(0, t)$. Since f is uniformely Lipschitz continuous we obtain

$$\begin{aligned} |\alpha'(t)| &= |g'(t) - \partial_t u(0, t)| = \big|f\big(g(t)\big) - f\big(u(0, t)\big)\big| \\ &\leqq c|g(t) - u(0, t)| = c\,|\alpha(t)|. \end{aligned} \tag{31}$$

For $t = 0$ we have $\alpha(0) = g(0) - u(0, 0) = 0$. From

$$\alpha(t) = \int_0^t \alpha'(\tau)\,\mathrm{d}\tau \quad \text{and} \quad |\alpha'(t)| \leqq c\,|\alpha(t)|$$

we obtain

$$|\alpha'(t)| = c \left| \int_0^t \alpha'(\tau)\, \mathrm{d}\tau \right| \leqq ct_0 \sup \{|\alpha'(t)| : 0 \leqq t \leqq t_0\}.$$

Hence

$$\sup \{|\alpha'(t)| : 0 \leqq t \leqq t_0\} \leqq ct_0 \sup \{|\alpha'(t)| : 0 \leqq t \leqq t_0\}$$

or

$$1 \leqq ct_0.$$

If $\alpha(t) \not\equiv 0$ and $t_0 < 1/c$ this is a contradiction. Therefore $\alpha(t) \equiv 0$ for $0 \leqq t \leqq t_0$. A similar proof holds if $\alpha(t) \not\equiv 0$ for $t_0 \leqq t \leqq 2t_0$ and so on. We have $\alpha(t) \equiv 0$ for $0 \leqq t$. Using this result we have proved that

$$u(0, t) = g(t).$$

The pair (u, f) is a solution of (13)–(16). ∎

Let $f \in \mathbf{C}^\alpha$. We define $\tilde{f}$ by

$$\tilde{f}(g(t)) := T_g[f](t). \tag{32}$$

This uniquely defines $\tilde{f}$ as long as g is monotone. For most inverse problems such a condition is important for uniqueness (see Theorem 1.4, Corollary 7.2).

Using the representation (21) Pilant and Rundell [5] proved the following important inequality:

$$\|\tilde{f}\|_\alpha \leqq \max \{1, (\inf |g'|)^{-\alpha}\} \|g' - \partial_t \psi\|_\alpha + \frac{C(\alpha)}{\alpha(1-\alpha)} \left(1 + \frac{1}{2} t^\alpha\right) \left(\frac{|\partial_x^2 u|}{\inf |g'|}\right)^\alpha |f|_\alpha. \tag{33}$$

If $\|f\|_\alpha \leqq E$, one can ensure that $\|\tilde{f}\|_\alpha \leqq E$, provided the quantities

$$\frac{C(\alpha)}{\alpha(1-\alpha)} \left(1 + \frac{1}{2} t^\alpha\right) \left(\frac{|\partial_x^2 u|}{\inf |g'|}\right)^\alpha \quad \text{and} \quad \|g' - \partial_t \psi\|_\alpha$$

are sufficiently small. The quantity

$$\lambda := \frac{|\partial_x^2 u|}{|g'|} \tag{34}$$

is the ratio of "unknown" to "known" information.

Theorem 7.3. *Under the assumptions on the data given by* A_1–A_4 *the transformation* $f \mapsto T_g[f]$ *maps a ball to a ball in* $\mathbf{C}^\alpha$.

One can find the complete proof in Pilant and Rundell [5]. In order to prove uniqueness we have to require an additional property of the unknown function f. We say that a function $f \in \mathrm{Lip}_1$ has "*Property S*" if for given functions $u, v \in \mathbf{C}^\alpha$ the mapping $u \mapsto f(u)$ satisfies

$$\|f(u) - f(v)\|_\alpha \leqq C_s \|u - v\|_\alpha.$$

The main theorem proved by Pilant and Rundell [5] is

Theorem 7.4. *If the overposed boundary-value problem possesses a g-fixed point and f has property S, then it is unique. Furthermore, the operator* $f \mapsto T_g[f]$ *is a contraction on* $\mathbf{C}^\alpha$ *for* $\alpha < 1$ *under the assumptions of Theorem* 7.3. *Then the sequence of iterates defined by* (24) *converges strongly in* $\mathbf{C}^\alpha$, $\alpha < 1$, *to the unique solution.*

8. Global uniqueness theorems for linear differential equations

In Theorem 3.14 we proved that the measure μ is uniquely determined if the potential $G\mu$ is known on the whole space. If $G\mu$ is not known on an open subset of the support supp μ then intrinsic ambiguities arise. In this chapter we shall prove a uniqueness theorem if the solutions of a linear differential equation of second order are known on the whole space. More general uniqueness theorems were proved by BUKHGEIM and KLIBANOV [1], see also BUKHGEIM [1].

In the following we consider a linear partial differential operator $P(x, \partial)\, u$ of second order of the form

$$P(x, \partial)\, u := \sum_{j,k=1}^{n} a_{jk}(x)\, \partial_{jk}^2 u + \sum_{j=1}^{n} a_j(x)\, \partial_j u + b(x)\, u \tag{1}$$

defined on an open set $\Omega_0 \subset \mathbb{R}^n$, $n \geqq 1$. The functions a_{jk}, a_j, b are real functions on Ω_0 satisfying $a_{jk} = a_{kj}$ and $b \leqq 0$. We denote by $\mathbf{C}^k(U)$ the linear space of k-times differentiable real functions on the open set $U \subset \Omega_0$. Defining

$$\mathscr{H}_P(U) := \{u \in \mathbf{C}^2(U) : P(x, \partial)\, u = 0 \text{ on } U\} \tag{2}$$

we obtain a harmonic sheaf $\mathscr{H}_P$ on Ω_0 (BLIEDTNER and HANSEN [1], MIRANDA [1]).

A relatively compact open subset $V \subset \Omega_0$ is called *regular* if any $h \in \mathbf{C}(\partial V)$ possesses a unique continuous extension H_h^V on $\overline{V}$ such that $H_h^V|_V \in H(V)$ and $H_h^V \geqq 0$ provided $h \geqq 0$. For any $x \in V$ the mapping $h \mapsto H_h^V(x)$, $h \in \mathbf{C}(\partial V)$, is a positive linear functional, hence a measure μ_x^V on ∂V is called the *harmonic measure* of x with respect to V (in the concrete case see (2.1.8) and (2.1.45)). Then

$$H_h^V(x) = \int h\, \mathrm{d}\mu_x^V, \qquad x \in V.$$

Let X be a linear space. A convex set $K \subset X$ is called a *convex cone with vertex at p*, if $p + x \in K$ implies $p + rx \in K$, $r \geqq 0$. For any subset U of Ω_0 let $^*\mathscr{H}(U)$ be the convex cone of *superharmonic functions* on U, i.e.,

$$^*\mathscr{H}(U) := \{u : U \mapsto\,]-\infty, \infty],\ u \text{ l.s.c.}, \int u\, \mathrm{d}\mu_x^V \leqq u(x) \text{ for each regular } V \subset \Omega_0\}. \tag{3}$$

In the case of the Laplace equation see Definition 3.8, Theorem 3.21 through Theorem 3.24. By means of the superharmonic functions a generalized Dirichlet problem can be studied. Let $U \subset \Omega_0$ be an open set. For every $h\colon \partial V \mapsto \overline{\mathbb{R}}$ we define the set

$$\mathcal{U}_h^U := \{u \in {}^*\mathcal{H}(U)\colon \liminf_{x \to z} u(x) \geqq h(z) \text{ on } \partial U\} \tag{4}$$

of all *upper functions* of h. The set of all *lower functions* of h is defined by

$$\mathcal{L}_h^U := -\mathcal{U}_{-h}^U. \tag{5}$$

Then

$$\overline{H}_h^U := \inf \mathcal{U}_h^U \text{ resp. } \mathbf{H}_h^U = \sup \mathcal{L}_h^U \tag{6}$$

is called the *upper* (*lower* resp.) *solution* of the *generalized Dirichlet problem*. The functions of

$$\mathcal{R}(U) := \{h\colon \partial U \mapsto \overline{\mathbb{R}}, \mathbf{H}_h^U = \overline{H}_h^U =: H_h^U \in \mathcal{H}(U)\} \tag{7}$$

are called *resolutive*. A boundary point $z \in \partial U$ is called *regular* (with respect to U) if for every $h \in \mathrm{C}(\partial U)$

$$\lim_{x \to z, x \in U} H_h^U(x) = h(z). \tag{8}$$

The boundary point $(0, 0, 0)$ of the Lebesgue spine is no regular boundary point (see Example 3.2).

Let U be open and $z \in \partial U$. For studying the behaviour of the Dirichlet problem for the Laplace equation at a boundary point $z \in \partial U$ the mathematicians H. Poincaré (1899) and H. Lebesgue (1912) introduced the notion of a barrier. A function w is called a *barrier* at $z \in \partial U$ with respect to U if there exists an open neighbourhood V of z and $w \in {}^*\mathcal{H}(U \cap V)$ such that $w > 0$ on $U \cap V$ and $\lim_{x \to z, x \in U} w(x) = 0$. It can be proved that $z \in \partial U$ is regular if and only if z has a barrier. The method for studying the generalized Dirichlet problem is called *Perron-Wiener-Brelot method* developed from 1923–1986. One can find classical results in Kellogg [1], Helms [1] and axiomatic approach in Bliedtner and Hansen [1], Doob [1].

The operator $P(x, \partial)\, u$ defined in (1) is called

a) *continuous* if $a_{jk}, a_j, b \in \mathrm{C}(\Omega_0)$ for all $j, k = 1, 2, \ldots, n$,

b) *smooth* if $a_{jk}, a_j, b \in \mathrm{C}^\infty(\Omega_0)$ for all $j, k = 1, 2, \ldots, n$,

c) *semi-elliptic* if for every $x \in \Omega_0$ the matrix $\big(a_{jk}(x)\big)$ is positive semidefinite and non-trivial, i. e.,

$$\sum_{j,k=1}^{n} a_{jk}(x)\, \xi_j \xi_k \geqq 0 \text{ for all } x \in \mathbb{R}^n \quad \text{and} \quad a_{jk}(x) \neq 0 \quad \text{for some } j, k.$$

The elliptic differential operators defined in (2.1.41) are special semi-elliptic differential operators. Further, the heat operator defined in (2.2.1) is a special semi-elliptic differential operator on $\mathbb{R}^{n+1}$.

Proposition 8.1. *Let $P(x, \partial)\, u$ be a semi-elliptic differential operator, $u \in \mathbf{C}^2(U)$ and $x \in U$ such that $u(x) = \inf \{u(y) : y \in U\} < 0$. Then $P(x, \partial)\, u \geqq 0$.*
Proof. By assumption, $\partial_j u(x) = 0$ for every $j = 1, \ldots, n$. From $u(y) \geqq u(x)$ it follows that the matrix

$$\left(\partial^2_{jk} u(x)\right)$$

is positive-semidefinite. Since $b \leqq 0$ and $u(x) < 0$ we get

$$P(x, \partial)\, u = \sum_{j,k=1}^{n} a_{jk}(x)\, \partial^2_{jk} u(x) + b(x)\, u \geqq b(x)\, u \geqq 0. \blacksquare$$

Definition 8.1. A relatively compact open set $U \subset \mathbb{R}^n$ is called *P-small*, if there exists a positive function $w \in \mathbf{C}^2(U)$ such that $P(x, \partial)\, w < 0$ and $\Delta w \leqq 0$ on U. We shall say that U is *P-regular*, if for every $z \in \partial U$ there exists a P-barrier at z, i.e., a function $w \in \mathbf{C}^\infty(U)$ such that $w(z) = 0$, $w > 0$ on $\overline{U} \setminus \{z\}$, $P(x, \partial)\, w < 0$ on U and $\Delta w \leqq 0$ on U.

Each relatively compact open set $U \subset \mathbb{R}^n$ is P-small relative to the Laplace operator Δu, since the function $x \mapsto w(x) = c - |x - x^0|^2$, $c > 0$, satisfies Definition 8.1. Further, each relatively compact open set $U \subset \mathbb{R}^{n+1}$ is P-small relative to the heat operator. In this case we consider $w(x) = c - (|x - x^0|^2 + nt)$, $c > 0$.

Maximum principles and minimum principles are basic for developing a potential theory in the sense of Perron-Wiener-Brelot.

Theorem 8.1 (Minimum principle). *Let $P(x, \partial)\, u$ be a continuous semi-elliptic differential operator on Ω_0, let U be a P-small open subset of Ω_0, and $c \in \mathbb{R}^+$. Then for every $u \in \mathbf{C}(\overline{U})$ such that* $\operatorname{rest}_U u \in \mathbf{C}^2(U)$, *$P(x, \partial)\, u \leqq 0$ and* $\operatorname{rest}_{\partial U} u \geqq -c$ *we have $u \geqq -c$ on U. In particular, $u = 0$ if $P(x, \partial)\, u = 0$ on U and $u = 0$ on ∂U.*
Proof. Let $x^0 \in U$ and suppose $u(x^0) < -c$. Choose a function $w \in \mathbf{C}^+(\overline{U})$ such that $\operatorname{rest}_U w \subset \mathbf{C}^2(U)$ and $P(x, \partial)\, w < 0$ on U. There exists $\varepsilon > 0$ such that the function $v := u + \varepsilon w$ satisfies $v(x^0) < -c$. Since $v \geqq u \geqq -c$ on ∂U, there is a point $x \in U$ such that $v(x) = \inf \{v(y), y \in U\}$ and hence $P(x, \partial)\, v(x) \geqq 0$ by Proposition 8.1. However, $P(x, \partial)\, v = \varepsilon P(x, \partial)\, w < 0$ on U. This is a contradiction. Therefore $u(x^0) > -c$. $\blacksquare$

In the following theorem (formula (10)) we shall construct special P-regular open subsets $U \subset \mathbb{R}^n$ for a given continuous semi-elliptic operator $P(x, \partial)\, u$. In the case of the elliptic differential operator (2.1.41) the balls $B(x^0, R)$ are further regular open subsets. The open subsets defined in (10) are also regular open subsets relative to the heat operator defined in $\mathbb{R}^{n+1}$.

Theorem 8.2. *Let $P(x, \partial)\, u$ be a continuous semi-elliptic differential operator on Ω_0. Then the P-regular subsets form a base of Ω_0, i.e., each open set $U \subset \Omega_0$ can be represented as union of P-regular sets.*
Proof. Let $y \in \Omega_0$ and let V be a neighbourhood of y. Since $P(x, \partial)\, u$ is semi-elliptic there exists $z^0 \in \mathbb{R}^n$ such that $|z^0| = 1$ and

$$\sum_{j,k=1}^{n} a_{jk}(x)\, z^0_j z^0_k > 0. \tag{9}$$

Further, since $P(x, \partial)\, u$ is continuous there exists a compact neighbourhood K of y and $0 < \varepsilon < 1$ such that $K \subset V$ and

$$\sum_{j,k=1}^{n} a_{jk}(x)\, z_j z_k > 0$$

for every $x \in K$ and every $z \in \overline{B}(z^0, \varepsilon)$. The distance between y and $y \pm z^0$ is equal to 1 (see Fig. 8.1).

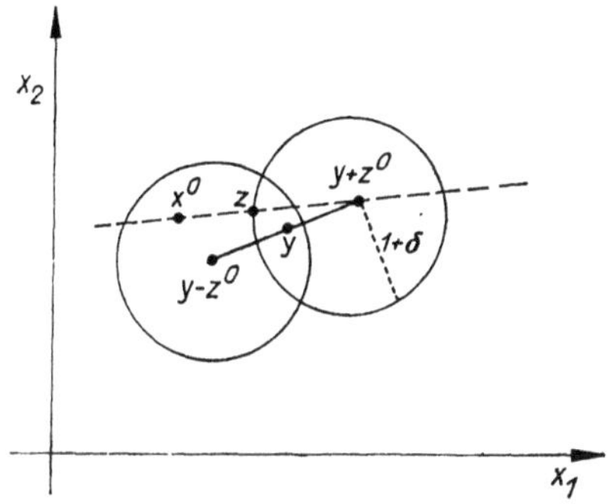

Fig. 8.1

For a sufficiently small $\delta > 0$ the *lenticular open neighbourhood*

$$U := B(y + z^0, 1 + \delta) \cap B(y - z^0, 1 + \delta) \tag{10}$$

of y is contained in $K \cap B(y, \varepsilon/4)$. In the following we shall prove that U is P-regular. Let $z \in \partial U \cap \partial B(y \pm z^0, 1 + \delta)$ and let x^0 be the point on the line joining $y \pm z^0$ and z such that (see Fig. 8.1)

$$\overline{B}\left(x^0, \frac{\varepsilon}{4}\right) \cap U = \{z\}. \tag{11}$$

If $x \in U$ we get

$$|x - x^0| = |z - x^0| \geqq \frac{\varepsilon}{4}$$

and

$$|x - x^0| \leqq |x - z| + |z - x^0| \leqq \frac{\varepsilon}{4} + \frac{\varepsilon}{4} = \frac{\varepsilon}{2},$$

i.e., $x - x^0 \in B(\pm z^0, \varepsilon)$. Hence

$$\sum_{j,k=1}^{n} a_{jk}(x)\,(x_j - x_j^0)\,(x_k - x_k^0) > 0 \tag{12}$$

for every $x \in U$. We define $w\colon \mathbb{R}^n \mapsto \mathbb{R}$ by

$$w(x) := \mathrm{e}^{-k\varepsilon/4} - \mathrm{e}^{-k|x-x^0|^2}. \tag{13}$$

Then $w \in \mathbf{C}^\infty(\mathbb{R}^n)$, $w(z) \geqq 0$ on $\overline{U} \setminus \{z\} \subset \mathbb{R}^n \setminus B(x^0, \varepsilon/4)$ and $w(z) = 0$. For every $x \in \mathbb{R}^n$ we get

$$P(x, \partial)\, w(x) = 2k\, \mathrm{e}^{-|x-x^0|^2}\Bigg[-2k \sum_{j,k=1}^{n} a_{jk}(x)\,(x_j - x_j^0)\,(x_k - x_k^0) + \sum_{j=1}^{n} \big(a_{jj}(x) + a_j(x)\,(x_j - x_j^0)\big)\Bigg] + b(x)\, w(x) \tag{14}$$

and

$$\Delta w(x) = 2k\, \mathrm{e}^{-|x-x^0|^2}[-2k\,|x - x^0|^2 + n], \tag{15}$$

where $b(x)\, w(x) \leqq 0$ if $x \in U$. Hence w is a P-barrier at z if k is sufficiently large.

Following Bliedtner and Hansen [1] we shall say that a *Bauer space* $(X, \mathscr{H})$ is associated with a continuous semi-elliptic differential operator $P(x, \partial)\, u$ on $X = \Omega_0$ if $\mathscr{H} \subset \mathscr{H}_P$, i.e., if $\mathscr{H}(V) \subset \mathscr{H}_P(V)$ for every open subset V of Ω_0. The minimum principle of Theorem 8.2 allows to prove the following important theorem.

Theorem 8.3. *Let $(X, \mathscr{H})$ be a Bauer space associated with a continuous semi-elliptic differential operator $P(x, \partial)\, u$ on X. Then, for every open subset U of X,*

$$\mathscr{H}(U) = \mathscr{H}_L(U) \text{ and } {}^*\mathscr{H}(U) \cap \mathbf{C}^2(U) = \{u \in \mathbf{C}^2(U)\colon P(x, \partial)\, u \leqq 0 \text{ on } U\}.$$

Proof. Let $h \in \mathscr{H}_P(U)$ and let V be a regular P-small set satisfying $\overline{V} \subset U$. Then $v = H_h^V \in \mathscr{H}(V) \subset \mathscr{H}_P(V)$, hence $P(x, \partial)\,(v - h) = 0$ on V and $v - h = 0$ on ∂V. By Theorem 8.1 $v = h$ on V and therefore $h \in \mathscr{H}(U)$ since $\mathscr{H}$ is a sheaf. Let $u \in \mathbf{C}^2(U)$ such that $P(x, \partial)\, u \leqq 0$ on U. Let V be a regular P-small set such that $\overline{V} \subset U$. We define $v := H_u^V$. Then $P(x, \partial)\,(u - v) \leqq 0$ on V and $u = v$ on ∂V. By Theorem 8.1 we conclude that $u - v \geqq 0$ on V, i.e., $u \geqq v = H_u^V$ or $u \in {}^*\mathscr{H}(U)$.

Conversely, let $u \in {}^*\mathscr{H}(U) \cap \mathbf{C}^2(U)$. Assume $P(x, \partial)\, u(x) > 0$ for some $x \in U$, hence $P(x, \partial)\, u > 0$ on some neighbourhood V of x. Using the above result, $-u$ is superharmonic on V. Thus $u \in \mathscr{H}(V)$, i.e., $P(x, \partial)\, u = 0$. That is a contradiction. ∎

Theorem 8.4. *Let $(X, \mathscr{H})$ be a normal Bauer space associated with continuous semi-elliptic differential operators $P(x, \partial)\, u$ and $M(x, \partial)\, u$. Then there exists $f \in \mathbf{C}^+(X)$ such that $P(x, \partial)\, u = fM(x, \partial)\, u$.*

Proof. Let $x \in X$ and define for $\alpha_{jk}, \alpha_j \in \mathbb{R}$, $j, k = 1, 2, \ldots, n$,

$$u(y) := \sum_{j,k=n}^{n} \alpha_{jk}(y_j - x_j)\,(y_k - x_k) + \sum_{j=1}^{n} \alpha_j(y_j - x_j), \qquad y \in \mathbb{R}^n.$$

If

$$2 \sum_{1 \leqq j \leqq k \leqq n} a_{jk}(x)\, \alpha_{jk} + \sum_{j=1}^{n} a_j(x)\, \alpha_j < 0$$

then $P(x, \partial)\, u(x) < 0$, hence $P(x, \partial)\, u < 0$ in a neighbourhood V of x. From

Theorem 8.3 it follows that $u \in {}^*\mathcal{H}(V) \cap \mathbf{C}^2(V)$ and $M(x, \partial)\, u \leqq 0$ on V. This implies

$$2 \sum_{1 \leqq j \leqq k \leqq n} b_{jk}(x)\, \alpha_{jk} + \sum_{j=1}^{n} b_j(x)\, \alpha_j \leqq 0,$$

where b_{jk} and b_j are the coefficients of M. Since x is fixed there exists a real number $f(x) > 0$ such that

$$f(x) \left(\sum_{1 \leqq j \leqq k \leqq n} b_{jk}(x)\, \alpha_{jk} + \sum_{j=1}^{n} b_j(x)\, \alpha_j \right) = \sum_{1 \leqq j \leqq k \leqq n} a_{jk}(x)\, \alpha_{jk} + \sum_{j=1}^{n} a_j(x)\, \alpha_j .$$

The continuity of $P(x, \partial)\, u$ and $M(x, \partial)\, u$ implies the continuity of f. ∎

Further, in Bliedtner and Hansen [1] it is proved, that every smooth Bauer space $(X, \mathcal{H})$ is associated with a semi-elliptic differential operator $Q(x, \partial)\, u$ on $X \subset \mathbb{R}^n$.

Potential theory has reached a high level. There are many deep results which can be applied to concrete problems in mathematical physics.

One can find the latest results on harmonic spaces with applications to the Helmholtz equation in Hansen and Hueber [1] and Sturm [1]. In these papers the equation $\Delta u + V(x)\, u = \lambda u$ is replaced by $\Delta u + \mu u = \lambda \nu u$, where μ is a signed measure and ν a positive measure, $\nu u(f) = \int f(y)\, u(y)\, \mathrm{d}\nu(y)$. These problems are completely solved. In future we have to apply such results to inverse problems of coupled fields in geophysics, medicine, and so on.

9. Some historical remarks

As we have seen in this book, the interpretation of nature in general is done by means of an equation of the first kind $Af = g$ (inverse problem). One of the first inverse problems was explicitly solved by N. H. ABEL [1, 2] in 1823. This problem consists in finding a curve $\mathscr{C}$ in a vertical plane where $\mathscr{C}$ is assumed to be a graph of an increasing function $x = \Phi(y)$ such that the falling time of a body (falling only under the influence of gravity) is a known function $t(y)$ of the height from which the body falls. This problem reduced to an Abel integral equation has a unique solution. If Φ is multi-valued then infinitely many solutions exist.

In 1867 G. G. Stokes for the first time raised the question of the relationship between the Newtonian potential at the exterior of a closed surface containing an attracting body, and the distribution of the body's density ϱ producing the assigned potential. In his lectures on Celestical Mechanics at the University of Pavia in 1875/76 the astronomer G. V. Schiaparelli showed how the internal density of a body can be altered without affecting the outer potential (see MARUSSI [1]). In 1889 J. BERTRAND [1] applied the notion '*mal posée*' in probability in the sence underdetermined.

In 1877 Lord RAYLEIGH [1] in his book '*The theory of sound*' briefly discussed the possibility of inferring the density distribution from the frequencies of vibration. Its modern analogue and generalization was given in a famous lecture by Marc Kac (1966) entitled '*Can one hear the shape of a drum?*' (see CHADAN and SABATIER [1]).

In 1905–1907 E. Wiechert, H. Benndorf and G. HERGLOTZ [1] studied the inverse kinematic problem for a spherical earth under the assumption that the propagation velocity increases monotonically with depth. Herglotz reduced this problem to an Abel integral equation. When the propagation velocity is not monotone, the inverse problem does not have a unique solution. Additional information is necessary to study the inverse problem in seismics.

In 1906 C. NEUMANN [1] studied a special inverse source problem relative to the logarithmic potential. It is well known that the external potential of a ball with constant density can be replaced by a point measure at the center of the ball. C. Neumann proved a similar result for an ellipse. The mass distribution is concentrated on the straight line connecting the two foci. C. Neumann writes in his paper the following interesting sentences:

‚*Bei wissenschaftlichen Arbeiten möchte man gerne in gleichmäßiger und ein-*

heitlicher Weise vorgehen. Das wird aber mehr oder weniger verhindert, wenn einzelne Punkte von hervorragender Schwierigkeit sich in den Weg stellen. Denn die Überwindung solcher Schwierigkeiten verlangt in der Regel die Entdeckung und Anwendung von neuen Methoden und verlangt namentlich auch, daß man das im Wege stehende Hindernis mit allen möglichen Mitteln und von ganz verschiedenen Seiten her in Angriff zu nehmen sucht.'

G. Herglotz [2] further developed C. Neumann's ideas in 1914. For similar problems see also A. Wangerin (1915, 1917) (see Lichtenstein [1], p. 209). Herglotz proved that the external logarithmic potential of a domain bounded by algebraic curves having constant density can be replaced by potentials whose mass distribution is concentrated on lines connecting the foci. Further, he proved similar results for rotating bodies in $\mathbb{R}^3$. In that case the external potential is uniquely determined by the special mass distributions considered. This result is the first uniqueness theorem for inverse source problems. Herglotz's paper which appeared during the First World War was forgotten.

Further, the results of the geodesist P. Pizetti and of G. Lauricella are not well known (see Marussi [1]). Pizetti introduced the concept of bodies of null attraction in 1909–1910. In his fundamental papers published in 1911–1912 G. Lauricella characterized any distribution of a body of null attraction by the relation

$$\int_{\Omega} \varrho_0(x)\, u(x)\, \mathrm{d}x = 0 \text{ for all } u \in \mathrm{C}(\bar{\Omega}) \text{ satisfying } \Delta_n u = 0 \text{ in } \Omega.$$

This was the first equivalence relation in inverse problems and fundamental in applying results of functional analysis (G. Anger [8]). In 1938 V. G. Novikov [1] used this equivalence relation for proving uniqueness theorems relative to Newtonian potentials, in 1956 V. K. Ivanov [1] studied the inverse problem of potential for a body closely approximated by a given one. Later similar methods were used by A. I. Prilepko [2, 3], V. I. Isakov [1], V. N. Strakhov and M. A. Brodsky [1, 2]. From a systematic point of view such equivalence relations can be studied only by using results of modern potential theory and functional analysis. It was about the year 1960 that this approach had become possible (Anger [6, 12]). In this way one gets new results which might have applications in physics like astronomy, solid-state physics, microelectronics, diffusion, and so on. Further remarks concerning the contributions of Italian scientists to inverse problems can be found in Marussi [1] and Oganesyan and Starostenko [1].

Before the year 1906, the Dutch physicist H. A. Lorentz studied the so-called Radon transform (computed tomography). He found a solution of the three-dimensional problem where a function is to be determined (recovered) from its integrals over plans. A complete solution was given by J. Radon in 1917. There are many important contributions to this problem, for instance, by F. John (1934–1955), V. Ambartsumian (1936), J. Szarski and T. Wazewski (1947), A. M. Cormack (1963–1982), G. N. Hounsfield (1973). The latter two scientists were awarded the Nobel price in 1979. There is a long time from the first idea to the concrete realization in computer tomography and other fields. Similar methods are used in differential geometry and for inverse problems in hyperbolic equations (Lavrent'ev et al. [1], Romanov [1, 2]). We remark that the Radon transform can be uniquely inverted. The inverse contains derivatives and therefore is an ill-posed problem. Further difficulties arise in discretization. Only finitely many measurements are possible. Computer tomography works. One sees that most difficulties could be clarified (see NATTERER [1]). But for most inverse problems like in geophysics and medicine the information content of an inverse problem is too weak, i.e., there exist infinitely many internal parameters which give the same measured data. Therefore a unique inversion is not possible. If all solutions have been characterized by mathematicians one can try to realize some of them by experiments.

In 1936 R. E. LANGER [1] studied a special identification problem. The conductivity to be determined depends only on one variable. The existence of the inner core of the Earth was proved by I. Lehmann in 1936. Identification problems relative to the heat equation are of special interest. Since 1962, such problems have been studied by J. R. Cannon, P. DuChateaux, J. R. Beck and many other scientists. One can find the history of inverse problems relative to the heat equation in BECK et al. [1] and CANNON [2]. Inverse problems for hyperbolic equations have a long tradition. One can find historical remarks in BALTES [1, 2], CASSINIS [1], COLTON [1], GRANT [1], LAVRENT'EV et al. [1], ROMANOV [1, 2].

Most papers were published on inverse spectral problems. An operator is to be determined by its spectrum. The first who proved a uniqueness theorem was V. AMBARTSUMIAN [1] in 1929. The specific problem of reconstructing the coefficients of a differential operator, or the boundary conditions, of a Sturm-Liouville system from its spectral data at first was seriously attacked by G. BORG [1] in 1946. It was then pursued in detail and in depth by N. Levison (1949). L. A. Chudov (1949, 1956), V. A. Marchenko (1950–1982), I. M. Gelfand and B. M. Levitan (1951), M. G. Krein (1951–1955), Ju. M. Berezanskij (1955, 1958), P. C. Sabatier (1965–1984), R. G. Newton (1955–1986), K. Chadan (1955–1976), L. D. Faddeev (1956–1984), O. E. Hald (1977), H. P. Baltes, and other scientists. For historical remarks see CHADAN and SABATIER [1], BALTES [1, 2], LEVITAN [1] and NEWTON [1]. For the inverse problem relative to the Schrödinger equation and the latest results as well as unsolved problems see R. G. NEWTON [1, 2]. Further applications can be found in ABLOWITZ and SEGUR [1], BEREZANSKIJ [1], MCLAUGHIN [1], PAO et al. [1], PÖSCHEL and TRUBOWITZ [1], TALENTI [1].

In 1923 J. HADAMARD [1] introduced the notion of a well-posed problem. In inverse theory often exist different solutions $f \neq 0$ satisfying $Af = g$. If A^{-1} is uniquely determined, then A^{-1} can be discontinuous like in computed tomography and potential theory. In the latter two cases A^{-1} contains derivatives. Let $A: X \mapsto Y$ be a compact one-to-one mapping on an infinite-dimensional space. Then A^{-1} is discontinuous. This result has many consequences in differential equations. In order to study ill-posed problems as well as their numerical aspect, different methods were developed by several authors, such as A. N. Tikhonov (1943, 1963), V. K. Ivanov (1962), M. M. Lavrent'ev (1962), R. Lattès and J. L. Lions (1969), V. G. Morozov (1966), M. Z. Nashed, and so on. For historical remarks see IVANOV et al. [1], TIKHONOV et al. [1], MOROZOV [1], HOFMANN [1], ZIELKE [1] and BAUMEISTER [1] and ENGL et al. [1].

Conclusions

The main objective of this book is to discuss the information content of equations of mathematical physics. Most problems in areas such as physics, chemistry, biology, medicine, technology, ecology and economy are often governed by a very complex system of (nonlinear) partial differential equations. The coefficients and the source term of the system are internal parameters having a special meaning. The problem mathematics has to solve is to identify those external parameters of the system to be measured, together with those subsets of $\mathbb{R}^3$, on which the measurements are to be made, so that several unknown coefficients or sources of the differential equations, or functional of these quantities inaccessible to measurement and depending continuously on the measured external parameters can be uniquely determined. In order to be able to make concrete statements about a system of partial differential equations one must often vastly simplify the system. This leads to a situation similar to that which arises when the system studied is modeled too coarsely, or the differential equations are discretized. Sometimes too few measured values or no measured values at all are available. The model simplification leads to intrinsic ambiguities, i.e., different values of the internal parameters can give the same external parameters (measured values). Such intrinsic ambiguities are typical for inverse problems. The following facts hold for inverse problems: (i) It is exceptional that an inverse problem can be uniquely solved. If such a problem has a unique solution, A^{-1} often is discontinuous. (ii) In general one can study the set of all solutions of an inverse problem as a first step. (iii) In a laboratory one can attempt to select a solution which is of interest in applications.

Most sophisticated methods of mathematics are necessary for studying the information content of an inverse problem. These unsolved problems are problems of pure mathematics.

By using the sense-organs one can to a certain degree determine the information content of electromagnetic waves, acoustic waves, chemical products, and so on. But it is not possible to look behind a hill or behind the moon. An inverse problem is either solvable or not solvable. For most physical problems we do not have sense-organs. Sometime we can decide inverse problems by means of mathematical models and experiments.

In science, technology and medicine most results were obtained in a laboratory. In a laboratory most internal parameters of a special process can be kept constant. By using such results, special physical apparatus or machines have been constructed. Most results obtained in mathematics are proved by using constructive methods. This corresponds to the deductive way of thinking of mathematicians. But solving inverse problems is much more difficult. If a problem has too large an information content, like in geophysics, soil mechanics, space research, meteorology, production of materials, biology, agriculture and in other fields, practical experiences often can only be applied thanks to the fact that in such a system many parameters may simultaneously change.

The practical importance of inverse and improperly posed problems is such that they may be considered among the pressing problems of current mathematical research. One can find unsolved problems in the theory of inverse problems in Anger [16].

References

ABEL, N. H.
[1] Solution de quelques problèmes à l'aide d'integrales definies. Magazin for Naturvidenskaberna, Aergang, Bind 2, Christiana, **11** (1823).
[2] Résolution d'un problème de mécanique. J. Reine Angew. Math. **1** (1826), 97–101.

ABLOWITZ, M. J., and H. SEGUR
[1] Solitons and the inverse scattering transform. SIAM Studies in Applied Mathematics **4**, Philadelphia, Pa., Soc. Ind. Appl. Mathem. (1981).

ABLOWITZ, M. J., B. FUCHSSTEINER and M. KRUSKAL (eds.)
[1] Conference on nonlinear evolution equations and the inverse scattering transform. World Scientific Publishing, Singapore 1987.

AIFANTIS, E. C.
[1] On the problem of diffusion in solids. Acta Mechanica **37** (1980), 265–296.

ALEKSEEV, A. S., and B. A. BUBNOV
[1] On a combined statement of inverse seismics and gravimetry problems. Dokl. Akad. Nauk SSSR **261** (1981) 5, 1086–1090 (Russian).
[2] Stability of the solution of an inverse problem of coupled seismology and gravimetry. Dokl. Akad. Nauk SSSR **275** (1984) 2, 332–335 (Russian).

ALEKSEEV, G. A., and A. YU. CHEBOTARËV
[1] Inverse problems of acoustic potentials. Zh. Vychisl. Mat. i Mat. Fiz. **25** (1985) 8, 1189–1199 (Russian).

ALEKSIDZE, M.
[1] Approximation methods of the solutions of direct and inverse problems in gravimetry. Nauka, Moscow 1987 (Russian).

ALESSANDRINI, G.
[1] On the identification of the leading coefficient of an elliptic equation. Boll. Un. Mat. Ital., Ser. VI, **4** (1985), 87–111.
[2] Stable determination of conductivity by boundary measurements. Appl. Anal. (to appear).

ALIFANOV, O. M., E. ARTJUKHIN and S. RUMYANTSEV
[1] Extremal methods for solving improperly posed problems. Nauka, Moscow 1988 (Russian).

AMBARZUMIAN, V.
[1] Über eine Frage der Eigenwerttheorie. Z. Physik **53** (1929), 690–695.

ANDERSSEN, R. S.
[1] The linear functional strategy for improperly posed problems. In: CANNON and HORNUNG [1], pp. 11–30.

ANDERSSEN, R. S., and G. N. NEWSAN
[1] Inverse problems. Papers from the workshop held in Canberra, February 3–7, 1986. Canberra: Centre for Mathematical Analysis. Australian National University, 1986.

ANGER, G.
[1] Stetige Potentiale und deren Verwendung für einen Neuaufbau der Potentialtheorie. Thesis, Technische Hochschule Dresden, Dresden 1957.
[2] Sur le rôle des potentiels continues dans les fondements de la théorie du potentiels. Séminaire de théorie du potentiel, t. 2 (1958), 30 pp., Institut H. Poincaré, Université de Paris.
[3] Die Entwicklung der Potentialtheorie im Hinblick auf ihre grundlegenden Existenzsätze. I, II. Jahresber. Deutsch. Math.-Verein. **64** (1961), 51–78, 100–134.
[4] Funktionalanalytische Betrachtungen bei Differentialgleichungen unter Verwendung von Methoden der Potentialtheorie. Akademie-Verlag, Berlin 1967.
[5] Eindeutigkeitssätze und Approximationssätze für Potentiale II. Math. Nachr. **50** (1971), 229–244.
[6] Direct and inverse problems in potential theory. In: Colloquium on Non-Linear Evolution Equations and Potential Theory. Proceedings of a summer school held in Podhradi (ČSSR) in September 1973, pp. 11–44. Akademia, Prague 1975.
[7] Die Rolle der modernen Potentialtheorie in der Theorie der inversen Aufgabenstellungen (transl. into English). Gerlands Beitr. Geophysik **85** (1976), 1–20.
[8] Uniquely determined mass distributions in inverse problems. Veröffentl. Zentralinstitut Phys. Erde **52** (1977) 2, 633–656, Akad. Wiss. DDR, Potsdam.
[9] Some remarks on inverse problems. In: Colloquium on Elliptic Differential Equations, pp. 31–51. Rostocker Mathematisches Kolloqium, Universität Rostock, 1979.
[10] Ein inverses Problem der Wärmeleitungsgleichung I. In: Lecture Notes in Mathem. **743** (1979), 485–507.
[11] Inverse and improperly posed problems in differential equations. Conference held at the Martin Luther University Halle–Wittenberg, May 29–June 2, 1979. Mathematical Research, vol. 1, Akademie-Verlag, Berlin 1979.
[12] Lectures on potential theory and inverse problems. In: ANGER and RÖSLER [1], pp. 15–95.
[13] A characterization of the inverse gravimetric source problem through extremal measures. Rev. Geophys. Space Phys. **19** (1981) 2, 299–306. Reprinted in GRAFAREND and RAPP [1].
[14] Einige Betrachtungen über inverse Probleme, Identifikationsprobleme und inkorrekt gestellte Probleme. In: Jahrbuch Überblicke Mathematik 1982, pp. 55–71. Bibliographisches Institut, Mannheim–Zürich.
[15] Inverse problems in mathematics, science and technology. Results, open problems and tendencies in the development. In: MICHEL [1], vol. 2, pp. 41–52.
[16] Unsolved problems in the theory of inverse problems. In: Partial Differential Equations, Conference held on the occasion of the 75th birthday of Academician S. L. Sobolev, Novosibirsk, October 10–14, 1983, pp. 17–31. Nauka, Novosibirsk

1986. English original Preprint 04/84. Institute of Mechanics, Academy of Sciences of the GDR, Karl-Marx-Stadt.

[17] On the relationship between mathematics and its applications: a critical analysis by means of inverse problems. Inverse Problems **1** (1985) 2, L7–L11.

[18] On inverse problems in differential equations: a critical analysis. In: 5th international symposium "Geodesy and Physics of the Earth" (Magdeburg 1984). Veröffentl. Zentralinstitut Phys. Erde **81** (1986) 1, 38–44, Akad. Wiss. DDR, Potsdam.

[19] Inverse problems: examples, counterexamples, consequences for the applications. In: SKORVANEK [1], pp. 7–20.

[20] On diagnostics in science, technology and medicine (German). In: MICHEL [1], vol. 38, pp. 14–29.

[21] The information content of an inverse problem: a theory of equivalence classes. Indianapolis J. Math. Phys. **3** (1984–1988), 57–83.

[22] Zur Interpretation von Meßergebnissen. Z. Anal. Anwendungen **7** (1988) 6, 483–493.

ANGER, B., and H. BAUER

[1] Mehrdimensionale Integration. Eine Einführung in die Lebesguesche Theorie. de Gruyter, Berlin–New York 1976.

ANGER, G., and R. CZERNER

[1] Solution of an inverse problem for the heat equation by methods of modern potential theory. In: ANGER [11], pp. 9–23.

[2] An inverse source problem for the heat conduction equation II, III. Math. Nachr. **105** (1982), 163–170; **108** (1982), 73–78.

ANGER, G., and E. KLEINE

[1] Some special inverse problems for the Laplace equation and Helmholtz equation. Z. Anal. Anwendungen **4** (1985) 6, 493–502.

ANGER, G., and R. RÖSLER (eds.)

[1] Die moderne Potentialtheorie als Grundlage des inversen Problems der Geophysik (Sommerschule der KAPG Arbeitsgruppe 5.2., Freiberg (Sachsen), 5.–16. Juni 1978). Geodätische und Geophysikalische Veröffentlichungen, Reihe III, Heft 45 (1980).

ATAKHODZHAEV, M. A.

[1] Improperly posed problems for the biharmonic equation. Fan, Tashkent: 1986 (Russian).

BACKUS, G. E.

[1] Inference from inadequate and inaccurate data. In: REID [1], vol. 2, pp. 1–105.

BAKUSHINSKIJ, A., and A. GONCHARSKIJ

[1] Iteration methods for solving improperly posed problems. Nauka, Moscow 1988 (Russian).

BALLANI, L.

[1] Inverse problems concerning the dynamics of the earth's rotation. In: HOLOTA [1], pp. 851–878.

BALTES, H. P. (ed.)

[1] Inverse source problems in optics. Topics in Current Physics 9. Springer-Verlag, Berlin–New York 1978.

[2] Inverse scattering problems. Topics in Current Physics 20. Springer-Verlag, Berlin — New York 1980.
[3] Forschung in theoretischer optischer Physik: ein Überblick. Landis & Gyr-Mitteilungen 1—80, pp. 7—11. Landis and Gyr, Zug (Switzerland) 1980.

BALZER, W.
[1] Theorie und Messung. Springer-Verlag, Berlin—New York: 1985.

BARCILON, V.
[1] Inverse problem for the vibrating beam in the free-clamped configuration. Phil. Trans. R. Soc. London **A 304** (1982), 211—251.
[2] Explicit solution of the inverse problem for a string. J. Math. Anal. Appl. **93** (1983), 222—231.
[3] Inverse eigenvalue problems. In: TALENTI [1], pp. 1—51.

BARCILON, V., and G. TURCHETTI
[1] Extremal solutions of inverse eigenvalue problems with finite spectral data. Wave Motion **2** (1980) 2, 139—148.

BARTHELMES, F.
[1] Untersuchungen zur Approximation des äußeren Gravitationsfeldes der Erde durch Punktmassen mit optimierten Positionen. Veröffentl. Zentralinstitut Phys. Erde **92** (1986), Akad. Wiss. DDR, Potsdam.

BAUER, H.
[1] Korovkin approximation in function spaces. Approximation theory and functional analysis (Proc. Intern. Sympos. Approximation Theory, Univ. Estadual de Campinas), Campinas 1977, pp. 19—36. North-Holland Publishing, Amsterdam—New York 1979.
[2] Probability theory and elements of measure theory. Academic Press, New York—London 1981.
[3] Harmonische Räume. In: Jahrbuch Überblicke Mathematik 1981, pp. 9—35. Bibliographisches Institut, Mannheim—Zürich—Wien 1981.
[4] Harmonic spaces, a survey. III Gruppo dei Seminari Matematici delle Università Italiane, Conference del Seminario di Matematica dell'Università di Bari, no. 197 (1984), 34 pp.

BAUMEISTER, J.
[1] Stable solution of inverse problems. Vieweg & Sohn, Braunschweig 1987.

BEALS, R.
[1] The inverse problem for ordinary differential operators on the line. Amer. J. Math. **107** (1985) 2, 281—366.

BECK, J. V., B. BLACKWELL and C. R. S. CLAIR
[1] Inverse heat conduction. John Wiley, Sussex 1985.

BEDNAR, J. B., J. REDNER and A. WEGLEIN (eds.)
[1] Inverse scattering: theory and applications. Soc. Ind. Appl. Math., Philadelphia, Pa. 1983.

BERDICHEVSKIJ, M. N. and M. S. ZHDANOV
[1] Interpretation of the variable electro-magnetic field of the Earth. Nedra, Moscow 1981 (Russian).

BEREZANSKIJ, YU. M. (ed.)
[1] Direct and inverse problems of spectral theory of differential operators. Akad. Nauk Ukrain. SSR, Inst. Mat., Kiev 1987 (Russian).

BERS, L.
[1] Mathematical aspects of subsonic and transonic gas dynamics. John Wiley, New York 1958.

BERTERO, M.
[1] Regularization methods for linear inverse problems. In: TALENTI [1], pp. 52–113.

BERTERO, M., C. DE MOL and E. R. PIKE
[1] Applied inverse problems in optics. In: ENGL and GROETSCH [1], pp. 291–314.

BERTRAND, J.
[1] Calcul des probabilités. Gauthier-Villars, Paris 1889.

BEZNOSHTSHENKO, N. JA.
[1] Determining the coefficient in a parabolic equation. Differencial'nye Uravnenija **10** (1974), 24–35 (Russian).

BLEISTEIN, N.
[1] Mathematical methods for wave phenomena. Academic Press, Orlando, Fla. 1984.

BLIEDTNER, J., and W. HANSEN
[1] Potential theory. An analytic and probabilistic approach. Springer-Verlag, Berlin–New York 1986.

BOCK, H. G.
[1] Recent advances in parameteridentification techniques for ODE. In: DEUFLHARD and HAIRER [1], pp. 95–121.

BOERNER, W. M., H. BRAND, L. A. CRAM, T. D. GJESSING, A. K. JORDAN, W. KEYDEL, G. SCHWIERZ, M. VOGEL (eds.)
[1] Inverse methods in electromagnetic imaging I, II. Reidel, Dordrecht 1985.

BOLT, B. A.
[1] What can inverse theory do for applied mathematics and its applications. Austral. Math. Soc. Gaz. **7** (1980), 69–78.

BORG, G.
[1] Eine Umkehrung der Sturm-Liouvillschen Eigenwertaufgabe. Acta Math. **78** (1946), 1–96.

v. BORZESKOWSKI, H.-H.
[1] Sind Gravitonen physikalisch real? Spectrum (1987) 2, 15–17.

BOURBAKI, N.
[1] Intégration, Chap. I–IV. Hermann, Paris, 1952.
[2] Espaces vectorielles topologiques, Chap. III–V. Hermann, Paris, 1955.

BRODSKY, M. A.
[1] On the uniqueness of the inverse potential problem for homogeneous polyhedrons. SIAM J. Appl. Math. **46** (1986) 2, 345–350.

BRODSKY, M. A. and STRAKHOV, V. N.
[1] Solution of the inverse potential problem for polyhedra with variable polynomial densities. Sov. Phys., Dokl. **32**, 209–210 (1987); transl. from Dokl. Akad. Nauk SSSR **293**, 336–339 (1987).

BUKHGEIM, A. L.
[1] Volterra equations and inverse problems. Nauka, Novosibirsk 1983 (Russian).
[2] Introduction to the theory of inverse problems. Nauka, Moscow 1988 (Russian).

BUKHGEIM, A. L., and M. V. KLIBANOV
[1] Uniqueness in the large of a class of multidimensional inverse problems. Dokl. Akad. Nauk SSSR **260** (1981), 269–272 (Russian).

CANNON, J. R.
[1] Determination of an unknown coefficient in a parabolic differential equation. Duke Math. J. **30** (1963) 2, 313–323.
[2] The one-dimensional heat equation. Encyclopedia of mathematics and its applications. 23. Addison-Wesley, Reading, Mass. 1984.

CANNON, J. R., and P. DUCHATEAU
[1] Determining unknown coefficients in a nonlinear heat conduction problem. SIAM J. Appl. Math. **24** (1973) 3, 298–314
[2] An inverse problem for a nonlinear diffusion equation. SIAM J. Appl. Math. **39** (1980), 272–289.
[3] An inverse problem for an unknown source term in a wave equation. SIAM J. Appl. Math. **43** (1983) 3, 553–564.

CANNON, J. R., and U. HORNUNG (eds.)
[1] Inverse problems. Proceedings of the Conference held at the Mathematical Research Institute at Oberwohlfach, May 18–24, 1986. Birkhäuser-Verlag, Basel–Boston, Mass. 1986.

CANNON, J. R., and S. PÉREZ-ESTEVA
[1] A note on an inverse problem related to the 3-D heat equation. In: CANNON and HORNUNG [1], 133–138.

CANNON, J. R., and W. RUNDELL
[1] An inverse problem for an elliptic partial differential equation. J. Math. Anal. Appl. **126** (1987), 329–340.

CARRION, P. M.
[1] Inverse problems and tomography in acoustics and seismology. Penn, Atlanta, GA 1987.

CARROLL, R.
[1] Transmutation theory and applications. Notas de Matématica, 105. North-Holland Publishing, Amsterdam–New York 1985.

CARROLL, R., and F. SANTOSA
[1] On complete recovery of geophysical data. Math. Methods Appl. Sci. **4** (1982), 33–73.

CASSINIS, R. (ed.)
[1] Solution of the inverse problem in geophysical interpretation. Ettore Majorana International Studies: Physical Sciences, 11. Plenum-Press, New York–London 1981.

CHADAN, K., and P. C. SABATIER
[1] Inverse problems in quantum scattering theory. Springer-Verlag, Berlin—New York 1977.

CHOQUET, G.
[1] Lectures on analysis I, II, III. W. A. Benjamin, New York—Amsterdam 1969.

COHEN, P.
[1] Set theory and the continuum hypothesis. W. A. Benjamin, New York—Amsterdam 1966.

COLTON, D.
[1] The inverse scattering problem for time-harmonic acoustic waves. SIAM Review **26** (1984) 3, 323—350.

COLTON, D., and R. KRESS
[1] Integral equation methods in scattering theory. John Wiley, New York 1983.

COLTON, D., P. MONK and F. SANTOSA
[1] Inverse problems in scattering theory. In: Oakland conference on partial differential equations and applied mathematics (Rochester, Mich. 1986), Pitman Res. Notes Math. Ser. 154, 39—73. Longman Sci. Techn., Harlow 1987.

CORMACK, A. M.
[1] Computed tomography: some history and recent developments. In: SHEPP [1], pp. 35—42.

DAHN, B. I.
[1] Zwei Wegbereiter der modernen Logik: Alfred Tarski (1901—1983) und Kurt Gödel (1906—1978). Wissenschaft und Fortschritt (1986) 12, 324—326.

DAY, W. A.
[1] Further remarks on a property of the equations of dynamic thermoelasticity. Arch. Rat. Mech. Analysis **84** (1984), 69—82.

DENY, J.
[1] Les potentiels d'énergie finie. Acta Math. **82** (1950), 107—183.

DEUFLHARD, P., and E. HAIRER
[1] Numerical treatment of inverse problems in differential equations. Proceedings of an international workshop held at Heidelberg, August 30—September 3, 1982. Progress in Scientific Computing, 2. Birkhäuser-Verlag, Basel—Boston, Mass. 1983.

DODD, R. K., J. C. EILBECK and J. D. GIBBON
[1] Solitons and nonlinear wave equations. Academic Press, New York—London 1982.

DONT, M.
[1] On the continuity of heat potentials. Casopis Pest. Mat. **106** (1981), 156—167.

DOOB, J. L.
[1] Classical potential theory and its probabilistic counterpart. Springer-Verlag, Berlin—New York 1984.

DOUGLAS, R. G.
[1] On the extremal measures and subspace density. Michigan Math. J. **11** (1964), 243—246.

DuChateau, P.
[1] Monotonicity and uniqueness results in identifying an unknown coefficient in a nonlinear diffusion equation. SIAM J. Appl. Math. **41** (1981), 310–323.
[2] Inverse problems for parabolic partial differential equations. In: Cannon and Hornung [1], pp. 51–59.

DuChateau, P., and W. Rundell
[1] Unicity in an inverse problem for an unknown reaction term in a reaction-diffusion equation. J. Diff. Equat. **59** (1985), 155–164.

Dümmel, S., and S. Handrock-Meyer
[1] Uniqueness of the solution of an inverse problem for a quasilinear parabolic equation in divergence form. Z. Anal. Anwendungen **7** (1988) 6, 241–246.

Dunford, N., and J. T. Schwartz
[1] Linear operators, I. General theory. Interscience Publishers, New York –London 1958.

Dupuis, M. (ed.)
[1] Supercomputer simulations in chemistry. Lecture Notes in Chemistry, 44. Springer Verlag, Berlin–New York 1986.

Engl, H. W., and C. W. Groetsch (eds.)
[1] Inverse and ill-posed problems. Academic Press, New York–London 1987.

Engl, H. W., K. Kunisch and A. Neubauer
[1] Tikhonov regularization for the solution of nonlinear ill-posed problems I. Inverse Problems (to appear).

Faddeev, L. D.
[1] The method of inverse scattering problem for the solution of evolution equations in mathematical physics. Materials of the VIIIth Far-Eastern Mathematical School held in Vladivostok, August 19–31, 1976. Preprint no 1. Khaborovsk. Kompleks. Nauchno-Issled. Inst., Khabarovsk 1977, 37 pp.

Fenyö, S., and H. W. Stolle
[1] Theorie und Praxis der linearen Integralgleichungen III. Deutscher Verlag der Wissenschaften, Berlin 1983.

Fitzgibbon, W. E.
[1] Mathematical and computational methods in seismic exploration and reservoir modeling. Soc. Ind. Appl. Math., Philadelphia, Pa. 1986.

Flügge, S. (ed.)
[1] Encyclopedia of Physics, vol. IX. Fluid Dynamics III. Springer-Verlag, Berlin–New York 1960.

Franek, A., I. Kratochvil and L. Trávníček
[1] Inhomogeneous inverse problem in finite elasticity. J. Elasticity **14** (1984) 4, 363 to 372.

Franzone, P. C.
[1] Some inverse problems in electrocardiology. In: Deuflhard and Hairer [1], pp. 180–205.

FREHZE, J.
[1] Capacity methods in the theory of partial differential equations. Jahresber. Deutsch. Math.-Verein. **84** (1982), 1–44.

FRIEDMAN, A.
[1] Partial differential equations of parabolic type. Prentice-Hall, Englewood Cliffs, N.J. 1964.

FRIEDMAN, A. and B. GUSTAFSSON
[1] Identification of the conductivity coefficient in an elliptic equation. SIAM J. Math. Anal. **18** (1987), 777–787.

GAJEWSKI, H.
[1] On existence, uniqueness and asymptotic behavior of solutions of the basic equations for carrier transport in semiconductors. ZAMM **65** (1985) 2, 101–108.

GALIULLIN, A. S.
[1] Methods for solving inverse problems of dynamics. Nauka, Moscow 1986 (Russian)

GILBARG, D.
[1] Jets and cavities. In: FLÜGGE [1], pp. 311–445.

GILBARG, D., and N. S. TRUDINGER
[1] Elliptic partial differential equations of second order. Second edition. Springer-Verlag, Berlin–New York 1983.

GILJAZOB, S.
[1] Methods for solving incorrectly posed problems. University Press, Moscow 1987 (Russian).

GLADWELL, G. M. L.
[1] Inverse problems in vibration. Martin Nijhoff Publishers, Dordrecht 1986.

GÖPFERT, A., G. WANKA and J. WANKA
[1] Approximation durch Lösungen partieller Differentialgleichungen. Z. Anal. Anwendungen **4** (1985) 4, 291–303.

GORENFLO, R.
[1] Inkorrekt gestellte Probleme I, II. Fachbereich Mathematik, Freie Universität, Berlin (West) 1977.
[2] Approximation of discrete probability distributions in spherical stereology. In: CANNON and HORNUNG [1], pp. 103–115.
[3] Abel integral equations: applications-motivated solution concepts. In: Methoden und Verfahren der mathematischen Physik (to appear).

GORENFLO, R., and S. VESSELLA
[1] Abel integral equations, an historico-bibliographical survey. Consiglio Nationale delle Ricerche, Istituto di Analisi Globale e Applicazioni, Firenze 1984.
[2] Abel integral equations: Applications and analytic properties. Fachbereich Mathematik, Freie Universität, Berlin (West), Preprint no. 236 (1986).
[3] Abel integral equations: Applications and analytic properties II. Fachbereich Mathematik, Freie Universität, Berlin (West), Preprint no. 245 (1987).

GRAFAREND, E. W., and R. H. RAPP (eds.)
[1] Advances in geodesy. Selected papers from Reviews of Geophysics and Space Physics. Amer. Geophys. Union, Washington 1984.

GRANT, F. S.
[1] In memoriam. Geoexploration **23** (1985), 443–444.

GRANT, F. S., and G. F. WEST
[1] Interpretation theory in applied geophysics. International Series in the Earth Sciences. McGraw-Hill, New York—London 1965.

GROETSCH, C. W.
[1] The theory of Tikhonov regularization for Fredholm equations of the first kind. Research Notes in Mathematics, 105. Pitman, London 1984.

GRÜNBAUM, F. A.
[1] Some mathematical problems motivated by medical imaging. In: TALENTI [1], pp. 114–147.

GRÜNDEMANN, H.
[1] Zur Entwicklung von Randintegralgleichungen für eine Aufgabenklasse der mathematischen Physik. Z. Anal. Anwendungen **7** (1988) 6, 495–501.

GÜNTHER, M.
[1] Über zwei spezielle inverse Probleme der Potentialtheorie. Z. Anal. Anwendungen **6** (1985) 5, 459–470.

HADAMARD, J.
[1] Lectures on Chauchy's problem in linear partial differential equations. Yale University Press, New Haven 1923.

HALD, O. E.
[1] Inverse eigenvalue problems for layered media. Comm. Pure Appl. Math. **30** (1977), 69–94.

HALMOS, P. R.
[1] Measure theory. Van Nostrand, New York—London 1950.

HAMANN, U.
[1] Approximation durch Lösungen elliptischer Randwertprobleme auf kompakten Mengen mit leerem Innern. Math. Nachr. **143** (1989), 143–156.

HÄMMERLIN, G., and K.-H. HOFFMANN, (eds.)
[1] Ill-posed problems and their numerical treatment. Intern. Ser. Numer. Math., 63, Birkhäuser-Verlag, Basel—Boston, Mass. 1983.

HANDROCK-MEYER, S.
[1] Inverse Probleme für die Wärmeleitungsgleichung. Z. Anal. Anwendungen **7** (1988) 6, 503–510.

HELMS, L. L.
[1] Introduction to potential theory. Wiley-Interscience, New York—London 1969.

HERGLOTZ, G.
[1] Über das Benndorfsche Problem der Fortpflanzungsgeschwindigkeit der Erdbebenstrahlen. Physik. Z. **8** (1907), 145–147.
[2] Über die analytische Fortsetzung des Potentials ins Innere der anziehenden Massen. Teubner-Verlag, Leipzig 1914.

HOFFMANN, K.-H., and J. SPREKELS
[1] On the identification of parameters in general variational inequalities by asymptotic regularization. SIAM J. Math. Anal. **17** (1986), 1198–1217.

HOFMANN, B.
[1] Regularization for applied inverse and ill-posed problems. Teubner-Texte zur Mathematik, 85. Teubner-Verlag, Leipzig 1986.

HOLOTA, P. (ed.)
[1] Figure and dynamics of the earth, moon and planets I, II, III. Proceedings of an international symposium held in Prague, September 15–20, 1986. Research Institute of Geodesy, Topography and Carthography, Prague 1987.

HORNUNG, U.
[1] Identification of nonlinear soil physical parameters from an input-output experiment. In: DEUFLHARD and HAIRER [1], pp. 227–245.

HÖRMANDER, L.
[1] The boundary problems of physical geodesy. Arch. Rat. Mech. Anal. **62** (1976), 1–52.
[2] The analysis of linear partial differential operators I, II. Springer-Verlag, Berlin–New York 1983.

ISAKOV, V. M.
[1] On uniqueness of solutions for inverse problems of potential theory. In: ANGER [11], pp. 135–139.

ISKENDEROV, A. D.
[1] Multidimensional inverse problems for linear and quasilinear parabolic equations. Dokl. Akad. Nauk SSSR **225** (1975), 1005–1008.

IVANOV, V. K.
[1] The inverse problem of potential for a body closely approximated by a given body. Izv. Akad. Nauk SSSR, Ser. Mat. **20** (1956), 793–818 (Russian).

IVANOV, V. K., V. V. VASIN and V. P. TANANA
[1] Linear ill-posed problems and their applications. Nauka, Moscow 1979 (Russian).

JAYNES, E. T.
[1] Prior information and ambiguity in inverse problems. In: McLAUGHIN [1], pp. 151 to 166.

JENTZSCH, L.
[1] Über einige neuere Probleme der Elastizitätstheorie stückweise homogener Körper. In MICHEL [1], vol. 2, pp. 79–107.

JEROME, J. W.
[1] Consistency of semiconductors modeling: an existence-stability analysis for the stationary van Roosbroeck system. SIAM J. Appl. Math. **45** (1985) 4, 565–590.

JOCHMANN, H.
[1] Interpretation of variation of the Earth's vector of rotating using inverse solution. Z. Anal. Anwendungen **7** (1988) 6, 519–529.

JOHN, F.
[1] Plane waves and spherical means: Applications to partial differential equations (1955). Reprint 1981. Springer-Verlag, Berlin—New York 1981.
[2] Collected papers, vol. 1. Birkhäuser-Verlag, Basel—Boston, Mass. 1985.

KAISER, H., and C.-H. KAISER
[1] Zur Identifikation von Interferenzschichten. Z. Anal. Anwendungen **7** (1988) 6, 531—556.

KAPANADZE, D. V.
[1] On a problem of G. Anger. Math. Nachr. **117** (1984), 247—251 (Russian).
[2] On uniqueness of solution of the inverse problem of potential theory. Math. Nachr. (to appear) (Russian).

KELLER, J. B.
[1] Inverse problems. Amer. Math. Monthly **83** (1976) 2, 107—118.

KELLER, W.
[1] On the treatment of the geodetic boundary value problem by contact transformations. Gerlands Beitr. Geophysik **96** (1987), 3—4, 186—196.

KELLOGG, O. D.
[1] Foundations of potential theory. Springer-Verlag, Berlin—Heidelberg 1929.

KLEINE, E.
[1] A special inverse source problem in potential theory. Math. Nachr. **126** (1986), 63—81.
[2] An inverse problem for biharmonic potentials. Math. Nachr. **128** (1986), 7—27.
[3] On the identification of the adiabatic equation of state in compressible gas flow. Z. Anal. Anwendungen **7** (1988) 6, 557—562.

KLUGE, R.
[1] Zur Parameterbestimmung in nichtlinearen Problemen. Teubner-Texte zur Mathematik 81, Leipzig (1985).

KNOPS, R. J.
[1] Instability and the ill-posed Chauchy problem. In: Mechanics of Solids — The Rodney Hill 60th Ann. Vol. (eds. H. G. HOPKINS and M. J. SEWELL). Pergamon Press, Oxford—New York 1982.

KOHN, R. V., and M. VOGELIUS
[1] Determining conductivity by boundary measurements II. Interior results. Comm. Pure Appl. Math. **38** (1985) 5, 643—667.

KOSAREV, G. L., E. Z. RIZAEV and K. SUDO
[1] Determination of the earthquake source parameters from the residual displacements of the Earth's surface. The optimal source model of the Gazli earthquakes of 1976. Dokl. Akad. Nauk SSSR **288** (1986) 4, 842—846 (Russian).

KÖTHE, G.
[1] Topological vector spaces I. Springer-Verlag, Berlin—New York 1971.

KRAVARIS, C., and J. H. SEINFELD
[1] Identificability of spatially-varying conductivity from point observation as an inverse Sturm-Liouville problem. SIAM J. Control and Optimization **24** (1986) 3, 522—542.

[2] Identification of spatially varying parameters in distributed systems by discrete regularization. J. Math. Anal. Appl. **119** (1986), 125–152.

LANDKOF, N. S.
[1] Foundations of modern potential theory. Springer-Verlag, Berlin–New York 1972.

LANGER, R. E.
[1] An inverse problem in differential equations. Bull. Amer. Math. Soc. **39** (1933), 814–820.

LASCH, J., and R. KUHNAU
[1] Evaluation of enzyme distribution in spherical supports by computed fluorescence tomography. Enzyme Microb. Technol. **8** (1986), 115–119.

LATTÈS, R., and J. L. LIONS
[1] Méthodes de quasi-reversibilité et applications. Dunod, Paris 1967.

LAVRENT'EV, M. M.
[1] Some improperly posed problems of mathematical physics. Springer-Verlag, Berlin–New York 1967 (transl. from Russian).

LAVRENT'EV, M. M., V. G. ROMANOV and S. P. SHISHATSKIJ
[1] Ill-posed problems of mathematical physics and analysis. Amer. Math. Soc., Providence, R. I. 1986 (transl. from Russian).

LAVRENT'EV, M. M., K. G. REZNITSKAYA and V. G. YAKHNO
[1] One-dimensional inverse problems of mathematical physics. Amer. Math. Soc., Providence, R. I. 1986 (transl. from Russian).

LEHMANN, I.
[1] P'. Publ. Bureau Central Seismol. Internat., Trav. Sci. Sér. A, **14** (1936), 87–115.

LEIS, R.
[1] Initial-boundary value problems in mathematical physics. Teubner-Verlag, Stuttgart 1986.

LEVITAN, B. M.
[1] Inverse Sturm-Liouville problems. Nauka, Moscow 1984 (Russian).

LICHTENSTEIN, L.
[1] Neuere Entwicklung der Potentialtheorie. Konforme Abbildung. Encykl. Math. Wissenschaften II, C 3, Teubner-Verlag, Leipzig 1918.

LIONS, J. L., and G. DUVAUT
[1] Inequalities in Mechanics and Physics. Springer-Verlag, Berlin–New York 1976.

LORENZI, A.
[1] An inverse problem for a quasilinear parabolic equation. Ann. Mat. Pura Appl. (4) **142** (1985), 145–169

LORENZI, A., and C. D. PAGANI
[1] An inverse problem in potential theory. Ann. Mat. Pura Appl. (4) **129** (1982), 281 to 303.

LOUIS, A.
[1] Inverse und schlecht gestellte Probleme. Teubner-Verlag, Stuttgart, 1988.

MADERNA, C., C. D. PAGANI and S. SALSA
[1] Existence results in an inverse problem of potential theory. Nonlinear Anal. **10** (1986) 3, 277—298.

MARCHENKO, V. A.
[1] Spectral theory of Sturm-Liouville operators. Naukova Dumka, Kiev 1972 (Russian).

MARUSSI, A.
[1] On the density distribution in bodies of assigned outer Newtonian attraction. Boll. Geof. Teor. Appl. **22** (1980), 83—94.

MAZYA, V. G., and T. O. SHAPOSHNIKOVA
[1] Theory of multipliers in spaces of differentiable functions. Pitman, Brooklyn—New York 1985 (transl. from Russian).

MCLAUGHIN, D. W.
[1] Inverse problems. Proceedings of the symposium held in New York, April 12—13, 1983. SIAM-AMS Proceedings, vol. 14. Amer. Math. Soc., Providence, R. I. 1984.

MCLAUGHIN, J. R., and W. RUNDELL
[1] A uniqueness theorem for an inverse Sturm-Liouville problem. J. Math. Physics. (to appear).

MICHEL, B. (ed.)
[1] Fracture mechanics, micromechanics, coupled fields, vol. 1—38 (1981—1987). Institute of Mechanics, Academy of Sciences of the GDR, Karl-Marx-Stadt.
[2] The equation of continuity — an important balance relation in solid mechanics and fluid mechanics (German). In: MICHEL [1], vol. 38, pp. 1—13.

MILITZER, H., and F. WEBER (eds.)
[1] Angewandte Geophysik, Bd. 1 Gravimetrie und Magnetik. Springer-Verlag, Wien—New York; Akademie-Verlag, Berlin 1984.

MILLER, M. A.
[1] Ambiguity of inverse problems in macroelectrodynamics. Spherical and toroidal sources of electromagnetic fields (A review). Radiofizika **29** (1986) 9, 991—1007 (Russian).

MIRANDA, C.
[1] Partial differential equations of elliptic type. Second edition. Springer-Verlag, Berlin—New York 1970.

MIZOHATA, S.
[1] On the Cauchy problem. Science Press, Beijing; Academic Press, Orlando 1985.

MORITZ, H.
[1] Advanced physical geodesy. Wichmann, Karlsruhe 1980.

MOROZOV, V. A.
[1] Methods for solving incorrectly posed problems. Springer-Verlag, Berlin—New York 1984 (transl. from Russian).

[2] Regularization methods for solutions of improperly posed problems. Nauka, Moscow 1987 (Russian).

Muzlev, N. V.
[1] Uniqueness theorems for some converse problems of heat conduction. Zh. Vych. Mat. i Mat. Fiz. **20** (1980), 388–400. (Russian).

Naats, I. E.
[1] The inverse method for problems of athmospherical optics. Nauka, Novosibirsk (1986) (Russian).

Nashed, M. Z. (ed.)
[1] Generalized inverses and applications. Academic Press, New York—London 1976.
[2] A new approach to classification and regularization of ill-posed operator equations. In: Engl and Groetzsch [1], pp. 53–76.

Natterer, F.
[1] The mathematics of computerized tomography. Teubner-Verlag, Stuttgart; John Wiley, Chichester 1986.

Netuka, I., and J. Vesely
[1] Harmonic continuation and removable singularities in axiomatic potential theory. Math. Ann. **234** (1978), 117–123.

Neubauer, A.
[1] Tikhonov regularization for the solution of nonlinear ill-posed problems II. (submitted).

Neumann, C.
[1] Über das logarithmische Potential. Ber. Abh. Königl. Sächs. Ges. d. Wiss. Leipzig **58** (1906), 482–559.

Newton, R. G.
[1] Scattering theory of waves and particles. Second edition. Springer-Verlag, Berlin—New York 1982.
[2] The Marchenko and Gel'fand-Levitan methods in the inverse scattering problem in one and three dimensions. In: Bednar et al. [1], pp. 1–74.

Novikov, P. S.
[1] On uniqueness of inverse potential problems. Dokl. Acad. Nauk SSR **18** (1938) 3, 165–168. (Russian)

Novikov, R. G., and G. M. Khenkin
[1] $\bar{\partial}$-equation in multidimensional inverse scattering problem. Russ. Math. Surveys **42** (1987), 93–152.

Nunez, P. L.
[1] Volume conducted fields in the brain: an overview. EEG Laboratory, Department of Neurosciences, University of California, San Diego, La Jolla, California, 1975.

Ochs, R. L., jr.
[1] The limited aperture problem in acoustic scattering: Dirichlet boundary conditions. SIAM J. Appl. Math. **47** (1987) 6, 1320–1431.

Oganesyan, S. M., and V. I. Starostenko
[1] Bodies of zero external gravity field potential: the forgotten ideas, and the current state of the theory. Izvestija Akad. Nauk SSSR, Fisika Zemli (1985) 3, 49–52 (Russian).

Pagani, C. D.
[1] Inverse problems for the volume potential. Portugal. Math. **41** (1984), 383–391.
[2] Analytic solutions to the inverse problem of the Newtonian potential. In: Cannon and Hornung [1], pp. 127–132.

Pao, Yih-Hoing, F. Santosa, W. W. Syms and Ch. Holland (eds.)
[1] Inverse problems of acoustic and elastic waves. Papers from the international conference held at Cornell University, Ithaca, N.Y., June 4–6, 1984. Soc. Ind. Appl. Math., Philadelphia, Pa. 1985.

Payne, L. E.
[1] Some improperly posed problems in partial differential equations. Regional Conference Series in Applied Mathematics. Soc. Ind. Appl. Mathem., Philadelphia, Pa. 1975.

Peitgen, H.-O., and P. H. Richter
[1] The beauty of fractals. Images of complex dynamical systems. Springer-Verlag, Berlin–New York 1986.

Pilant, M., and W. Rundell
[1] An inverse problem for a nonlinear parabolic equation. Comm. Part. Diff. Equations **11** (1986) 4, 445–457.
[2] Undetermined coefficient problems for nonlinear elliptic and parabolic equations. In: Cannon and Hornung [1], pp. 139–154.
[3] An inverse problem for a nonlinear elliptic differential equation. SIAM J. Math. Anal. **18** (1987) 6, 1801–1809.
[4] A uniqueness theorem for determining conductivity from overspecified boundary data. J. Math. Anal. Appl. **136** (1988), 20–28.
[5] Fixed point methods for a nonlinear parabolic inverse coefficient problem. Comm. Part. Diff. Equations **13** (1988), 469–493.

Plonsey, R.
[1] Bioelectric phenomina. McGraw-Hill, New York–London 1969.

Polking, J.
[1] An interview. Notices Amer. Math. Soc. **34** (1987) 5, 731–736.

Pöschel, J., and E. Trubowitz
[1] Inverse spectral theory. Pure and Applied Mathematics, 130. Academic Press, Boston, Mass. 1987.

de la Pradelle, A.
[1] Approximation et caractère de quasi-analyticité dans la théorie axiomatique des fonctions harmoniques. Ann. Inst. Fourier **17** (1967) 1, 383–399.

Prilepko, A. I.
[1] Mixed inverse potential problems for star bodies. Sib. Mat. Zh. **12** (1971), 1341–1353.
[2] Über die Existenz und Eindeutigkeit von Lösungen inverser Probleme. Math. Nachr. **63** (1974), 135–153.

[3] Selected questions of the exterior inverse problem of potential theory. In: Partial differential equations with free boundary (ed. I. V. Skrypnik), pp. 93–97, Naukova Dumka, Kiev 1983 (Russian).

Radon, J.
[1] Über die Bestimmung von Funktionen durch ihre Integralwerte längs gewisser Mannigfaltigkeiten. Ber. Verh. Königl. Sächs. Ges. d. Wiss. Leipzig **69** (1917), 262–277.

Ramm, A. G.
[1] Scattering by obstacles. Reidel Publishing Company, Dordrecht–Boston, Mass. 1986.

Rapoport, I. M.
[1] On the stability of inverse potential problems. Dokl. Acad. Nauk SSSR **31** (1940) 4, 303–306 (Russian).

Lord Rayleigh, J. W. S.
[1] The theory of sound (1877). Reprinted by Dover Publications, New York 1945.

Reid, W. H. (ed.)
[1] Mathematical problems in the geophysical sciences. 1. Geophysical fluid dynamics; 2. Inverse problems, dynamo theory and tides. Lecture in Applied Mathematics, 13, 14. Amer. Math. Soc., Providence, R. I. 1971.

Richter, G.
[1] An inverse problem for the steady state diffusion equation. SIAM J. Appl. Math. **41** (1981), 210–221.
[2] Numerical identification of a spatially varying diffusion coefficient. Math. Computation **36** (1981), 375–386.

Riesz, F.
[1] Sur les fonctions subharmoniques et leur rapport à la théorie du potentiel I, II. Acta Math. **48** (1926), 329–343; Acta Math. **54** (1930), 321–360.

Romanov, V. G.
[1] Integral geometry and inverse problems for hyperbolic equations. Springer-Verlag, Berlin–New York 1974 (transl. from Russian).
[2] Inverse problems of mathematical physics. VNU Science Press, Utrecht 1987 (transl. from Russian).

Romanov, V. G., S. I. Kabanikhin and I. P. Pukhnacheva
[1] Inverse problems in electrodynamics. Computing Center, Acad. of Sciences of the USSR, Novosibirsk 1984 (Russian).

Rösler, R.
[1] Theoretische Grundlagen der angewandten Gravimetrie und Magnetik. In: Militzer and Weber [1], pp. 13–63.

Rossi Casta, L.
[1] Problemi inversi per l'equazione del calcore con conduttivitá dipendente dal tempo. Publicazione Serie III. Istituto per le Applicazioni del calcolo „Mauro Picone", Rome 1983.

RUNDELL, W.
[1] The determination of a parabolic equation from initial and final data. Proc. Amer. Math. Soc. **99** (1987), 637–642.
[2] Some inverse problems for elliptic equations. Appl. Anal. **28** (1988), 67–78.

SABATIER, P. C. (ed.)
[1] Inverse problems, non linear evolution. Éditions du CNRS, Paris 1980.
[2] Critical analysis of the mathematical methods used in electromagnet inverse theories: a quest for new routes. In: BOERNER et al. [1]. pp. 43–64.
[3] Formation of waves by ground motion. In: Encyclopedia of Fluid Mechanics, pp. 723–759. Gulf, Houston, Texas 1986.
[4] Inverse problems. Papers presented at the RCP meeting on interdisciplinary study of inverse problems held in Montpellier, Nov. 26–30, 1985. Université des Sciences et Technique du Languedoc, Montpellier 1986.
[5] A few geometrical features of inverse and ill-posed problems. In: ENGL and GROETZSCH [1], pp. 1–18.
[6] Algorithmic approaches to sets of "good answers" in inverse problems. In: TALENTI [1], pp. 312–323.

SALEM, R.
[1] On some singular monotonic functions which are strictly increasing. Trans. Amer. Math. Soc. **53** (1943), 427–439.

SANSO, F., R. BARZAGHI and C. C. TSCHERNING
[1] Choise of norm for the density distribution of the earth. Geophys. J. Royal Astr. Soc. (to appear).

SCHACHTZABEL, H.
[1] Phasenrekonstruktion für Stapel homogener dielektrischer Schichten gleicher optischer Dicke. Z. Anal. Anwendungen **7** (1988) 6, 563–569.

SCHIFFER, M.
[1] Analytical theory of subsonic and supersonic flow. In: FLÜGGE [1], pp. 1–161.

SCHULZE, B.-W., and G. WILDENHAIN
[1] Methoden der Potentialtheorie für elliptische Differentialgleichungen beliebiger Ordnung. Birkhäuser, Basel–Boston, Mass.; Akademie-Verlag, Berlin 1977.

SEIDMAN, T. I.
[1] Nonconvergence results for the applications of least-squares estimation to ill-posed problems. J. Optim. Theory Applications **30** (1980) 4, 535–547.
[2] Coefficient identification for a parabolic problem. In: TALENTI [1], pp. 340–351.
[3] The method of generalized interpolation for approximate solutions of ill-posed problems. In: CANNON and HORNUNG [1], pp. 155–162.

SHEPP, L. A.
[1] Computed tomography. Proc. Symp. Appl. Math. 27. Amer. Math. Soc., Providence, R. I. 1982.

SHITZER, A., and R. C. EBERHART
[1] Heat transfer in medicine and biology. Analysis and Applications I, II. Plenum Press, New York–London 1985.

ŠKORVANEK, M. (ed.)
[1] Mathematical methods of solution of inverse problems of geophysical fields. Bratislava (ČSSR): Slovak Academy of Sciences, Geophysical Institute, 1987.

SMITH, R. A.
[1] A uniqueness theorem concerning gravity fields. Proc. Cambridge Philos. Soc. **57** (1961), 865–870.

STEPHAN, E. P.
[1] Boundary integral equations for mixed boundary value problems in $\mathbb{R}^3$. Math. Nachr. **134** (1987), 21–53.

STRAKHOV, V. N., and M. A. BRODSKY
[1] On the uniqueness of the inverse logarithmic potential problem. SIAM J. Appl. Math. **46** (1986) 2, 324–344.
[2] On uniqueness of the determination of a polyhedron by its edges and its applications to the inverse problem of potential theory. Sibirsk. Mat. Zh. (to appear) (Russian).

STROMEYER, D., and L. BALLANI
[1] Uniqueness of the inverse gravimetric problem for point mass models. Manuscripta Geodetica **9** (1984) 1/2, 125–136.

SUZUKI, T.
[1] Deformation formulas and their applications to spectral and evolutional inverse problems. In: Nonlinear partial differential equations in applied science (eds. H. FUJITA, P. D. LAX, G. STRON) pp. 289–311, North-Holland Math. Stud., 81, North-Holland Publishing, Amsterdam–New York 1983.
[2] Uniqueness and nonuniqueness in an inverse problem for the parabolic equation. J. Diff. Equations **47** (1983) 2, 296–316.
[3] On a multidimensional inverse parabolic problem. Proc. Japan Acad. Ser. A, **62** (1986) 3, 83–86.
[4] Gel'fand-Levitan's theory and related inverse problems. In: CANNON and HORNUNG [1], pp. 163–172.

SUZUKI, T., and MURAYAMA
[1] A uniqueness theorem in an identification problem for coefficients of a parabolic equation. Proc. Japan Acad. Ser. A, **56** (1980), 259–263.

SYLVESTER, J., and G. UHLMANN
[1] A uniqueness theorem for an inverse boundary value problem in electrical prospection. Comm. Pure Appl. Math. **39** (1986), 91–112.
[2] A global uniqueness theorem for an inverse boundary value problem. Ann. Math. **125** (1987), 153–169.

TALENTI, G. (ed.)
[1] Inverse problems. Papers from the 1st 1986 session of the Centro Internazionale Matematico Estivo held in Montecatini Terme, May 28–June 5, 1986. Lecture Notes in Mathem., 1225. Springer-Verlag, Berlin–New York 1987.

TARANTOLA, A.
[1] Inverse problem theory. Methods for data fitting and model parameter estimation. Elsevier, Amsterdam–New York, 1987.

TIKHONOV, A. N., and J. A. SAMARSKIJ
[1] Equations of mathematical physics. Pergamon Press, London—New York 1963 (transl. from Russian).

TIKHONOV, A. N., V. K. IVANOV and M. M. LAVRENT'EV
[1] Ill-posed problems. In: Partial differential equations. (ed. A. V. BITSADSE), pp. 224 to 239. Nauka, Moscow 1970 (Russian).

TIKHONOV, A., and V. ARSENIN
[1] Solutions of ill-posed problems. Winston, Washington, D.C.; John Wiley, New York—London 1977 (transl. from Russian).

TIKHONOV, A., V. ARSENIN, A. TIMONOV
[1] Mathematical problems of computed tomography. Nauka, Moscow 1987 (Russian).

TIKHONOV, A., and A. GONCHARSKIJ
[1] Improperly posed problems in natural sciences. Mir, Moscow 1987 (English).

TONELLI, L.
[1] Su un problema di Abel. Math. Ann. **99** (1928), 183—199.

TREDER, H.-J.
[1] Gauß und die Gravitationstheorie. Die Sterne **53** (1977), 9—14.

TRÈVES, F.
[1] Basic linear partial differential equations. Academic Press, New York—London 1975.

VOGEL, C. R.
[1] An overview of numerical methods for nonlinear ill-posed problems. In: ENGL and GROETSCH [1], pp. 231—246.

WANG, CHENG T.
[1] A new set of semiconductor equations for computer simulation of submicron devices. Solid-State Electronics **28** (1985) 8, 783—788.

WECK, N.
[1] Inverse problems of potential theory. Appl. Analysis **2**, (1972), 195—204.

WILDENHAIN, G.
[1] Uniform approximation by solutions of general boundary value problems for elliptic equations of arbitrary order I. Z. Anal. Anwendungen **2** (1983) 6, 511—521.

YAKHNO, V.
[1] Inverse problems for differential equations in elasticity. Nauka, Moscow 1989 (Russian).

ZAKHARIEV, B. N., and A. A. SUZKO
[1] Potentials in quantum scattering. Direct and inverse problems. Springer-Verlag, Berlin—New York 1988 (transl. from Russian).

ZAKHAROV, V. E., S. V. MANAKOV, S. P. NOVIKOV and L. P. PITAEVSKIJ
[1] Theory of solitons, the method of inverse problems. Nauka, Moscow 1980 (Russian).

ZEIDLER, E.
[1] Lectures on nonlinear functional analysis and its applications. I—V. Springer-Verlag Berlin—New York.
Part I: Fixed-point theorems (1986)
Part II: Monotone operators (1987)
Part III: Variational methods and optimization (1985)
Part IV, V: Applications to mathematical physics (to appear).

ZHDANOV, M. S.
[1] Integral transforms in geophysics. Springer-Verlag, Berlin—New York 1988. (transl. from Russian).

ZIDAROV, D.
[1] On the solutions of some potential field inverse problems and its applications in geophysics. Publishing House of the Bulgarien Academy of Sciences, Sofia **1968** (Russian with English Summary).
[2] Direct, characteristic and derivative solutions of the inverse gravimetric and magnetic problems. In: ANGER and RÖSLER [1], pp. 219—234.
[3] Inverse gravimetric problem in geoprospecting and geodesy. Publishing House of the Bulgarian Academy of Sciences, Sofia **1984** (Bulgarian with Russian and English summaries).

ZIELKE, G.
[1] Verallgemeinerte inverse Matrizen. In: Jahrbuch Überblicke Mathematik **1983**. Bibliographisches Institut, Mannheim—Zürich 1983, pp. 95—116.

Added in proof

ANDERSEN, R. S., and G. N. NEWSAN
[2] Special program on inverse problems. Proceedings of the Centre for Mathematical Analysis, Australian National University, **17** (1988).

ANGER, G., and J. GOTTLIEB
[1] On coupled fields in geophysics (to appear).

BOCK, H. G.
[2] Randwertprobleme zur Parameteridentifizierung in Systemen nichtlinearer Differentialgleichungen. Bonner Math. Schriften **183** (1987).

CANNON, J. R., J. V. KOEBBE and R. M. SCHMIDT
[1] The determination of rate constants from known linear combinations of concentrations in a chemical reaction. IMA J. Appl. Math. **41** (1988), 67—84.

FOKAS, A. S., and V. PAPAGEORGIN
[1] Inverse problems and a unified approach to integrability in 1, $1 + 1$ and $2 + 1$ dimensions. In: Solitons (Tipuchirapalli, 1987), 63—85. Springer Ser. Nonlinear Dynamics, Springer-Verlag, Berlin—New York 1988.

GERVER, M. L., and V. M. MARKUSHEVICH
[1] Hodograph properties of a surface source. In: Some direct and inverse problems in seismology, pp. 15—63. Published in: Vychislitel'naya Seismologiya, vol. **4**, Nauka, Moscow 1968 (Russian).

GILBERT, R. P., and Y. XU
[1] Starting fields and far fields in ocean acoustics. Wave Motion (to appear).

GORENFLO, R., and S. VESSELLA
[4] Abel integral equations: Analysis and applications. To appear in: Lecture Notes in Mathem., Springer-Verlag, Berlin—New York 1989.

HAARIA, H. (ed.)
[1] Theory and applications of inverse problems. Pitman Res. Notes Mathem. Ser. **167**. Longman Sci. Techn., Harlow 1988.

HANSEN, W., and H. HUEBER
[1] Eigenvalues in potential theory. J. Differential Equations **73** (1988), 133—152.

ISAKOV, V. M.
[2] Inverse source problems (to appear).

KELLER, W.
[2] An overdetermined boundary value problem for a nonspherical boundary. Veröff. Zentralinstitut Phys. Erde **102** (1989) Teil II, 229—254, Akad. Wiss. DDR, Potsdam.

KOHN, R. V., and B. D. LÖWE
[1] A variational method for parameter identification. Math. Mod. Num. Anal. **22** (1988), 119—158.

KOHN, R. V., and M. VOGELIUS
[2] Relaxation of a variational method for impedance computed tomography. Comm. Pure Appl. Math. **40** (1987), 745—777.

KOUNCHEV, O.
[1] Extremal problems for the distributed moment problem. In: Potential theory (Prague, 1987), pp. 185—195. Plenum, New York—London 1988.

LEVITAN, B. M., and I. S. SARGSYAN
[1] Sturm-Liouville and Dirac operators. Nauka, Moscow 1988 (Russian).

LORENZI, A., and E. PAPARONI
[1] An identification problem for a semilinear parabolic equation. Ann. Mat. Pura Appl., IV Ser. **151** (1988), 263—287.

MADERNA, C., C. D. PAGANI and S. SALSA
[2] Analytic solutions to the inverse problem of the Newtonian potential. J. Math. Anal. Appl. **129** (1988), 71—89.

MARCHENKO, V. A.
[2] Sturm-Liouville operators and applications. Birkhäuser-Verlag, Basel—Boston, Mass., 1987.

MORITZ, H.
[2] Equilibrium figures in geodesy and geophysics. Veröff. Zentralinstitut Phys. Erde **102** (1989) Teil II, 58—97, Akad. Wiss. DDR, Potsdam.

NACHMAN, A. I.
[1] Reconstruction from boundary measurements. Ann. Math. **128** (1988), 531—576.

NACHMAN, A., J. SYLVESTER and G. UHLMAN
[1] An n-dimensional Borg-Levison theorem. Comm. Math. Phys. **115** (1988), 595–605.

NIZHNIK, L.
[1] Inverse scattering for hyperbolic equations. Naukova Dumka, Kiev 1990 (Russian).

POGORELOV, A. G.
[1] Inverse problems of nonstationary chemical kinetics: A system approach. Nauka, Moscow 1988 (Russian).

PRILEPKO, A. I., and D. G. ORLOVSKIJ
[1] On the determination of the parameters of an evolution equation and inverse problems of mathematical physics II, III. Diff. Urav. **21** (1985), 694–701; **23** (1987), 910–918 (English).

PTASHNIK, B. I.
[1] Ill-posed boundary value problems for partial differential equations. Naukova Dumka, Kiev 1984 (Russian).

RAKESH
[1] A linearized inverse problem for the wave equation. Comm. Partial Differ. Equations **13** (1988), 573–601.

RAMM, A. G.
[2] Multidimensional inverse scattering problems and completeness of the products of solutions to homogeneous PDE. ZAMM **69** (1989) 4, T13–T22.

SABATIER, P. C. (ed.)
[7] Basic methods of tomography and inverse problems. Malvern Physics Series, Adam Hilger, Bristol 1987.
[8] Some topics on inverse problems. World Scientific Publishing, Singapore 1988.

SCHLÖDER, J. P.
[1] Numerische Methoden zur Behandlung hochdimensionaler Aufgaben der Parameteridentifizierung. Bonner Math. Schriften **187** (1988).

STURM, K.-TH.
[1] Perturbation of Markov processes with applications to Schrödinger operators. To appear in: Lecture Notes in Mathem., Springer-Verlag, Berlin–New York 1990.

VRAGOV, V. N. (ed.)
[1] Boundary value problems for nonclassical equations of mathematical physics. Mathematical Institute of the Academy of Sciences of the USSR, Novosibirsk 1987.

Symbol Index

$[a, b]$ 15

$]a, b[$ 21

a_α 38

A 9, 15, 27, 37, 112, 125, 176

A^{-1} 16, 23, 37

A^* 111, 125, 179

$\mathfrak{B}(\mu)$ 114, 124, 132, 144

ex $\mathfrak{B}(\mu)$ 125, 133, 137, 146

$\mathfrak{B}_{\mathrm{H}}(\mu)$ 177

cap (K) 98, 100

cap $(K; G)$ 100

$\mathrm{cap}_{\mathrm{e}}$ $(B; G)$ 100

$\mathrm{cap}_{\mathrm{i}}$ $(B; G)$ 100

$\mathbb{C}$ 16

$\mathbf{C}([a, b])$ 27, 85

$\mathbf{C}(K)$ 48, 67, 77, 85, 104, 145

$\mathbf{C}^*(K)$ 104, 145

$\mathbf{C}^1([a, b])$ 27

$\mathbf{C}^1(K)$ 125

$\mathbf{C}^1(\bar{\Omega})$ 68, 125

$\mathbf{C}^k(U)$ 216

$\mathbf{C}^\alpha$ 210

$\mathbf{C}_\omega(\mathbb{R}^n)$ 105, 176

$\mathbf{C}_\omega^*(\mathbb{R}^n)$ 105

$\mathbf{C}_0^\infty(\mathbb{R})$ 29

$\mathbf{C}_0(\mathbb{R}^n)$ 66

$\mathbf{C}_0^\infty(\mathbb{R}^n)$ 43

$D(G)$ 96

$D(K; G)$ 96

$D_K(G)$ 96

Δu 40, 76, 114

$\Delta^2 u$ 41, 125, 161

$\partial u = \mathrm{d}u/\mathrm{d}x$ 26

$\partial_j u = \partial u/\partial x_j$ 38

$\partial^\alpha u$ 38

$\partial_{\mathfrak{n}} u = \partial u/\partial \mathfrak{n}$ 41

$\mathrm{d}s$ 17, 66

$\mathrm{d}S$ 41, 66

$\mathrm{d}y = \mathrm{d}y_1 \ldots \mathrm{d}y_n$ 43, 66

δ_x 43

$\delta_{x^0, R}$ 68

$\delta_{\overline{0t}}$ 163

$E^+(G)$ 96, 98

$E(G)$ 96

$f(u)$ 210

$\|f\|_{\mathbf{C}}$ 27

$\|f\|_{\mathbf{L}^p}$ 27

$\|f\|_{\mathbf{L}^1(\nu)}$ 146

$\|f\|_{\mathbf{L}^\infty}$ 146

$\|f\|_{\mathbf{C}^1}$ 27, 125

$\|f\|_\alpha$ 210

$|f|_\alpha$ 210

$\langle f, f^* \rangle = f^*(f)$ 125, 178

$F^+(G)$ 96

$F(G)$ 96

$F^+(K; G)$ 96

$F(K; G)$ 96

grad $G_{\mathrm{L}}\mu$ 44

G 68, 92

G^+, G^- 92

G^* 92, 102

G_{D} 60, 92, 99

G_{H} 50, 101, 175

G_{L} 43

G_{S} 46

G_{W} 60

G_{L}^{Ω} 33, 78, 109, 117, 121

$G_L m_{x^0,R}$	69
$G_L \delta_{x^0,R}$	69
$G\mu$	92
$G^*\mu$	92
$G_L\mu_\varrho$	43, 66
$G_L\mu_\sigma$	66
$G_L\mu_\tau$	66
$G_L\lambda$	96
H	27
$\mathfrak{H}^0$	89
$\mathfrak{H}^u$	89
$\mathcal{H}$	48, 216
$H_G(\bar{\Omega})$	146
$\mathcal{H}(U)$	48, 220
$\mathcal{H}_P(U)$	216
$H_h(x)$	42, 49, 103, 216
$J(\partial\Omega)$	87, 110, 179
$\ker A$	33
K_B, K_D	49
l.s.c.	88
λ	96
λ_0	98, 145
$\mathbf{L}^1(\nu)$	90
$\mathbf{L}^p([a, b])$	15, 21, 77
$\mathbf{L}^\infty(\nu)$	146
Lip_1	210
$\max A$	14
$m_{x^0,R}$	68
μ	67, 81
μ_x	103, 109
$\mu_{(x,t),\tau}$	175
$\mu(f)$	67, 81
μ^+, μ^-	87
$\Vert\mu\Vert$	87
$\int f(x)\,d\mu(x)$	81
$\mu(B)$	81, 90
μ_K	97
μ_ϱ	66
μ_σ	66, 107
$\mu_{\delta,\varrho}$	67
$\mu_{\delta,o}$	67
$\langle f, \mu\rangle = \int f\,d\mu$	111
(μ, ν)	99
$\mathbb{N}$	14
ω	105
Ω	32, 41, 42, 49, 66
Ω_T	50
Ω_T^*	55, 174
$\partial\Omega_T$	50
$\partial\Omega_{[0,T]}$	50
$\partial\Omega_T^*$	55, 174, 179
$P(\partial)u$	32, 38
$P(x, \partial)u$	35, 38, 216
$P(x, y)$	41, 47, 103
$\Pi_{\partial\Omega}\mu$	112, 137, 139
$Q(y, t)$	162
$\mathbb{R}$	14
$\mathbb{R}$	15
$\mathbb{R}^+$	15
$\mathbb{R}^n$	37
$\mathrm{rest}_{\partial\Omega}\, u = u\vert_{\partial\Omega}$	67, 87, 97
$\sup \mathcal{A}$	14
$\sup_k f_k$	48, 88
$\mathrm{supp}\,\mu$	87
$\mathrm{supp}\,\varphi$	29
$\sigma(y)$	107
$T\psi$	207
u.s.c.	88
U, V	48, 216
V_n	41, 103
$W(x)$	44, 67
$x = (x_1, \ldots, x_n)$	37
$\langle x, y\rangle$	59, 111
$\vert x - y\vert$	37
$\tilde{x} = (x, t)$	175
X	110, 125
X^*	125, 130
X_1, X_2	110, 125
X_1^0, X_2^0	113
χ	83, 97
$\int f\,d\chi$	85
$y = (y_1, \ldots, y_n)$	37
$\tilde{y} = (y, \tau)$	175
$\nearrow$	69
$\searrow$	69
1_B	86, 97, 99

Subject Index

Abel integral equation 14, 21, 23
abstract balayage principle 127
adjoint heat equation 55, 175, 178
– heat potential 175
– kernel 92
– mapping 112, 125, 179
– potential 92, 175
– projection 112
analytic continuation 30, 47, 74, 121, 123, 124, 163, 174, 178, 180
– function 30, 34, 151, 174, 180
analytical set 100
approximation properties 42, 81, 89, 93, 95, 115, 120, 146, 178
Archimedian property 14

balayage principle 109, 130, 145, 179
Banach space 27, 37, 125
Banach-Schauder theorem 129
barotropic fluid 195
barrier 217
base of regular subsets 48, 218
Bauer harmonic structure 49
Bauer space 49, 220
Bessel function 187
biharmonic equation 41, 64, 161
bilinear form 111
bioelectric phenomena 44, 61
boundary-flux measurement 56
boundary-value problem 33, 42, 45, 48, 64, 217
brain research 44, 80
Brelot harmonic structure 49
Brelot space 49
Brelot's convergence axiom 49

Cantor set 83, 101
Cantor's Continuum Hypothesis 13, 16
capacitable set 100
capacity 98, 99, 100, 102
Cauchy problem; adjoint heat equation 175
– –; heat equation 50
– –; Laplace equation 42
Cauchy-Riemann system 64
characteristic function 86, 147
Choquet capacity 101
compact mapping 16
– operator 16
– set 15
computed tomography 19
conductivity 47, 63, 204
conductor potential 98
cone 108, 116, 119, 123, 124, 216
continuation of a function 87, 126
– property 115, 119, 122, 123
continuity equation 196
– principle 97, 98
continuous potential 98, 99
control problem 57
convergence theorems for integrals 91
convex set 76, 132, 145, 177, 217
convolution integral 29

decomposition theorem of superharmonic functions 104
denumerable 14
diffusion coefficient 11, 52, 64, 204
dipole 67, 75, 80
Dirac measure 43, 67, 90, 103
direct problem 9, 13, 17, 49, 51, 61, 204
– sum 110
Dirichlet problem 41, 64, 112
– –, generalized 217
– –; instability 42
– –; stability 42

distribution of finite order 60, 79
domination principle 107
Doob's convergence axiom 49

eigenfunction 46, 61, 109, 221
eigenvalue 35, 36, 61
elastic tensor 60
elastodynamic equation 59
electric conductivity 61
— permeability 61
electrocardiology 44
electroencephalographic research 44
electromagnetic field 44, 60
electron charge 64
electrostatic field 11, 47
— potential 40, 64
elliptic coordinates 71
— differential operator 47, 116, 217
— — — of higher order 99, 103, 121
energy integral 44, 80, 96, 99, 105, 161
equation of the first kind 9, 16, 36
— of the second kind 10, 16
equilibrium distribution 135, 145
— potential 98, 145
equivalence relation 106, 107, 112, 114, 136, 223
Euclidean space 37
Euler equations of motion 195
Evans-Vasilesco theorem 97, 117
exterior capacity 100, 102
external parameter 39
extremal element 77, 125, 132, 136, 138, 141, 145, 149, 174, 175

far-field pattern 59
fixed point 213
Fourier transform 169
free boundary-value problem 45
Fubini theorem 91
functional 39, 66, 67
fundamental solution 30, 32, 43, 46, 50, 60, 62

generalized Dirichlet problem 217
gravimetry 44
gravitational field 10, 41, 65, 144
Green function 32, 51, 62, 104, 108, 109, 116, 117, 162, 207, 211

Hahn-Banach theorem 105, 117, 119, 120, 122, 123, 131, 146
harmonic function 41, 48, 95, 110, 114, 124, 216
— measure 103, 107, 114, 116, 175, 216, 223
— polynomial 122
— sheaf 48, 216
— structure 48, 49
Harnack's first theorem on convergence 106
Harnack's second theorem on convergence 48
heat equation 49, 190, 210
Heaviside function 27, 82
Helmholtz equation 46, 58, 59, 109
Hilbert transform operator 207
hodographic plane 198
hyperbolic equation 58, 63

ideal 87
identification problem 11, 40
ill-posed problem 37
improperly posed problem 16, 37, 42
indicator function 86
information content 10, 12, 26, 38, 144
initial-value problem 34, 36, 42
inner capacity 100
instability of the Dirichlet problem 42
instable mapping 16, 23, 25, 40
integrable function 90
— set 90
internal parameter 39
interval, closed 15
—, open 21
—, semi-closed 84
intrinsic ambiguity 38
inverse problem 9, 10, 18, 37, 38, 43, 45, 47, 57, 59, 65, 132, 144, 191, 204, 215, 220, 224
— source problem 10, 33, 37, 40, 43, 57, 65, 144, 160, 174, 210
— — — in gravimetry 44, 173
— spectral problem 10, 35, 36, 61
irregular boundary point 70
isotropic medium 47

kernel 15, 92
Korovkin approximation 47
Krein-Milman-Choquet theorem 132

Lamé coefficient 60, 63
Laplace equation 40

Lebesgue integrable function 90
Lebesgue measure zero 83, 84
Lebesgue spine 70, 106, 108, 117
lemma of Novikov 77
lenticular neighbourhood 219
line distribution 66, 70
linear form 67, 85, 86, 106, 175
linearized inverse problem 62
Lipschitz condition 210
logarithmic kernel 33, 43
lower function 217
— semicontinuous (l.s.c.) function 88
— solution 217

magnetic field 44, 67, 79, 80
— permeability 61
— resonance imaging 21
Mal posée 222
mapping 9, 14, 16, 37, 111, 125
mass density 10, 44, 56, 59, 63
maximum 14
— principle of the heat equation 50, 297
— — of the Laplace equation 47, 100, 109, 130, 207, 218
Maxwell equations 61
measure 86
— of a set 90
metric space 37, 115
minimum principle for superharmonic functions 104, 218
mixed boundary-value problem 204
mobile electron density 64
— hole density 64
Molodensky's problem 45
monotone functions in inverse problems 9, 18, 134, 190, 211
multipole 80
multi-valued mapping 18, 114

negligible set 98, 101, 102, 146
Neumann function 212
Neumann problem 41, 204
Newtonian kernel 33, 43, 100
Newtonian potential 66, 92, 105, 163
non-denumerable 14, 25, 34, 83
non-radiating 12, 161
norm 27, 28, 125, 146
normal derivative 41

one-dimensional inverse problem 33, 35, 36, 58, 132, 160, 190, 194
operator *see* mapping
oscillation of a particle 18
order-completeness 14
overspecification 52, 53, 54, 55, 56, 58, 63

parabolic equation 51, 145
permeability 47, 61
Perron-Wiener-Brelot method 217
Poisson equation 32
Poisson integral 41, 103, 106
Poisson kernel 41, 65, 103, 162
potential 18, 47, 57, 61, 80, 92, 97, 105, 144, 160, 175
— of a ball 68, 89, 162
— of a double layer 44, 67
— of an ellipse 70
— of a line distribution 66, 70, 106, 117
— of a measure 80, 92, 97, 121, 175
— of a polygon 79
— of a polyhedron 79
— of a surface distribution 66, 132
— of a volume distribution 43, 66, 89, 95, 160
— plane 197
pressure 47, 195
product measure 91
projection 111, 112
— theorem 112
proper data 51, 52, 54, 55, 58
P-small set 218

quantum theory 46, 61

radiation pattern 59
Radon transform 20
reaction diffusion equation 210
recombination rate 64
regular boundary point 217
— set 48, 218
relatively compact set 15
resolutive function 217
restriction of a function 67, 87, 97
— of a measure 97
Riemann-Stieltjes integral 85
Riesz decomposition theorem 104
van Roosbrock system 63
Runge approximation theorem 124

Salem's function 84
scalar product 99
scattering theory 59

Schrödinger equation 61
seismic field 19, 59, 63
semicontinuous function 88
semi-elliptic differential operator 217
sheaf 48
singular part of a function 83
Sommerfeld radiation condition 59
source term 39, 43, 44
specific heat 56, 63
stability 37, 39
stable solution of an inverse problem 16, 37, 39
— solution of the Dirichlet problem 42
steady flow 195
step function 82
stream function 196
subharmonic function 104, 216
subsonic flow 196, 202
superharmonic function 104, 216
supersonic flow 196
support of a function 29
— of a measure 87
supremum 14
surface distribution 66
swept-out measure 107, 179

temperature 47, 56, 64
thermal conductivity 47, 56, 63, 191, 192
thermoelastic slab 63
Tietze's theorem 87, 126
Tschebicheff polynomial 20

uniqueness 37, 39
upper function 217
— semicontinuous (u.s.c.) function 88
— solution 217

velocity potential 196
volume distribution 43, 66, 163

wave equation 58, 63
— propagation 46
weak convergence 115
— limit 85, 115
— topology 115
weakly compact set 145
well-posed problem 37
w^*-topology 115

9 780306 431647

MIX
Papier aus verantwortungsvollen Quellen
Paper from responsible sources
FSC® C105338

If you have any concerns about our products,
you can contact us on
ProductSafety@springernature.com

In case Publisher is established outside the EU,
the EU authorized representative is:
Springer Nature Customer Service Center GmbH
Europaplatz 3, 69115 Heidelberg, Germany

Printed by Libri Plureos GmbH
in Hamburg, Germany